CAMBRIDGE EARTH SCIENCE SERIES

Editors:
A.H.Cook, W.B.Harland, N.F.Hughes, J.G.Sclater

T0296053

The Great Tolbachik Fissure Eruption:
geological and geophysical data
1975–1976

In this series:

PALAEOMAGNETISM AND PLATE TECTONICS
M. W. McElhinny

JURASSIC ENVIRONMENTS
A. Hallam

PALAEOBIOLOGY OF ANGIOSPERM ORIGINS
N. F. Hughes

MESOZOIC AND CENOZOIC PALEOCONTINENTAL MAPS
A. G. Smith and J. C. Briden

MINERAL CHEMISTRY OF METAL SULFIDES
David J. Vaughan and James R. Craig

PHANEROZOIC PALEOCONTINENTAL WORLD MAPS
A. G. Smith, A. M. Hurley and J. C. Briden

EARTHQUAKE MECHANICS
K. Kasahara

EARTH'S PRE-PLEISTOCENE GLACIAL RECORD
M. J. Hambrey and W. B. Harland

A GEOLOGIC TIME SCALE
W. B. Harland, A. Cox, P. G. Llewellyn, C. A. G. Pickton,
A. G. Smith and R. Walters

THE GREAT TOLBACHIK FISSURE ERUPTION

geological and geophysical data 1975–1976

EDITORS-IN-CHIEF

S.A.FEDOTOV & YE.K.MARKHININ

*Translated by J. E. Agrell, M. R. Colenso, F. Cooley, A. Lee,
D. McDuff, P. Miles and A. R. Prowse
(Cambridge Arctic Shelf Programme translation team, under the
leadership of P. Miles)*

CAMBRIDGE UNIVERSITY PRESS

Cambridge

London   New York   New Rochelle

Melbourne   Sydney

CAMBRIDGE UNIVERSITY PRESS
Cambridge, New York, Melbourne, Madrid, Cape Town, Singapore,
São Paulo, Delhi, Dubai, Tokyo, Mexico City

Cambridge University Press
The Edinburgh Building, Cambridge CB2 8RU, UK

Published in the United States of America by Cambridge University Press, New York

www.cambridge.org
Information on this title: www.cambridge.org/9780521158893

Originally published in Russian as *Geologicheskiye i geofizicheskiye dannyye o Bol'shom treschinnom izverzhenii 1975-1976 gg* by Nauka Press, Moscow and © Izdatel'stvo "Nauka" 1978 g.

First published in English by Cambridge University Press as *The Great Tolbachik Fissure Eruption: geological and geophysical data 1975- 1976*

English translation © Cambridge University Press 1983

First paperback edition 2010

A catalogue record for this publication is available from the British Library

Library of Congress Catalogue Card Number: 82-9586

ISBN 978-0-521-24345-2 Hardback
ISBN 978-0-521-15889-3 Paperback

Contents

Acronyms and abbreviations used in this book

AN SSSR Academy of Sciences of the USSR
DVNTs Far Eastern Scientific Centre of the Academy
GSZ Geo-seismic sounding
GTFE Great Tolbachik Fissure Eruption
IV Institute of Volcanology of the DVNTs
L Leningrad
M Moscow
SakhKNII Sakhalin Multi-Disciplinary Scientific Research Institute of the Academy
SO Siberian Division of the Academy
SVKNII North-eastern Multi-Disciplinary Scientific Research Institute of the Academy

Preface

The Ploskiy (= flat) Tolbachik volcano and its vicinity had always been one of the most intriguing volcanic regions of Kamchatka. On the flat top of the three kilometre basalt volcano, at the bottom of a well-like crater, a lava lake could be seen. For tens of kilometres to the north-east and south-west extended series of cinder cones and fissures; wide areas were covered with lavas of the Hawaiian type; and the landscape was reminiscent of the moon.

At the end of June 1975 to the south of Ploskiy Tolbachik a large earthquake swarm commenced. Here on 2 July occurred the most violent volcanic earthquake yet recorded in the Klyuchi group of volcanoes. Seismological data had indicated an imminent eruption. The prediction made by P. I. Tokarev of when the eruption would begin, had proven correct.

On 6 July, 18 km south of Ploskiy Tolbachik, an eruption began whose violence and dimensions were such that after only a few months it was called the Great Tolbachik Fissure Eruption (GTFE). It lasted more than one and a half years and ended on 10 December 1976. The active area of the eruption extended 30 km from the summit of Ploskiy Tolbachik across the Northern Breakthrough of the eruption (where it had started) to its Southern Breakthrough, in which from 18 September 1975 onwards the eruptive activity was concentrated. In the course of the eruption four new monogenic volcanoes – cinder cones of up to 340 m – appeared, an eruptive cloud rose to a height of more than 13 km, basalt rivers flowed out, sheets of lava were formed covering an area of around 40 km² and with a thickness of up to 80 m, and the top of Ploskiy Tolbachik collapsed to form a new summit caldera with a diameter of 1700 m and a depth of more than half a kilometre.* The volume of extruded lava, pyroclastics and ash reached 2 km³. This was the largest basalt eruption of recorded time in the Kuril–Kamchatka belt, and one of the most interesting volcanic eruptions known.

Thanks to the successful prediction of the time and place of the eruption, the first group of scientists from the Institute of Volcanology of the Far Eastern Scientific Centre of the AN SSSR were in place in the area of the eruption and began observations several days before it commenced. In order to carry out a comprehensive study of the eruption, the Institute of Volcanology formed the large Tolbachik Expedition, whose organiser and scientific leader was the present author. It included a seismological expedition (headed by V. V. Stepanov), a geodetic one (headed by V. B. Enman), a geophysical (headed by S. T. Balesta) and a geological/geo-chemical (headed by A. M. Chirkov and Ye. K. Markhinin). The scientists in these expeditions carried out observations on the eruption from its first until its last day, they collected samples of pyroclastics, lava, ash, and magmatic and fumarole gases, measured earthquakes, volcanic tremor, the growth of the new volcanic cones, the deformations of the land surface associated with the injection of feeder dykes and the displacement of basalt magma within its deep-seated chambers, made a seismic survey of the eruption area, studied its deep structure and carried out various other investigations.

Important work was done by the General Directorate of Geodesy and Cartography. In 1976, at the request of the Institute of Volcanology, it carried out repeated geodetic measurements in an extensive area that included the region of the eruption and Ploskiy Tolbachik volcano itself.

Numerous analyses were made in the V. I. Vernadskiy Institute of Geochemistry and Analytical Chemistry of the AN SSSR, the Institute of Geology and Geophysics, the Institute of Geochemistry of the Siberian Division of the AN SSSR and other institutions.

The results of the studies carried out in 1975 and 1976 were presented at two sessions of the Academic Council of the Institute of Volcanology in the spring of 1976 and 1977. The present collection of articles was compiled from the papers and reports read at these sessions. The first

---

* At the present author's suggestion, the Academic Council of the Institute of Volcanology requested that the first and largest of the new cones of this eruption be called the G. S. Gorshkov volcano in memory of the late Georgiy Stepanovich Gorshkov, a well-known volcanologist, former President of the International Association of Volcanology and Chemistry of the Earth's Resources, and former Director of the Institute of Volcanology.

session was held in honour of the fortieth anniversary of the founding of the Kamchatka (Klyuchi) volcanic station of the AN SSSR, which marked the beginning of continuous volcanological observations in our country. Thus the two parts of this collection of papers are preceded not only by this preface but also by B. V. Ivanov's article on the history of the Kamchatka volcanic station and its role in the development of Soviet volcanology.

The first part of the collection contains fourteen papers and short reports describing the course of the eruption and presenting data from geological and geochemical studies. A number of results are exceptionally interesting. A description is given of the activity of the Southern Breakthrough, where the development of a major eruption of the Hawaiian type and the formation of basalt lava sheets were observed in all their detail for the first time in our country. Age classifications and petrochemical characteristics are given for previous eruptions of this type in the Tolbachik Valley, since they are necessary for a comparative assessment of the GTFE of 1975–76 and to throw light on the origin of its basalts. One of the most remarkable features of the eruption was the abrupt changes in the latter's composition. Detailed petrogeochemical data are presented concerning this phenomenon. During the eruption the rare opportunity occurred to take a large number of samples of magmatic gases from the basalt rivers. Important data on the chemistry and metal content of the magmatic gases were obtained as a result.

The second half of this volume contains twelve papers and short reports setting out the results of the various geophysical and geodetic studies. For instance, the seismological studies provided key data on the source of the basalts and the mechanism and development of the eruption. They showed, in particular, that for nineteen days preceding the eruption basalts were being injected and raised from the lower layers of the crust or from the transitional layer between crust and mantle, at an average rate of about 100 m/h. Geodetic observations showed the rapid development of deformations associated with the injection of the feeder dykes of the new volcanoes, and revealed an extensive region of vertical and horizontal deformations of the surface. This region evidently came about as a result of the movement of basalts in deep magma chambers during the eruption. Seismic methods were used to determine the structure of the upper horizons of the crust on the site of the eruption and seismic shadows were discovered, the source of which could be newly-formed sills.

The papers and short reports in this collection, together with the series of communications about the eruption, published in *Doklady AN SSSR* 1976, and Fedotov's paper on the mechanism of fissure basalt eruptions (*Izvestiya AN SSSR*, Seriya geol., 1976, No. 10), give a fairly complete account of the GTFE of 1975–76 and the investigations carried out in the course of it. The overall results of these interdisciplinary studies is that this eruption has become the most studied large volcanic eruption ever. It may become one of those eruptions from which data will be used by generations of scientists. Many of the results obtained are as unique as the eruption itself. They will contribute to a deeper understanding of the mechanism of volcanic and magmatic processes, the geodynamics of volcanically active regions, sources of ores, the formation of geothermal deposits and many other problems in Earth Sciences.

In conclusion, it should be said that the observations were carried out in difficult circumstances. The camps and equipment were constantly being bombarded with cinders and had to be moved from place to place before the advancing lava flows. Many members of the scientific expeditions displayed great courage and self-sacrifice in their work on the Tolbachik eruption – as was noted in a Resolution of the Presidium of the AN SSSR of 20 May 1976. The scientific results presented in this volume are the product of the intense, combined efforts of more than 150 scientists, engineers, laboratory assistants and auxiliary service workers.* To them, his comrades in the great Tolbachik epic, the author expresses his profound gratitude.

S. A. Fedotov, Associate Member of the AN SSSR.

* Among the authors of this volume are sixty-one scientists of the Institute of Volcanology of the DVNTs AN SSSR and thirteen from other institutions.

The Kamchatka Volcanological Station and its part in the development of Soviet volcanology

B. V. Ivanov

Forty-three years ago, the Kamchatka Volcanological Station was established. It was the first volcanological centre in the country for the systematic study of contemporary volcanism, in the first instance for the detailed study of the Klyuchi group of volcanoes on Kamchatka.

The first years of Soviet power saw an intensive development of geological studies. The founding of the volcanological station was preceded by a number of measures, of which the most important were the decision of the Geological Committee and the Academy of Sciences in 1929 to make a systematic study of Kamchatka, and the resolution of the Pacific Ocean Committee of the AN SSSR concerning the organisation of permanent observations of the activity of the largest volcano in Eurasia – Klyuchevskaya Sopka.

Soviet volcanology originated basically as a branch of magmatic geology. The entire experience of geological science and, in particular, magmatic geology, had pointed to the exceptionally large role played by volcanism in the history of the Earth: volcanoes, after all, were the most important source of information about the composition of the Earth's depths and the processes that go on inside them. It was also significant that volcanic eruptions are one of the most terrifying and dangerous natural phenomena known to man. It was necessary to learn how to predict them and thereby protect lives and property from eruptions. All this contributed towards the decision to set up a volcanological station in Kamchatka.

The founding fathers of Soviet volcanology included such luminaries of geological science as F. Yu. Levinson-Lessing, A. N. Zavaritskiy and D. S. Belyankin. On the initiative of Academician F. Yu. Levinson-Lessing, who was at that time Director of the Petrographical Institute, a volcano-

1

logical station was organised on 1 September 1935 in the settlement of Klyuchi. This date marks the beginning of the activity of the station that is known today as the F. Yu. Levinson-Lessing Kamchatka Volcanological Station. Immediately, work began on the sustained, systematic study of the active volcanoes, which has continued to this day. The tasks put before the volcanologists of the Station were as follows: to collect data on the morphology and structure of the volcanoes, the nature of volcanic activity and the types of eruptions; to study the composition of the volcanic products – gases, sublimates, volcanic bombs and other pyroclastic ejecta; to investigate the chemical composition of lavas and the laws governing their variation; to appraise the possibilities of exploiting volcanic products in the national economy; and to make systematic observations of the nature of volcanic activity (in each individual active volcano) with a view to predicting eruptions.

For the purpose of acquainting the scientific community with the work of the Station it was decided to publish the *Bulletin of the Kamchatka Volcanological Station*, the first issue of which was published in 1937 in English and Russian.

The first head of the Station was V. I. Vlodavets, today an Honoured Scientist of the RSFSR, Doctor of Geological and Mineralogical Sciences and Professor, to whose lot fell the great task of organising the construction and scientific work of the Station. The research collective in those far-off times was made up of only five people: V. I. Vlodavets and V. F. Popkov (geologists), I. Z. Ivanov and N. N. Shavrova (chemists) and A. M. D'yakonov (topographer). At that time the Station was just a one-storey house made of logs, accommodating both working and living quarters.

The scientific work was carried out by groups of researchers who were sent out on official missions from Moscow to Kamchatka for two to three years' 'winter camp'. In 1936 S. I. Naboko and A. A. Menyailov arrived at the Station. Volcanological studies were extended and intermediate bases organised at the foot of the Klyuchi volcano. Regular trips were made to the subsidiary Kirgurich, Tuyla and Biokos' craters that had formed in 1932 and produced lava flows, and systematic observation was organised of the temperature and composition of the volcanic gases.

Within a year or two, the scientific data amassed by the Station needed to be synthesised, and by 1937 it was clear that there was a need for monograph publications in which such syntheses could be presented. Thus in 1940 the first volume appeared of the Transactions of the Kamchatka Volcanological Station, in the form of V. I. Vlodavets' monograph *The*

Klyuchi group of volcanoes. In this work, on the basis of a detailed description of the petrography and chemical properties of the products of these volcanoes, Vlodavets identified the variation in the composition of the lavas for each volcano and came to the conclusion that all the volcanic rocks in the region had been produced from basaltic magma by normal differentiation.

The eruption of Klyuchevskaya Sopka that began in April 1937 and continued until spring 1939 was studied comprehensively by Menyaylov and Naboko. The extensive and interesting material which they obtained and then published constituted the first detailed Soviet study of the dynamics of a complete cycle of volcanic activity.

In the period preceding the Second World War the area covered by the Station's researches was already expanding: as well as the Klyuchi group, the Avacha, Mutnovskiy, Karymskiy and other volcanoes were being studied, and the nature and state of their activity reported in the station's *Bulletin.*

In the difficult conditions of wartime the scientists of the Station continued their volcanological research. From 1944 to 1945 B. I. Piyp made a detailed study of the eruption of the Klyuchi volcano, and this material subsequently formed the basis of his classic monograph which appeared in 1956. In 1944 A. N. Zavaritskiy published *An introduction to the petrochemistry of igneous rocks.* This book by the distinguished volcanologist initiated a new branch of science – petrochemistry. It should be added that this period of research was characterised by volcanological theory of a general nature and a predominance of descriptive petrography.

The years immediately following the war saw a new stage in the development of volcanological investigations. Both in organisational and scientific terms the Kamchatka Station emerged as the founder of Soviet volcanology. The station formed the basis of the Laboratory of Volcanology, which was organised in Moscow in 1945 under the aegis of the AN SSSR with Academician Zavaritskiy as its director.

In December 1946 a seismic department was opened at the Station, the Klyuchi seismic station was set up, and slightly later the station at the settlement of Kozyrevsk. Geophysical investigations of the active volcanoes were begun. In the same year an aerial photo survey was used for the first time to study the morphology of the volcanoes. The first aero-volcanological studies led to the publication of the *Atlas of the volcanoes of the USSR* edited by A. Ye. Svyatlovskiy.

The post-war period was an extremely productive one in the scientific life of the Station. A remarkable team of scientists worked in the Laboratory and at the Station under Zavaritskiy's leadership. They included an older generation headed by Vlodavets – B. I. Piyp, S. I. Naboko, A. A. Menyaylov, A. Ye. Svyatlovskiy and L. A. Basharina, and a younger, but already fairly experienced one consisting of G. S. Gorshkov, Ye. K. Markhinin, P. I. Tokarev, I. I. Gushchenko, G. Ye. Bogoyavlenskaya, A. N. Sirin, K. M. Timerbayeva and others. Wide-ranging regional volcanological studies were undertaken in this period. Volcanic activity began to be regarded as an intermediate stage in the general tectono–magmatic cycle of the region, whilst the extensive factual material collected indicated that the structural geology of the volcanic centres had a considerable influence on the nature of constituent vulcanites. The early 1950s saw the beginning of studies in practical volcanology, such as volcanic regionalisation and the exploitation of thermal waters for the national economy.

A number of fundamental works appeared at this time, of which the foremost were Piyp's monograph *Klyuchevskaya Sopka and its eruptions of 1944–45 and in the past* (1956); Vlodavets' and Piyp's *Catalogue of active volcanoes of the USSR* (1957); Gorshkov's original and daring work (1956) in which he determined the dimensions of the magma chamber beneath the Klyuchi volcano using body waves from remote earthquakes; and *Volcanic emanations and their reaction products* by Naboko and others.

The logical culmination of this period was the First All-Union Volcanological Conference called on the initiative of the scientists of the Station in 1959, and the publication of its transactions *Voprosy vulkanizma* (*Questions of volcanism*).

The next period in the development of volcanology was connected with an event of enormous importance – the amalgamation of the Laboratory of Volcanology, the Kamchatka Volcanological Station and the Kamchatka Geological and Geophysical Observatory into the Institute of Volcanology. Understandably, the first few years of the Institute's existence were essentially a continuation of many of the studies carried out by the Station. Works on the volcanism of the Klyuchi group continued to come out – in 1965 Gorshkov and Bogoyavlenskaya's *Bezymyannyy Volcano and features of its last eruption (1955–1963)* and Gushchenko's *The ashes of Northern Kamchatka and the conditions of their formation* and works by Tokarev, Markhinin, Timerbayeva and Sirin.

The transfer of leading volcanologists to the Institute played a positive part in terms of educating a talented new constellation of young volcan-

ologists. Most of the volcanologists of the second generation –
I. T. Kirsanov, I. A. Menyaylov, A. A. Vazheyevskaya, V. I. Gorel'chik,
V. A. Yermakov, Yu. M. Dubik, L. P. Nikitina, Ye. K. Serafimova, B. V.
Ivanov and many others – served their volcanological apprenticeship at
the Station.

At the present time most of the Station's attention is being directed to
developing combined geological, geophysical and geochemical studies,
which are carried out in collaboration with many of the laboratories of
the Institute of Volcanology and other institutes. The first experiment in
such work was a combination of studies of the Klyuchi eruption in 1966,
carried out under the leadership of I. T. Kirsanov. In 1970–71, jointly with
SakhKNII AN SSSR and the Laboratory of Regional Geophysics of the
Institute of Volcanology of the DVNTs AN SSSR, deep seismic sounding
was carried out for the first time in the USSR on the Klyuchi group of
volcanoes. This work, by S. T. Balesta, V. K. Utnasin, B. V. Ivanov,
Ye. K. Markhinin, V. A. Yermakov and A. I. Farberov, has yielded
unique data on the deep structure of the region.

Together with the V. I. Vernadskiy Institute of Geochemistry and
Analytical Chemistry (GEOKhI) of the AN SSSR, research is being
carried out on a physical and chemical appraisal of the behaviour of water
in the volcanic process (this group includes Associate Member of the AN
SSSR N. I. Khitarov, A. A. Kadik, B. V. Ivanov and A. P. Maksimov).
Firm scientific links exist with the Siberian Division of the AN SSSR,
especially with the laboratory headed by Academician V. A. Kuznetsov
(B. V. Ivanov and V. I. Sotnikov). The purpose of these joint researches
is to study the behaviour of elements of the petrogenic and metallic groups
in the volcanic process. An interesting way of scientifically summarising
the work of the Station over the past five years, and to celebrate the
Station's fortieth anniversary, proved to be the preparation of a collection
of papers entitled *The deep structure, seismicity and present-day activity of
the Klyuchi group of volcanoes*, edited by B. V. Ivanov and S. T. Balesta.
This collection sets out to combine geophysical, geological and geochemical
information into a model of crustal structure in a region of active
volcanism, and to achieve an understanding of the volcanic process.

One of the principal tasks of the Station is to devise a reliable method
of predicting eruptions using a combination of methods. Here the Station
has made conspicuous progress. A wide range of seismic studies, coupled
with the correct methodological approach, have enabled the Station to
make fairly reliable seismological forecasts of eruptions of various types

of volcanoes. P. I. Tokarev had in fact earlier demonstrated from the Bezymyannyy and Shiveluch volcanoes and the vents of subsidiary craters of the Klyuchi volcano that it was possible to predict eruptions. His method has now been further developed and improved in works by V. I. Gorel'chik and V. V. Stepanov. The lateral eruption of the Klyuchi volcano in 1974 and the Great Tolbachik Fissure Eruption (GTFE) of 1975 were both predicted by the Station's seismic service.

To carry out combined studies of this kind demands a considerable physical effort from everyone who works at the Station. The team acquits itself honourably. At the moment it includes Candidate of Geological–Mineralogical Sciences V. I. Gorel'chik, Junior Research Workers A. P. Maksimov and V. V. Stepanov, and Senior Laboratory Assistants V. T. Garbuzova, V. V. Stepanova, V. P. Khanzutin, K. S. Kirishev and V. V. Ushakov.

I should like to mention briefly some of the most important results of the studies carried out by the Kamchatka Volcanological Station over the past forty years, as well as the tasks that still confront it. The Station's research has enriched many branches of modern volcanology. It has been shown that in the twentieth century 1044 eruptions have occurred from 320 volcanoes, i.e. 54% of the volcanoes on our globe have been active. The majority of these eruptions were of an explosive character. In the nineteenth century the volcano with the largest number of eruptions was Mt Etna, whilst in the twentieth century it is Karymskiy in Kamchatka (V. I. Vlodavets).

The total quantity of ejected material has been calculated as 3–6 milliard t/a, and it has been postulated that volcanism plays a leading role in the formation of the crust, hydrosphere and atmosphere (Ye. K. Markhinin).

Seismic methods have been used to determine the depth of the magma chamber beneath the Klyuchi volcano (G. S. Gorshkov).

The study of eruptions of the directed blast type has led to the development of a classification scheme for them (G. S. Gorshkov, G. Ye. Bogoyavlenskaya).

The local and regional factors governing the eruption of the highest volcano in Eurasia – Klyuchevskaya Sopka – have been elucidated by Piyp, Vlodavets and Naboko.

The role of pyroclastics from volcanoes in the unconsolidated continental sediments has been assessed. On average 7% of the pyroclastics go into the formation of coarsely fragmented facies in the volcanic area, 43% into the unconsolidated continental deposits and 50% of them are in the waters of the Sea of Okhotsk and the Pacific Ocean (I. I. Gushchenko).

G. S. Gorshkov has proposed that volcanism is predominantly a mantle phenomenon and may consequently be considered an indicator of the state and composition of the material of the upper mantle. This idea has proved to be very fruitful and has found a wide response among both Soviet and foreign specialists.

Through their combined efforts, the scientists of the Station have identified the degree to which volcanism and petrochemical features are dependent on the type of tectonic regime within the various areas of island arcs, and the petrochemical differences between volcanoes that are fed from the mantle and those fed from the crust (K. M. Timerbayeva, A. N. Sirin, E. N. Erlikh, V. A. Yermakov, B. V. Ivanov, I. T. Kirsanov). By geophysical methods (seismic sounding, magnetotelluric probes, etc.) the thickness of the crust in the area of the Klyuchi volcano group has been determined, the deep structure of the volcanic systems and the relation of volcanism to the deep structure of the crust have been demonstrated, and a model of the crust has been constructed for the whole of the region of the Klyuchi group (S. T. Balesta, B. V. Ivanov). Seismological methods have been used as a basis for reliable predictions of eruptions in the case of volcanoes producing basaltic and andesitic-basalt compositions (P. I. Tokarev, V. I. Gorel'chik, V. V. Stepanov).

What are the tasks that face the Station in the forthcoming five-year plan? First and foremost is that of further improving the quality and standard of volcanological investigations of the multi-disciplinary kind. The Station must be further equipped with the most up-to-date geophysical apparatus and instruments for the study of material composition. The Station will continue to broaden the scope of its seismological studies by constructing new seismic stations around the Klyuchi volcanoes. Evidently the time has also come to establish constant geodetic monitoring of the Shiveluch and Bezymyannyy volcanoes, whose activity is increasing and, to judge from the cycles of eruptions, will have reached its maximum by 1980. There is also the possibility of drilling boreholes to a depth of 300–400 m in the region of the subsidiary cones of the Klyuchi group for the purpose of determining the extreme parameters of the volcanic process.

The fortieth anniversary of the Kamchatka Volcanological Station is being celebrated at a significant time for our country. Not long ago Soviet Science celebrated its 250th birthday. The Resolution of the Central

Committee of the Communist Party of the Soviet Union, *On the 250th Anniversary of the Founding of the Academy of Sciences*, paid high tribute to its achievements. To these achievements the work of the scientists at the Kamchatka Volcanological Station has certainly contributed. One recalls with affection those who are no longer within our ranks and wishes those who have worked and currently are working at the Station, or have assisted in its development, every success in the work they are doing to further our science.

Part I

Description of the eruption, geological and geochemical data

Chronology and features of the Southern Breakthrough of the Great Tolbachik Fissure Eruption 1975–1976

S. A. Fedotov, G. N. Kovalev, Ye. K. Markhinin,
Yu. B. Slezin, A. I. Tsyurupa, N. A. Gusev, V. I.
Andreyev, V. L. Leonov and A. A. Ovsyannikov

Introduction

The Great Tolbachik Fissure Eruption (GTFE) occurred, as was reported earlier (Fedotov *et al.*, 1976), to the south of the Ploskiy Tolbachik volcano, within the Southern Tolbachik zone of cinder cones that was named thus and first described by B. I. Piyp. Morphologically speaking, this is a gently sloping lava shield resting up against the Ploskiy Tolbachik stratocone and falling away at an average angle of 3–4° to the west, south and south-east, towards a large meander in the River Tolbachik. The shield has an area of about 800 km² (Piyp, 1956). Within it are situated several dozen monogenic eruptive systems, mainly cinder cones in varying states of preservation, with less frequent lava domes. The volcanic systems are often combined into groups and chains, but also occur singly. They are considerably closer together in the axial band of the shield than on the flanks. The eruptions of 1975–76 were also confined to the axial band of the shield and were evidently typical in all their features of the Southern Tolbachik zone.* As a result of the eruption at the Northern Breakthrough (Fedotov *et al.*, 1976) a chain of three cinder cones formed on top of the already existing chain of cones from earlier eruptions, while at the Southern Breakthrough, 10 km south-south-west of the Northern Breakthrough, a single cone emerged. In both regions extensive lava fields formed, and cinder and ash covered an enormous area.

During the eruption the summit caldera of Ploskiy Tolbachik subsided. This must have begun early in August 1975. In September–October of that

---

* In addition to the present authors' observations, this description is based on data collected by V. A. Andreyev, V. A. Budnikov, A. L. Budnikov, Yu. V. Vande-Kirkov, V. A. Dvigalo, A. P. Ivanov, N. N. Litasov, I. A. Menyaylov, N. V. Ogorodov, V. S. Petrov, A. A. Razina and A. Ye. Shantser.

11

year a warm-water lake appeared in the pit, fed mainly by a glacier that had fallen into it. Evidently the subsidence practically stopped at the end of 1975. In August 1976 the level of the water in the lake remained approximately the same as in April, but after the eruption at the Southern Breakthrough had stopped it fell considerably: according to observations of 29 March 1977 the surface of the lake had contracted, exposing screes at the foot of the sheer walls of the caldera.

The eruption at the Southern Breakthrough, whose description is the object of this paper, had a number of distinguishing features. It was almost exclusively effusive, its lavas were liquid and it was closer to the Hawaiian type than any other eruption observed in the Kuril–Kamchatka arc in historical times. Unlike the Northern Breakthrough, at the Southern the lavas are of the subalkaline basalt type that is already known in the region of the Klyuchi group (Volynets *et al.*, 1976a, b), and in texture they are megaplagiophyric (Yermakov, 1971).

A short chronology of the activity of the Southern Breakthrough

On 18 September 1975 a fissure of north–south strike opened, the northern end of which curved slightly north-north-westwards. Lava fountains appeared the entire length of the fissure, followed immediately by lava flows moving westwards and south-westwards in the direction of the general slope of the locality. The length of the fountaining part of the fissure grew at first from 200 m to 600 m, but subsequently decreased rapidly.

On the second day of the eruption, 19 September, the cinder–agglutinate embankment that had arisen along the fissure (Fig. 1) already began to change into a horseshoe-shaped cone and a single vent to form. The cinder–agglutinate masses that were accumulating to the west of the fissure were dispersed in various directions by the lava flows for a distance of up to several hundred metres; as a result, to the west of the cone arose an area of lava–agglutinate hills several tens of metres thick. This area, bounded by clearly-expressed ledges, retained its morphological individuality throughout the eruption (Fig. 2). The formation of the horseshoe-shaped cone was complete by 30 September. Lava poured from its mouth (open to the south-west) at an average rate of 25 m³/s until 7 November 1975.

Flows of cinders and blocks poured out of the horseshoe-shaped cone during the entire period of its activity.

On 8 November the cone closed up and thenceforth the discharge of lava occurred only through boccas that kept opening around the foot of the cone. From 8 November until 28 December 1975 they opened almost exclusively to the west of the cone, along the perimeter of the area of cinder–agglutinate hills. Each bocca gave rise to a lava river, which flowed continuously for as much as several weeks. It was at this time that the longest lava flows were formed, which remained in existence throughout the eruption. The lowest viscosities also occurred then (up to 1×10^4 poise). The cone was operating at a steady rate of 10–12 ejections a minute and with average height of up to 150–200 m.

Fig. 1. Stages in the development of a cone.
1, fissure of the initial breakthrough; 2, build up of loose material; 3, continuous cover of cinders and agglutinates on fresh lavas; 4, liquid, glowing lavas; 5, dark lavas, including stationary and chilled lavas; 6, elevation caused by lava being expressed from under a cone; 7, axis of a crest; 8, edge of the scarp of an intra-crater terrace; 9, morphological or geological boundary; 10, fissure.

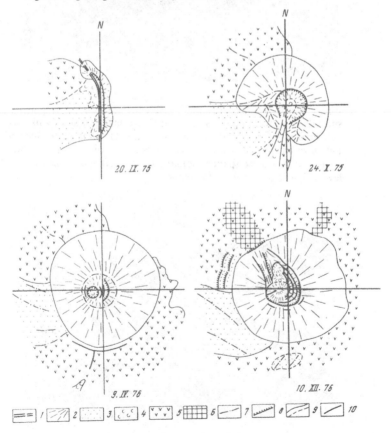

From 27 December 1975 until the beginning of April 1976, the discharge of lava occurred only to the south of the cone. Boccas opened mainly along the perimeter of the cinder–agglutinate hills and the lava plateaux adjoining the cone; these plateaux underwent extensive deformation (Fig. 3) in the

Fig. 2. Cinder cones and cinder–agglutinate hillocks. View from NNW in June 1976. (Photograph: A. I. Tsyurupa.)

Fig. 3. Scarp of a lava plateau. The blocks are tumbled fragments of smooth lavas forming the surface of a lava lake ponded by its own marginal ramparts. At the lower right, tube lava has been squeezed out through the base of the marginal rampart; on the left a hollow 'finger' of undulose (pahoehoe) lava has flowed across the crest of the levée. (Photograph: A. I. Tsyurupa.)

form of bulging and pulling apart at depth. Each bocca functioned for not more than several days, although any three or four might be working at once. Here hardly any continuous lava rivers were observed as in the preceding period, and the distribution of the lava sheets was more even. The average discharge of lava up to 9 April grew less and less, and the boccas functioned more and more erratically; brief spurts occurred followed by intervals of almost a week when hardly any lava was observed to flow. The activity of the crater fell steadily: the number of ejections per minute declined from 10–15 to 1, and the average height from 150–200 m to 5–10 m. The viscosity of the lavas throughout remained low, within 10^4–10^5 poise, and their surface was of pahoehoe type.

On 6 April there was a sudden increase in explosive activity, and on 9 April in effusive activity as well. Ejection frequency increased to 20 per minute, while the average height increased to 200–300 m. The discharge of lava, which on the first day attained 80 m³/s, grew gradually weaker in the course of a month, but remained above 20 m³/s. Discharge occurred in three main zones: to the east, south-south-east and north-west of the cone.

The activisation of the eruption was also accompanied by a series of qualitative changes. The viscosity of the lavas jumped by approximately an order. In terms of surface, all the flows of this period should be classed as the aa type. As at the end of 1975, steady rivers of lava occurred, but the flows formed by them were considerably shorter than those of December 1975 and had the characteristic aa form.

A distinguishing feature of the April 1976 activity was the sudden increase in deformations of the cone and of the zone of lava–cinder hills in areas where lava rivers had broken through. One could observe large fragments emerging from the foot of the cone, with simultaneous subsidence and slumping of its slopes, as well as internal collapse accompanied by the ejection of large masses of black ash from the crater. The largest fragment was detached from the southern part of the cone on 19 April, when the south-south-western lava river burst through. This fragment, which represented a whole segment of the structure and was initially almost as high as the cone itself, travelled about 100 m. However, the cone was not completely destroyed, and after a month intensive ejections of pyroclastics led to the 'wound' on the cone healing. Thousands of square metres of the cinder–agglutinate hills and lava plateaux were torn off and carried away by the lava, whilst the surface of the hills themselves became a network of wide and deep fissures. Moreover, the hill zone shifted from the cone as a whole (Fig. 4). Subsequently, although the general nature of the eruption remained the same, the discharge of lava and the intensity of deformation continued to diminish. The functioning of the cone was unaffected.

On 9 July a sudden increase occurred in the discharge of lava, similar to that of 9 April. At the same time, intensive deformation of the plateau around the cone occurred, parts of it were torn off and wrenched aside, and some areas were pushed up or subsided. The main events, as in April, occurred to the south of the cone, although a lava bocca to the west was also operating.

However, unlike its April counterpart the July activity was not accompanied by noticeable changes in the properties of the lava and the flow types. The activity of the cone remained at approximately the same level, but its nature changed slightly: whereas until 9 July bombs were ejected mainly in a fan, the ejecta were now more precisely directed and arrowlike,

Fig. 4. Sickle-shaped depression between the foot of the cone and the rear of the cinder-agglutinate hillocks. The depression was formed when these hillocks slid along the lava substrate. In the background are the summits of the largest of these separated parts of a cone, formed on 19 April 1976. Photographed at the beginning of June. (Photograph: A. I. Tsyurupa.)

and the discharge of black ash intensified. These changes indicated a considerable lowering of the lava level in the conduit, accompanied by collapse of its walls.

Subsequently the discharge of lava fell fairly quickly to an average level, and stable activity with a slight tendency to slackening continued until the beginning of November. Until 27 August the lava discharges took place mainly in the west and south-west. After that date the most important discharges occurred to the north of the cone, where practically all the effusive activity was concentrated after 18 October.

From 8 November 1976 onwards, up to three eccentric but similar conduits at a time began to form periodically in the northern segment of the cone. They produced mainly ash, but sometimes also ejected bombs. The lavas were discharged, as before, mainly through the northern boccas.

In the concluding phase, from the beginning of October until 10 December 1976, the activity of the Southern Breakthrough showed some special features. The increase and decrease in activity, as determined by the lava discharges and the frequency and height of the ejections of pyroclastic material, were clearly cyclical in character, with a periodicity that gradually decreased from 14 to 8 days. Moreover, each successive maximum of lava discharge was usually lower than the preceding ones, which implied a steady slackening in volcanic activity. The only exception was the last few days of the breakthrough's activity, when the visible lavas increased to twice the size of any others observed since October 1976.

Another peculiarity of the concluding phase was the fact that the orientation of the discharging sector of the cone (335° NW, based on the position of the boccas and eccentric conduits, orientation of the subsidences) was the same as that of the northern end of the original fissure. The possibility cannot be excluded that movements were continuing along this fault.

Characteristics of explosive activity

During the eruption, explosive activity was concentrated in the crater of the cone and took the form of a sequence of explosions whose characteristics are given in Fig. 5. Plot (*a*) shows the average height of ejections above the rim in a twenty-four hour period, and plot (*b*) shows the average frequency of the ejections. Some bombs were projected considerably further than the average height, attaining 400 m.

The average heights given in the plot were defined as the averaged maximum heights for several dozen ejections. In the majority of cases measurements were carried out on continuous series of ejections by comparing their height with that of the cone using a scale projected onto the object. The error might be as much as 10–15%. Some of the measurements were carried out using a seconds theodolite from a base 2300 m away, but because of the narrow field of vision and the loss of time in taking the reading, in these cases the averaging was done on random samples.

The average frequency was determined by counting every case when new pieces of incandescent material appeared above the rim of the crater over a particular interval of time.

All the measurements were carried out daily after dark, the weather and thick ash-showers permitting.

Apart from the frequency and height of the ejections, the explosions also differed in terms of the form of material ejected and the sound effects. All degrees of intensity and sharpness of sound were observed, from the bubbling 'blow-out' effect to an unusually piercing, powerful blast painful to the ears. These piercing reports were accompanied by the ejection in a symmetrical fan of a relatively small number of very large bright bombs. Until the fresh activity in April these occurred three to four times every

Fig. 5. Characteristics of eruption. (a) Average discharge calculated from volume of lava field. (b) Results of daily measurement of lava discharge through boccas (m³/s). (c) Height of the cone.

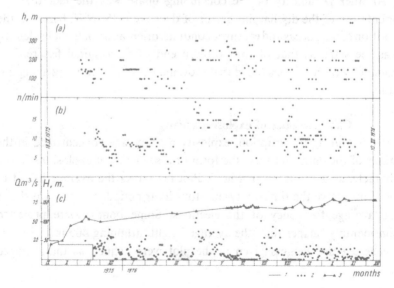

twenty-four hours, but after 6 April no less than once or twice an hour. The commonest of them were loud, but not very piercing explosions with an abundant ejection of small fragments of lava in the form of an absolutely vertical fountain.

Characteristics of the effusive activity

The lavas had the following physical properties. The temperature, measured at the sources of the lava rivers, fluctuated between 1050 and 1070 °C. Temperature was measured using a chrome–aluminium thermo-couple with a cold weld, placed in melting ice with the thermal-e.m.f. regis-tered on a PP-1 potentiometer. For the qualitative control of the upper limit of the temperature, samples of pure copper (melting point 1083 °C) were inserted into the lava. In no case was the copper observed to melt.

The effective viscosity of the flowing lava was calculated according to the formula for a plane layer: $\eta = (\zeta g h^2 \sin \alpha)/2v$, where ζ = density; η = effective viscosity; h = depth of flow; α = angle of incline (in fact it was the angle of incline of the flow's surface that was observed, the latter being taken as parallel to its base); v = maximum velocity on the surface of the flow and g = free fall acceleration.

Fig. 6. Volume weight of massive samples of lava.

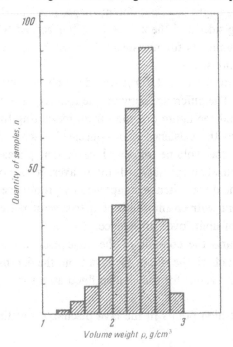

An alignment was chosen for these measurements which would give the most favourable conditions for applying the above formula.

It was established that the value η at the sources of the lava flows near the cone fluctuated between 10^4 and 10^5 poise. The secondary breakthroughs 2–3 km from the cone gave flows of viscosity 10^6–10^7 poise. A calculation of the density from massive samples of lava gave an asymmetrical distribution with a considerable scatter (Fig. 6).

From the second half of November 1975 onwards, in addition to estimates of the average discharge of lava calculated from the increase in area of the lava field using aerial photographs, daily measurements were made of the visible lava discharge (see Fig. 5, plot (*b*)). The visible discharge (R) was calculated according to the formula $R = hdv$, where d = width of flow; v = average velocity of flow; h = depth of flow.

For this purpose the average velocity of flow on the surface of the lava was used, as this was close to the average velocity of the entire lava flow, assuming the latter behaves essentially as a Bingham liquid. This assumption is based on the properties of the lava and is confirmed by a number of observations carried out by us at the Southern Breakthrough and by Hulme (1974) on the basalt flows of Etna. It was also assumed that the bed of the lava river was rectangular in section, as has been generally confirmed from observations of lava tubes left empty once feeding had ceased. In some cases the tube might have a different shape, particularly where there were overhanging sides and the average width exceeded the visible width. In selecting alignments for measuring lava discharge, we attempted to avoid such locations.

Lava rivers with an average discharge of 11 m³/s would supply 0.42 km³ over the time of the eruption. The difference between the total volume of the lava field (6.5–7.0 km³) and the figure obtained from measuring the discharges evidently points to the existence of a concealed discharge.

Throughout the eruption the volume of the lava–cinder plateau adjoining the cone grew as it slowly crept outwards (at an average rate of 1 cm/h). This outward movement was often accompanied by pronounced differentiated vertical movements with an amplitude of up to several metres a week and a predominance of uplift over subsidence.

The lava–cinder plateaux and the conduit of the cone itself act as accumulation-reservoirs which check the rise of the lava from the depths, and this is one of the possible reasons for the sudden fluctuations in the visible discharge.

However, the three distinct high points in the lava discharge – at the beginning of the eruption, and in April and July 1976 (see Fig. 5, plot (*c*)), were evidently caused by a change in the way lava was entering the conduit. The April peak is particularly interesting, as the change in the discharge was accompanied by a change in the properties of the lava and the activity of the crater.

The escaping lava formed a cover whose pattern of growth is shown in Fig. 7, whilst the rates of increase in area are given in Table 1. Whereas

Fig. 7. Diagrammatic map of areal growth of lava field.
1, by 20 Sept. 75; 2, by 4 Oct. 75; 3, by 22 Oct. 75; 4, by 26 Nov. 75; 5, by 18 Dec. 75; 6, by 9 April 76; 7, by 7 Sept. 76; 8, by the end of the eruption (survey of 22 Dec. 76); 9, lavas hidden by a continuous cinder–agglutinate cover; 10, cinder cones from old eruptions; 11, cone of the 1975–76 eruptions; 12, secondary bocca, fed from the beginning from horizontal lava conduits; 13, primary bocca, the result of a new breakthrough on the cone; 14, direction of the main systems of lava ducts with supply direct from the cone area. The outlines of the successive growth steps of the lava field were obtained by aerial photography; the survey and preliminary deciphering were carried out by N. A. Gusev and V. A. Dvigalo. (Compiled by A. I. Tsyurupa.)

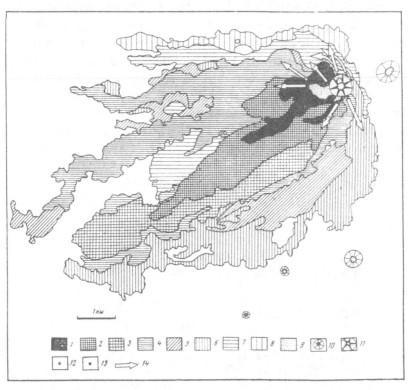

at the beginning of the eruption it was almost exclusively an increase in area that occurred, in the last few months no more than 10–15% of the newly-appearing lava was expended in increasing the area.

A lava field is the result of the superimposition of numerous 'single' flows (Walker, 1972) issuing from groups of intermittently functioning boccas. The longest of these flows (up to 9 km) were emitted at the end of 1975. In general the field has a stepped surface. There are various reasons for this. Some of the steps represent the front of flows emitted from relatively narrow feeder-boccas functioning for not more than a week. When the discharges are small, around 10–20 m^3/s, several steps may be formed on one flow, as lava breaks through secondary vents at the front of the practically stationary flow. This effect is caused by changes in the overall flow regime and may be satisfactorily described in terms of a rheological model that takes into account the fact that the lava has a yield point but can also flow slowly at stresses below this limit (Slezin, 1972; Hulme, 1974). The spreading of the lava was facilitated by the formation of lava tubes similar to those described on the Hawaiian volcanoes (MacDonald & Abbott, 1970). Areas where lava flowed along concealed tube conduits of up to several hundred metres' length, occurred all over the lava sheet within the individual steps. When lava had stopped flowing, these lava conduits were partially emptied, but their rather complex structure can be seen in places where the roof has collapsed. Fig. 8 is a plan of a cave discovered by F. A. Fedotov which is a typical empty conduit in a lava sheet that was formed when a neighbouring cone erupted not more than a few hundred years ago. The height of the roof in the individual passages varies from 0.5–4.5 m and is on average 1.5–2 m. The

Table 1. *The growth in area of the lava field*

Period	Complete increase (km^2)	Rate (km/24 h)
18 Sept. 1975 – 20 Sept. 1975	1.7	0.57
20 Sept. 1975 – 4 Oct. 1975	4.4	0.31
4 Oct. 1975 – 24 Oct. 1975	4.1	0.20
24 Oct. 1975 – 26 Nov. 1975	9.2	0.28
26 Nov. 1975 – 18 Dec. 1975	4.0	0.18
18 Dec. 1975 – 8 April 1976	5.6	0.05
8 April 1976 – 7 Sept. 1976	3.5	0.023
7 Sept. 1976 – 10 Dec. 1976	1.8	0.019

roof is vaulted, has sagged and melted in places and has lava icicles up to 15 cm long and 0.5–1 cm in diameter. The floor is flat, with a predominantly ropy surface. Judging by the thickness of the roof (in places as much as 2–3 m), this lava conduit came into operation after a fairly long interval – probably of some weeks. The distension and cracking of the roof indicate the hydraulic retention that existed during this intermission.

Fig. 8. Plan and sections of lava conduit cavern.
1, flow direction of lava (direction of curves on ropy surface); 2, steps or 'lava falls'; 3, roof collapse; 4, piles of fallen blocks; 5, lava 'plug'. (Compiled by Yu. B. Slezin.)

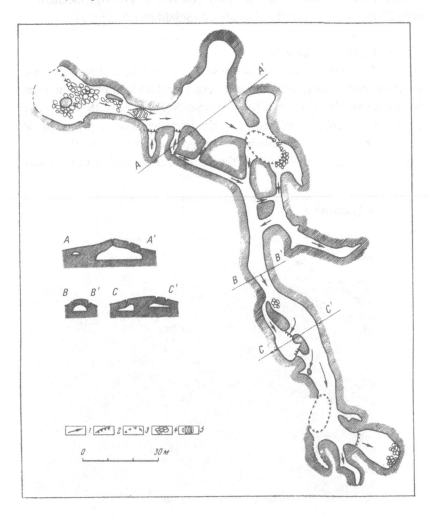

The surface of the lava field is complicated by numerous structural elements characteristic of mobile basalts: gently sloping domes (tumuli) and longitudinal levées, strips of smooth lavas reminiscent of airport runways and extending tens and hundreds of metres, pit-craters, empty lava channels and tunnels, hornitos, lava fingers and blisters of various shapes and sizes. The main morphological lava types are cinder-blocks (aa) and flat-blocks with smooth surfaces merging into hummocky ones. Regularly, but less frequently, one comes across ropy (Fig. 9), thickly coiled entrail-like (see Fig. 3) and pancake lavas. Some parts of the embankments look like large-blocked lavas; in reality they do not reflect the morphology of the moving flow, but have been created by a mechanical mobilisation of material that has already solidified (see Fig. 3).

Pyroclastic material
The pyroclastic material consisted of crystalline vitroclastic ash, vesicular black cinder, scoriaceous 'fracture' and nodular bombs, and less frequently bombs of various rotatory shapes. From May 1976 the cone collapsed noticeably more frequently and its material was flung out again in the form of fine re-explosive ash. In the lapilli fraction of the ejecta, especially during the first months of the eruption, abundant flat, occasionally

Fig. 9. Ropy lava with spheroid extrusions. (Photograph: N. L. Smelov.)

cruciform and spherical aggregates of plagioclase crystals were present. In the first few months fragments of white and light-grey pumice were found in small quantities, though fairly regularly.

Geological and energy aspects

The result of the eruption, which lasted 450 days, was a cone with a final height of 160 m and a volume of 0.018 km^3. In addition, 0.022 km^3 of the pyroclastic material was dispersed over the surface of the lava flows, while 0.025 km^3 of it was ejected beyond the cone during explosions. The area of the lava field is 34.3 ± 0.6 km^2, while its volume, according to preliminary estimates of the distribution of thicknesses, is 0.5–0.7 km^3. The explosion index has been reckoned at 4% (weight), taking the average density of the lavas as 2.2 g/cm^3, and of the pyroclastics as 1.2 g/cm^3. The thermal energy of the eruption amounted to 1.3×10^{18} J, and the energy of the volcanic explosions 3×10^{14} J. This thermal energy, then, was equivalent to 3.3×10^{10} W.

Conclusion

The basic distinguishing feature of the eruption at the Southern Breakthrough was its unmistakably effusive character compared with the Northern. Its other characteristics were all in one way or another connected with this, being either its causes (physical and petrochemical characteristics of the lava), or effects (style of eruption, predominant types and morphology of the accumulative formations).

The closest analogies with the Southern Tolbachik zone and the eruptions that occurred in it in 1975/76 may be found in the structure and activity of the rift zones of the Hawaiian shield volcanoes. From this point of view the events at the Southern Breakthrough are of the greatest interest. Lava sheets formed by eruptions of a similar type are often found in various regions of ancient and more recent volcanism. A study of them in the actual process of formation provides invaluable insights into the geology of these regions. The eruption in the Tolbachik region in 1975/76 was the first of this type in historical times on USSR territory which scientists could directly observe and study in all its detail.

26   *Fedotov, Kovalev* et al.

References

Fedotov, S. A., Khrenov, A. P. & Chirkov, A. M. (1976). Bol'shoye treschchinnoye Tolbachinskoye izverzheniye 1975 g. na Kamchatke. (The Great Tolbachik Fissure Eruption of 1975 in Kamchatka.) *Dokl. AN SSSR.* **228**(5), 1193–6.

Hulme, G. (1974). The interpretation of lava flow morphology. *J. Geophys. R. Astr. Soc.*, **39**, 361–83.

Macdonald, G. A. & Abbott, A. T. (1970). *Volcanoes in the sea.* Honolulu, University of Hawaii Press, 417 pp.

Piyp, B. I. (1956). *Klyuchevskaya sopka i yeye izverzheniya v 1944–1945 gg. i v proshlom. (Klyuchevskaya Sopka and its eruptions of 1944–45 and in the past.)* Trudy Labor. vulkanol., Issue 11, 310 pp.

Slezin, Yu. B. (1972). O vozmozhnoy svyazi dinamiki vulkanicheskikh izverzheniy s reologicheskimi svoystvami magmy. (Possible links between the dynamics of volcanic eruptions and the rheological properties of magma.) *Geologiya i geofizika*, No. 10, 17–22.

Volynets, O. N., Yermakov, V. A., Kirsanov, I. T. & Dubik, Yu. M. (1976a). Petrokhimicheskiye tipy chetvertichnykh bazal'tov Kamchatki i ikh geologicheskoye polozheniye. (Petrochemical types of Quaternary basalts from Kamchatka and their geological position.) *Byul. vulkanol. stantsiy*, No. 52, 115–26.

Volynets, O. N., Flerov, G. B., Khrenov, A. P. & Yermakov, V. A. (1976b). Petrologiya vulkanicheskikh porod Tolbachinskogo izverzheniya 1975 g. (The petrology of volcanic rocks of the Tolbachik eruption, 1975). *Dokl. AN SSSR*, **228**(6), 1419–22.

Walker, G. P. L. (1972). Compound and simple lava flows and flood basalts. *Bull. Volcanol.*, **35**, 579–90.

Yermakov, V. A. (1971). Megaplagiofirovyye lavy Kamchatki – veroyatnyy analog anortozitovykh porod. (Megaplagiophyre lavas of Kamchatka – a probable analogue of anorthosite rocks.) *Izv. AN SSSR*, Ser. geol., No. 10.

2 Calculation of the magma discharge, growth in the height of the cone and dimensions of the feeder channel of Crater I in the Great Tolbachik Fissure Eruption, July 1975

P. I. Tokarev

In studying the volcanic process it is necessary to know the quantitative characteristics of the individual volcanic phenomena for the entire period of activity of the volcano, and especially for the initial period of eruptions. For this reason observations should be carried out using a wide range of geophysical, geological and geochemical methods. An opportunity to do precisely this presented itself during the Great Tolbachik Fissure Eruption (GTFE) in 1975.

The time and place of the commencement of the GTFE in July 1975 were predicted from seismological data (Tokarev, 1976). This enabled us to organise observations even before the eruption began, and to monitor it from its very first days. According to observations by A. I. Farberov, the eruption began on 6 July at 09.45 local time (5 July at 21.45 GMT). On the evening of 6 July a group of scientists from the Klyuchi volcanological station headed by V. V. Stepanov flew around the site of the eruption in a helicopter and obtained the first photographs of the newly formed cones. On the morning of 7 July V. A. Droznin took a photograph from which we were able to determine the dimensions of the first new cinder cone. On 8 July a group of scientists from the Institute of Volcanology made up of V. A. Droznin, A. I. Farberov, V. V. Stepanov, O. B. Selyangin, P. I. Tokarev, Yu. M. Dubik and A. M. Rozhkov began a range of regular observations from a camp situated 1700 m from the centre of the eruption. From 9 to 11 July repeated measurements of the dimensions of the new cone were made. From 16 July measurements of the cone were carried out by M. A. Magus'kin using a phototheodolite.

From the measurements of the size and shape of the cinder cone in the initial period, the author determined the discharge of magma during the eruption, obtained a formula to calculate the subsequent growth in the

27

height of the cone, and estimated the size of the magma conduit, the pressure in the magma chamber and the velocity at which the magma was moving along the conduit.

Characteristics of the eruption. A new crater was being formed 18 km to the south-west of the central crater of Ploskiy Tolbachik, and according to visual and seismological observations (Tokarev, 1976) three stages can be distinguished in its activity.

Stage 1 (09.45–12.44 (local time) on 6 July): gentle discharge of gases and steam from fissures that had appeared on the site of the breakthrough of the new crater.

Stage 2 (12.44, 6 July – 10.23, 29 July): explosive activity.

Stage 3 (10.23, 29 July – end of July and subsequently): uninterrupted effusion of lava.

For the first three hours after the beginning of the eruption there was a comparatively mild discharge of gases, subsequently replaced (second stage) by explosions. The seismograms recorded volcanic earthquakes of the explosive type and constant volcanic tremor. By 18.00 hours three cinder cones covering a small area in the region of the eruption and with a height of up to 50 m had appeared, in the craters of which explosions occurred every two to three seconds which ejected incandescent bombs to a height of 300–400 m. In them a fourth conduit began to function and by 7 July had already become the main vent. Around it a cone quickly grew, which gradually covered the earlier formed ones.

During the observations of 8–11 July and up to 28 July the eruption proceeded from a single vent, around which a cinder cone literally grew before our very eyes. On 8 and 9 July the structures formed earlier still showed through its slopes, but by 10 July a non-stop vertical jet of incandescent gases, ash, cinder and volcanic bombs was bursting noisily from the crater of the cone to a height of 1–1.5 km, while above it, to a height of 6–8 km, rose a billowing cloud of ash blown sideways by the wind and forming a long trail which in the photographs taken by weather satellites could be traced for up to 300 km. Sand and ash showered constantly from this cloud, and scoriae anything up to 5 cm in diameter fell at a distance of 5–6 km. The jet bursting from the crater pulsated slightly, but its intensity varied little. It shone like a torch, even by day. At moments when the eruption intensified, the stream of gases, ash and visible bombs rose as much as 2 km above the crater. Some bombs, although only comparatively rarely, fell 1–1.5 km from the crater, setting bushes alight. The greater portion of the incandescent material fell close to the crater, and in the first few days of the eruption not less than 80% of it was deposited on the cone. The eruption continued with unremitting intensity until 23 July; from 23 to 29 July it alternated with short intervals. On 29 July at 10.23 uninterrupted effusion of lava began.

The most characteristic feature of the Tolbachik eruption of 8–11 July 1975 was the fact that incandescent material was discharged from the crater in an uninterrupted stream rather than as individual explosions, although pulsations were observed in the intensity of the jet. For the greater part of the time, the stream remained vertical and only occasionally inclined briefly at a small angle. Some rare ejection peaks were inclined at not more than 30° from the vertical. The maximum height above the crater to which some bombs were projected was 2000 m. The loose material that fell in the first days of the eruption 2–3 km from the crater consisted of cinder and sand of fresh lava without any ash. The diameter of the crater at the top of the cone increased as the cone grew in height, and the shape of the crater varied depending on the angle of the pyroclastic-laden gas stream and the direction of the wind – the leeward side of the crater being higher. The maximum duration of the flight of bombs was 50 s. When the wind speed was less than 5 m/s they were projected no more than 250 m.

Until 11 July the boundary of the cone showed clearly on the Earth's surface from the leeward side and the generator of the cone was a straight line.

Dimensions of the cone. To determine the dimensions of the cone measurements were taken of the vertical and horizontal angles of the individual elements of the cone from a distance of 1700 m. The point of measurement was situated at an azimuth of 90° from the new cone. The angles to the south-eastern and north-western portions of the foot (base) of the cone were measured, as well as the south-eastern and north-western rims of the summit crater and other elements. In order to fix the locality, measurements of angles to the old, immobile cone were used. Characteristically, the centre of the crater did not shift throughout the observations.

The results of the measurements, evaluated in linear dimensions, are given in Table 1. This records the date and time of measurements, height H of the south-eastern (H_{SE}) and north-western (H_{NW}) edges of the crater, radius of the crater r, radii of the base R of the south-eastern (R_{SE}) and north-western (R_{NW}) edges of the base, diameter of the base D, volume of the cone V_0, and rates of accumulation of material within the cone α_0.

Table 1. Characteristics of cinder cone growth

No. of measurement	Date	Time (hours)	Dimensions of cone (m)							V_0 (10^6 m^3)	a_0 (m^3/s)
			H_{SE}	H_{NW}	H_{mean}	r	R_{SE}	R_{NW}	D		
1	7 July	07.00	90	79	85	56	197	241	438	4.7	63
2	9 July	08.00	130	120	125	65	240	350	600	10.1	41
3	9 July	22.00	140	130	135	75	250	450	700	12.2	41
4	10 July	08.00	150	140	145	105	270	480	750	17.1	51
5	11 July	10.00	180	150	165	127	390	510	900	37.7	87

The first measurement was made from V. A. Droznin's photograph, the others directly by the author. The second and third measurements, performed with the aid of plumb-line and compass, are less exact than the last two, which were made with a theodolite. The errors in measuring the linear dimensions of the cone are not more than 10%.

Shape of the cone. The dimensions of the cone were continually changing, but its shape varied comparatively little in general terms. Fig. 1(*a*) shows the sequence of changes in the dimensions and shape of the cone from 7 to 11 July. The asymmetry of the base of the cone is connected with the latter's 'creeping' north-westwards on 8 and 9 July, which was evidently caused by the injection of magma from fissures beneath the cone. The base of the cone expanded north-westwards during this time by more than 100 m, as was noticeable to the eye. It appeared that a substantial lava flow was on the move.

The growth of the cone was accompanied by an increase in the crater's diameter, as Fig. 1(*a*) shows. The steepness of the south-eastern outer slope β was $35° \pm 5°$, and that of the inner slope about $55°$. On the basis of these data the cinder cone may be approximated to a frustum of a circular cone, a schematic diagram of which is shown in Fig. 1(*b*). The fact that the angle

Fig. 1. Dynamics of growth of the explosive cone of the Great Tolbachik eruption, 6–11 July 1975. (*a*) Change in form and dimensions of cone from observations. (*b*) Approximation: frustrum cone, *R* and *r* – radii of base and crater, *d* – diameter of conduit, β – angle of outer slope of cone. (*c*) Alteration of cone height with time; 1, from observations; 2, calculated from formula $H = 95t$; 3, author's measurements used for calculation; 4, subsequent measurements by M. A. Magus'kin. Time (*t*) is expressed as intervals of 24 hours.

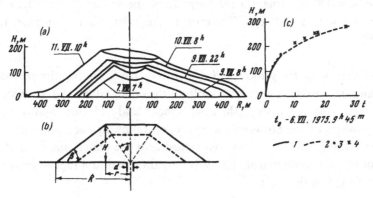

of inclination of the crater interior remained constant despite the increase
in height of the cone, indicates that the mouth of the conduit was always
on a level with the Earth's surface, and did not rise as the cone grew and
the crater filled with products of the eruption.

The volume of the cone was determined from the formula
$V_0 = \frac{1}{3}\pi H(R^2 + r^2 + Rr)$; the height of the cone was taken to be the mean
arithmetical value of the heights of the north-western and south-eastern
margins of the crater, i.e. $H = \frac{1}{2}(H_{SE} + H_{NW})$, and the radius of the base
to be the value $R = R_{SE}$. It was assumed that the low value obtained by
not taking into account the extra volume of the north-western side of the
cone was justified by making no allowance for the volume of the crater.
The error in the calculation of the volume does not exceed 30%. The values
of the calculated volumes of the cone are given in Table 1.

Discharge of magma. Using the change in volume of the cone ($V_0 t$) and
the time of the eruption, the rate of accumulation of material $\alpha_0 = V_0 t/t$
was calculated. The start of time calculations was taken as 1000 hours on
6 July. The individual measurements are given in Table 1. The mean
arithmetical value from the first, fourth and fifth measurements was taken
for the rate of accumulation of material α_0, as these were the most reliable.
This gave $\alpha_0 = 5.8 \times 10^6$ m³/24 h = 67 m³/s, while the standard deviations
were 1.3×10^6 m³/24 h and 15 m³/s respectively. Taking the density of the
material of the cone as $\rho_0 = 1.5 \times 10^3 \pm 0.3 \times 10^3$ kg/m³, we obtain α_0 in
units of mass: $\alpha_0 = 8.7 \times 10^3$ kg/24 h = 1.0×10^5 kg/s. The discharge of the
volcano $\alpha = (1/n) \times \alpha_0$, where n is that part of the mass of the material
which is deposited during the eruption on the cone of the volcano.
Assuming that 20% of the mass of the material is ejected beyond the cone
(i.e. $n = 0.8$) we obtain a discharge from the volcano of $\alpha = 1.1 \times 10^{11}$
kg/24 h = 1.25×10^5 kg/s. The error in this calculation is not more than
50%.

Calculation of the growth in height of the cone. During the period of
continuous extrusion of material from the crater either as a jet made up
of a mixture of gases, ash, sand, cinder and volcanic bombs, or as
explosions of the Strombolian type often following each other in quick
succession, most of the material settled near the crater, and around it a
cinder cone accumulated. This was bound to lead to a regular change in
the height of the cone.

Having determined the rate of accumulation of material α_0 and the shape of the cone during the initial period of the eruption, and assuming the discharge from the crater and the nature of the eruption were constant, we can calculate the height of the cone at any subsequent moment. On these assumptions the volume of the cone at moment t is equal to $V_0 t = \alpha_0 t$. On the other hand, the volume of the cone is equal to $V_0 = \frac{1}{3}\pi H(R^2 + r^2 + Rr)$. According to (*b*) in Fig. 1, $r = H \tan \beta$. After some simple conversions, we arrive at $R = 2H/\sin 2\beta$. Substituting for R and r in the general expression for the volume their value expressed through H and β and carrying out some simple conversions, we obtain $V_0 = \frac{1}{3}\pi H^3$ $(3 \tan^2 \beta + \cot^2 \beta + 3)$. Equating the right-hand terms of the two expressions for the volume of the cone at time t, we obtain $V_0 t = \alpha_0 t = \frac{1}{3}\pi H_t^3$ $(3 \tan^2 \beta + \cot^2 \beta + 3)$. Hence $H_t = \sqrt[3]{[3\alpha_0 t/(3 \tan^2 \beta + \cot^2 \beta + 3)]}$. Thus, assuming the discharge from the volcano and the nature of the eruption are constant, then the height of the cinder cone will increase proportionately to the cube root of the time of the eruption.

In the first five days of the New Tolbachik eruption the rate of accumulation of material and the shape of the cone were determined, and five measurements made of its height. The results obtained were $\alpha_0 = 5.8 \times 10^6$ m^3/24 h, $\beta = 35°$. Putting these values into the above formula and making the necessary calculations, we obtain $H_t = 95\sqrt[3]{t}$, where H is expressed in metres and the time t in 24 h periods. On the graph (Fig. 1(*c*)) a continuous line denotes the increase in height of the cone according to data from observations marked by large dots. The dashes show the growth in height calculated from the above formula, and the crosses show the height of the cone according to phototheodolite observations carried out later.

As Fig. 1(*c*) shows, the calculated height of the cone coincides fairly well with data from observations. The divergence is not more than 10% – which is within the precision limits of the initial measurements. From this we may draw two conclusions: (1) the method of calculating the increase in the height of the cone has been correctly chosen, (2) the rate of accumulation of material, and consequently of discharge of magma as well, remained constant during the eruption from 6 to 27 July.

The initial rate of discharge of material from the crater V_0 may be determined both directly from the cine-survey data and (less accurately) from observations of the height to which bombs were flung. In the initial period of the eruption the maximum height to which material was ejected, judging from angle measurements, was approximately 2000 m. Hence the initial rate V_0 was 200 m/s. On the other hand, the maximum measured flight-time of an individual bomb was 22 s. Consequently $V_0 = 220$ m/s. This is not taking into account air resistance or the acceleration of the bombs by the jet. Both these factors affect the velocity of the bombs in opposing directions and partially compensate each other. If the initial velocity of the discharge of material from the crater is 200 m/s, we have an error, evidently, of not more than 20%.

The pressure P_0 in the explosion chamber of the gases in the conduit may be determined from the formula $P_0 = \frac{1}{2}\rho\, V_0^2$. Taking $V_0 = 200$ m/s and $\rho = 2.5 \times 10^3$ kg/cm^3, we obtain $P_0 = 5 \times 10^7$ Pa.

Dimensions of the conduit. From observations in the first few days of the eruption and from the shape of the cone, it follows that the upper portion of the conduit was vertical, while its mouth remained on a level with the earth's surface. After the formation of the fissures and independent vents at the beginning of the eruption, by the end of the first day the main channel had formed from which the eruption proceeded for the next three weeks. Judging by the shape of the cone and the jet that burst from the crater, the upper part of the conduit was a right circular cylinder. One can assume that the conduit was cylindrical in shape up to the level where the discharge of gases from the magma (magma dissociation) begins, i.e. as far as the point at which the movement of the mixture of gases and magmatic material began to accelerate.

To estimate the diameter of the mouth of the conduit and the velocity of the magma along it, let us make the following simplifying assumptions: the conduit is a circular vertical cylinder extending to the depth where dissociation occurs in the magma, and having diameter d. The magma contains 3.0% by weight of gas, mostly water vapour (Markhinin, 1967). The density of the magma before bubbles of gas form in it is $\rho = 2.5 \times 10^3$ kg/m^3. At the mouth of the conduit the temperature of the gas is 1000 °C; its pressure is 1×10^5 Pa. According to these definitions, the discharge of the magma during the eruption is $\alpha = 1.25 \times 10^5$ kg/s, while the velocity of the discharge of the jet from the mouth of the conduit is $V_0 = 200$ m/s.

With the values given above, 1 m^3 of undissociated magma contains a mass of gas $m_g = 2.5 \times 10^3$ kg $\times 0.03 = 75$ kg. According to Vukalovich's tables (1965), at a temperature of 1000 °C and atmospheric pressure, the specific volume of vapour $V_s = 5.989$ m^3/kg; hence at the mouth of the conduit the volume of 75 kg of vapour will be 449 m³. The volume of the solid products occupies approximately 1 m³. Thus as it rises from the bottom of the conduit to its mouth, the volume of the magma increases 450 times. This means that with a discharge of magma $\alpha = 1.25 \times 10^5$ kg/s the volume of the mixture of gas and lava material discharged will be $\alpha_m = (\alpha/\rho) \times 450 = 2.25 \times 10^4$ m³/s. With a velocity of discharge of the jet $V_0 = 200$ m/s, the area of the section of the mouth of the conduit will be $S = \alpha_m/V_0 = 110$ m², and the diameter of the conduit $\alpha = \sqrt{(4S/\pi)} = 12$ m. Hence the velocity of the undissociated magma along the conduit at a given depth is $V_m = \alpha/\rho S = 0.5$ m/s.

The greatest uncertainty is in the estimates of gas content in the magma and its degree of expansion. However, these estimates do not exceed 0.3 of an order, and therefore it is unlikely that the error in determining the diameter of the conduit is greater than 0.3. Thus we may assume that the diameter of the conduit lies between 6 and 24 m.

Conclusion. According to observation data, during the first few days of the GTFE (6–11 July 1975) the initial rate of emission of magmatic material from the crater was 200 m/s, the rate of discharge from the volcano was 1.25×10^5 kg/s, and the maximum height to which volcanic bombs were thrown was 2000 m. It has been established that the height of the cone grew according to the law $H = 95 \sqrt[3]{t}$ m, where time t is expressed in 24 hour periods. The heights for the further growth of the cone as calculated from this formula, coincided with the data from subsequent observations; this indicates that over the period from 6 to 28 July the discharge from the volcano remained constant. According to calculations, the diameter of the feeder conduit was 12 m, and the rate of movement of the magma in the deep portions of the conduit about 0.5 m/s.

References

Markhinin, Ye. K. (1967). *Rol' vulkanizma v formirovanii zemnoy kory.* (*The role of volcanism in the formation of the Earth's crust.*) M., Nauka, 256 pp.

Tokarev, P. I. (1976). Predskazaniye mesta i vremeni Bol'shogo Tolbachinskogo izverzheniya v iyule 1975 goda. (Predicting the time and place of the Great Tolbachik eruption of July 1975.) *Dokl. AN SSSR*, **229**(2), 439–42.

Vukalovich, M. I. (1965). *Tablitsy termodinamicheskikh svoystv vody i vodyanogo para.* (*Tables of the thermodynamic properties of water and water vapour.*) M.-L., Energiya, 408 pp.

Activity of the Ploskiy Tolbachik volcano, June–July 1975

A. I. Farberov

During the period preceding the Tolbachik eruption, the Ploskiy Tolbachik volcano had for a number of years been in a relatively quiet state and discharged only water vapour and gases. The volcano began to become active in the second half of June 1975. At this time the moderate fumarole activity was replaced by an intensive discharge of gas. On 28 June at 13.00 hours (here and subsequently, local time) the first ejections of ashy material above the Ploskiy Tolbachik crater were observed from Apakhonchich seismic station. Periodic ejections of ash, as well as discharges of gas with an admixture of ash, were also observed over the following days right up to the morning of 7 July.

The activity of Ploskiy Tolbachik not only began before the Tolbachik eruption but continued throughout its duration. This could be constantly ascertained from ejections of ash above the crater. These occurred, for example, on 27–29 July, and 2, 9 and 17 August, i.e. at turning points in the course of the eruption of the Northern Breakthrough, corresponding to the opening of the fissure on the slope of Gora 1004, the formation of the northern bocca of Cone I, and the birth of Cones II and III.

The state of the crater

Observations in the crater of Ploskiy Tolbachik showed that its activity was different at the beginning of July from at the end. During the night of 2/3 July a concentrated stream of gas with diameter ≈ 6–8 m burst from an aperture in the bottom of the well-shaped crater pit. The gas was lit up from below and had a pinkish-red tinge. In spite of the high pressure, it rose practically without noise. There was an isometric aperture with transverse measurements ≈ 30 × 40 m in the southern area of the bottom of the pit. Where the gas emerged the edges of the aperture were heated

36

to a yellow colour, and the visible portions of its walls to a dark-red colour. During our observations ($\approx$ 1 hour) there were no signs of fresh lava in the aperture. Probably melting was occurring at a somewhat deeper point – beneath the south-western part of the aperture, from which a slightly inclined, high-temperature gas stream was gushing. The presence of fresh lava beneath the bottom of the crater was also indicated by a profusion of freshly-ejected material in the snow at the bottom of the caldera, on the southern slope and at the foot of the volcano. The approximate outline of the area where this material fell is shown by the broken line in Fig. 1.

On 28 and 29 July, when the eruption of Cone I had already been going on for four weeks, the nature of the activity in the Ploskiy Tolbachik crater changed. Gas began to be discharged over the whole section of the well-shaped pit in the crater, the bottom of which was no longer visible. Periodically there was an increase in the gas pressure, accompanied by subterranean rumbling lasting up to 30–40 s. No reflections of fiery-red colour were visible on the lower surface of the gas discharges, either by day or by night, which indicated that there was no fresh lava at the bottom of the well. At moments when the gas discharge from the pit intensified, gas was also emitted from a number of reactivated fissures at the bottom of the crater to the south of the pit itself. Previously occasional faint water vapour fumaroles had occurred along these fissures (Menyaylov, 1953; Alypova, 1964). Apart from fragments of pyroclastic formations of juvenile appearance, the bottom of the crater also contained numerous fragments of resurgent material. Evidently at the end of July the bottom of the crater began to shatter; this process was being accompanied by the destruction and ejection of material making up the bottom of the crater pit, which had begun at the end of June.

Products of the eruption of the volcano

A study was made of the solid products ejected from the Ploskiy Tolbachik crater in the period 28–30 June 1975. They consisted of fragments of cinder, aggregates of plagioclase, volcanic sand, ash and 'Pele's hair'. The pieces of light, highly porous cinders of juvenile appearance and bluish-black, brown and greenish-yellow colour, were up to 10–15 cm in size. The cinders were found to contain isometric tabular crystals of plagioclase 1.0–1.5 cm in size. The cavities in the cinders contained thread-like brown and dark-green formations of the 'Pele's hair' type.

Microscopic investigation shows the cinders to be made up mainly of transparent brownish-green glass. This accords with the results of chemical analysis (see Table 1) – the composition of the cinders and the glass derived from them is similar. In the glass from the cinders one finds isolated phenocrysts of plagioclase up to 1.5 mm in size, and of olivine up to 0.5 mm.

Volcanic sand and ash were collected from névés on the southern slope of the volcano. The medium and fine-grained varieties contain fragments

Fig. 1. Sketch map showing deposits of pyroclastic material ejected 28–30 June 1975, from the crater of Ploskiy Tolbachik.
1, approximate outline of the fallout zone of cinder and ash; 2, site of cinder sampling; 3, site of ash sampling; 4, eruption aperture in the bottom of the well-like crater pit, 3 July 1975.

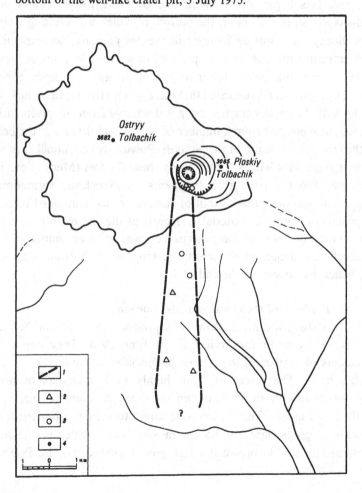

of black and light-brown glass, crystals of various minerals, and pieces of the groundmass of lava of a basaltic composition. The latter consist of dark-coloured minerals (around 50%, fine microlites of plagioclase and glass. The ash has a predominance (up to 60–70%) of fragments of light-brown, as a rule transparent, uncrystallised glass. The fragments of the crystals are made up of plagioclase (around 50% of all the mineral fragments), olivine and a lesser quantity of monoclinic pyroxene. The large-grained varieties contain resurgent material which probably resulted from the shattering of the rocks at the bottom of the crater pit as the eruptive aperture was opening in it.

Conclusion

The data presented above indicate that the Ploskiy Tolbachik volcano was reactivated in the incipient stages of the Tolbachik eruption

Table 1. *The chemical composition of ejection products of Ploskiy Tolbachik in June 1975*

Sample	SiO_2	TiO_2	Al_2O_3	Fe_2O_3	FeO	MnO	MgO
1. Cinder	51.02	2.26	14.64	3.69	8.30	0.18	4.63
2. Glass from cinder	51.06	2.12	14.49	4.39	8.44	0.21	4.53
3. Ash	51.82	1.92	16.07	5.60	5.83	0.16	5.05
				1.76			
4. Plagioclase from lapilli	51.27	not determined	—	—	—	—	—

Sample	CaO	Na_2O	K_2O	H_2O^-	H_2O^+	P_2O_5	Σ
1. Cinder	8.35	3.47	2.29	0.22	H/o	0.43	99.48
2. Glass from cinder	7.90	3.60	2.40	0.44	H/o	0.75	100.03
3. Ash	7.52	3.40	2.10	0.20	0.31	0.28	100.26
				Molecular %			
				Ab	An	Or	
4. Plagioclase from lapilli	13.18	3.86	0.74	33.2	63.0	3.8	

Analysis of samples: 1, G. P. Novoseletskaya; 2 and 4, T. V. Dolgova; 3, T. G. Osetrova.

proper. The commencement and duration of the two processes – the activity of the volcano and the first earthquake swarm presaging the formation of Cone I (Fedotov *et al.*, 1976) – coincide within 12–24 hours. We may suppose that the intensely compressed, high-temperature gas phase which formed at considerable depths in the course of the preparation of the eruption was partially 'tapped off' from 28 June to 7 July along the line of least resistance presented by the conduit of the active volcano. The solid juvenile material ejected in this process is typical basalt material from the uppermost portion of the magma column of Ploskiy Tolbachik, related chemically and mineralogically to the products of the previous historical eruptions of this volcano (Kirsanov & Ponomarev, 1974). It is completely different in its parameters from the products of the eruption of Cone I of the Northern Breakthrough (Volynets *et al.*, 1976; see Budnikov *et al.*, this volume), which were brought to the surface from depths of several tens of kilometres (Fedotov *et al.*, 1976). Apparently, at such depths the entire zone enclosing the fissure along which the cinder cones are aligned and within which the Ploskiy Tolbachik volcano is situated, has a single source of supply, as was suggested by B. I. Piyp (1954). If this is so, then the activisation of the deep-seated processes under any part of the zone must have its 'echoes' in the volcanic apparatus of an active volcano, as was repeatedly observed in the course of the Tolbachik eruption.

The author is grateful to Yu. V. Vande-Kirkov, O. N. Volynets and A. P. Khrenov for their assistance in analysing the material obtained and for many useful discussions.

References

Alypova, O. M. (1964). O magnitnom pole v kratere vulkana Ploskiy Tolbachik. (The magnetic field in the crater of the Ploskiy Tolbachik volcano.) *Byul. vulkanol. stantsiy*, No. 38, 62–5.

Fedotov, S. A., Gorel'chik, V. I. & Stepanov, V. V. (1976). Seysmicheskiye dannyye o magmaticheskikh ochagakh, mekhanizme i razvitii bazal'tovogo treshchinnogo Tolbachinskogo izverzheniya v 1975 g. na Kamchatke. (Seismic data on the magmatic chambers, mechanism and development of the basalt Tolbachik fissure eruption of 1975 in Kamchatka.) *Dokl. AN SSSR*, **228**(6), 1407–10.

Kirsanov, I. T. & Ponomarev, G. P. (1974). Izverzheniya vulkana Ploskiy Tolbachik i nekotoryye osobennosti ikh produktov. (Eruptions of the Ploskiy Tolbachik volcano and some features of their products.) *Byul. vulkanol. stantsiy*, No. 50. 53–63.

Menyaylov, A. A. (1953). Sostoyaniye vulkana Tolbachik v 1945–48 gg. (The state of the Tolbachik volcano 1945–48.) *Byul. vulkanol. stantsii*, No. 17, 41–5.

Piyp, B. I. (1954). Vulkan Tolbachik. (The Tolbachik volcano.) *Byul. vulkanol. stantsii*, No. 20, 69–71.

Volynets, O. N., Flerov, G. B., Khrenov, A. P. & Yermakov, V. A. (1976). Petrologiya vulkanicheskikh porod treshchinnogo Tolbachinskogo izverzheniya 1975 g. (Petrology of the volcanic rocks of the Tolbachik fissure eruption of 1975.) *Dokl. AN SSSR*, **228**(6), 1419–22.

The quantity, distribution and petrochemical features of pyroclastics of the Great Tolbachik Fissure Eruption

V. A. Budnikov, Ye. K. Markhinin and
A. A. Ovsyannikov

Course of the eruption and formation of the pyroclastics

Several distinct periods were observed in the eruption of the Northern Cones. In the initial period, from 6 to 9 July 1975, the eruption was purely explosive in character and was of the Strombolian type, which is characterised by ejections of mainly plastic lava material to a height of up to 200–300 m, with a gradual transition over the next few days to the Vulcanian–Strombolian type. A feature of this period was the distinctive gas blast, which flung a mass of cinders and volcanic bombs onto the surface. Thus, during the period from 09.45 on 6 July to 02.00 on 9 July a cinder cone was formed with a height of up to 150 m, a crater diameter of about 100 m and a base diameter of about 500 m.

Characteristic of this period of the eruption were cinder showers which spread in the direction of the prevailing winds and thinned out in proportion to the distance from the cone. The axis of the ash-fall in the first few days of the eruption was west-south-west (250°), and the total thickness of the cinder and ash 1 km from the centre of the eruption was 20 cm. The average intensity of the ejection of material in the initial period fell within the following parameters: average altitude of ejection of bombs – 0.3–0.5 km, average height of the ash column – 5 km, maximum radius of flight of bombs – 700 m.

Over the next few days powerful explosions occurred against a background of rhythmic fountaining of pyroclastic material. The height to which bombs were flung and the radius of their flight increased noticeably. The eruption changed to a Plinian type of activity, which is characterised by the steady, non-stop emission of an enormous volume of cinders, bombs and ash in a mighty gas jet. The incandescent pyroclastic material formed a fiery torch up to 1–2 km high. A constantly spreading gas–ash eruptive

41

column rose higher up to 5–6 km, while the inversion cloud developed higher still, at 8–10 km. The discharge of material occurred along a vertical or nearly vertical trajectory. The flight of the bombs did not exceed 1 km. Against a background of non-stop blast, isolated explosions occurred, but more often series of explosions accompanied by sharp increases in the discharge of cinder and ash and the ejection of isolated bombs to a height of up to 2.5 km; after these the intensity of the blast increased abruptly, but gradually diminished again to the previous level. This activity was characteristic of Cone I and did not alter essentially until 23 July.

From 23 July onwards pauses occurred in the eruption of varying duration (from a few seconds to several hours). During these there was either a complete absence of discharge of material from the crater, or there were gentle puffs of ash and blue gas. Between the pauses the jets and fountaining continued unabated, with occasional powerful explosions.

From 27 July the dynamics of the eruption changed radically and lava bubbles were observed periodically appearing above the rim of the crater and splashing out material beyond it. The first instalment of the lava appeared at 16.30 hours in the form of a spot on the pass between Cone I and elevation 1004. On 29 July the first lava flow began. On 2 August the effusion of the second lava flow commenced, accompanied by fountaining to a height of 150 m. The lava continued to flow until 8 August.

A characteristic period in the formation of the pyroclastic products of Cone I was the eruption of a pale, almost white powdery ash in the night of 8/9 August. This event was preceded by considerable pauses in the activity of Cone I. Thus on 4, 5, and 6 August periods of rhythmic fountaining alternated with prolonged pauses (from a few minutes to several hours). On 7 August the pause lasted nine hours (from 23.00 on 7 August to 08.18 on 8 August); on 8 August at 17.50 white powdery ash began to fall and at times there was a strong smell of sulphur. The fall of white ash turned overnight into an ash storm which lasted until 06.00 on 9 August. During this storm sounds of powerful blasts came from the volcano. In the course of 12 hours, according to our calculations, around 7 million m^3 of white ash was emitted. It was ejected to the west of Cone I ($\approx$ 5 million m^3), as well as to the south and south-east of it ($\approx$ 2 million m^3) as a result of the wind direction at the time (Fig. 1). This ash was various shades of pale grey and differed completely from the other kinds. The thickness of the layer of pale ash was 6–7 cm at a distance of 2.5 km from Cone I and up to 0.5 cm at a distance of 8 km. It contained predominantly fine and medium psammitic and pelitic material (Table 1).

Apart from its purely external features (colour, dimensions), the pale ash was also chemically and mineralogically quite different from other ash, as will be described below.

At 18.50 hours on 9 August a new breakthrough opened and Cone II began to form. A jet of incandescent pyroclastics shot up to a height of 1.5 km and simultaneously a thick flow (up to 40 m) of large blocks of lava began to pour out. A distinguishing feature of the activity of Cone II was the formation of peculiar ellipsoid and spherical bombs which looked as though they had been rolled into these shapes and which contained xenoliths of basement rocks. Bombs of this shape were probably formed in the gas–ash stream at high temperatures (up to 1400 °C, according to G. N. Kovalev's calculations).

Subsequently, the eruption (formation of vent III on 17 August, of vent IV on 22 August, and of the fissure vent on 23 August) took the form of pulsating blasts with the ejection of pyroclastic material (ash, cinder and

Table 1. *Granulometric composition of the pale ash*

No. of sample	Fraction sizes (mm), amount (%)						
	2.0	2.0–1.0	1.0–0.5	0.5–0.25	0.25–0.1	0.1–0.01	< 0.01
216	—	3.0	12.3	48.5	4.15	1.8	30.25
III-6	7.05	19.15	16.3	18.75	17.30	11.25	10.20
III7/3	1.5	4.05	3.85	30.65	33.70	15.3	10.95

Fig. 1. Wind roses for the area of Tolbachik volcano (August, 1975).

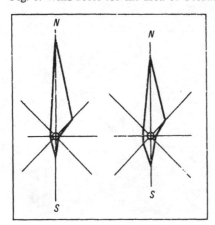

bombs) and the simultaneous effusion of lava. Only the intensity of the ejection of pyroclastic material and the duration of the activity of the new vents changed. The entire activity of the Northern Breakthrough ended on 15 September with the termination of the effusive–explosive eruptions of Cone II.

On 18 September at 08.00–09.00 8 km south of the Northern Breakthrough, liquid lava began to fountain to a height of 40 m along the entire length ($\approx$ 200 m) of a fissure of north-western strike, and pour out. A pulsating form of activity was observed, with periodic blasts from one vent and the ejection of small quantities of ash. In other words, from the very start the eruption was essentially effusive.

The main types of pyroclasts formed during the eruption

Volcanic bombs. These make up the cones and cover an area up to 1.5 km in radius around them. Bombs vary from a few centimetres to 1–1.50 m (rarely 2 m) in size (Fig. 2). Two types are characteristic: mainly spherical or ellipsoidal forms made up of massive basalts, and flattened ones of highly porous basalts. On Cone I the bombs constituted 10–15% of the overall volume of the pyroclastics; on Cone II their volume increased to 25–35%. Moreover, on Cone II there were many bombs containing xenoliths of various basement rocks: sandstones, mudstones and effusives. These are mostly 10–15 cm, rarely 20–25 cm, in size.

Fig. 2. Volcanic bombs at Cone II.

Scoriae. These are fragments of porous and vesicular lava of varying dimensions – from a few millimetres to 10 cm; sometimes it is difficult to draw a clear line between scoriae and porous cinder bombs. The scoriae of the Northern Breakthrough may, and indeed must, be considered the richest occurrence of them in Kamchatka. Over an area bounded by the 50 cm isopach they comprise 170 million m³. The dirt road from Kozyrevsk to the region of the Northern Cones leads as far as here.

Ash. According to Rittmann's definition, ashes are 'friable materials similar to dust or sand, made up of sprayed-out magma (vitreous ash) or finely comminuted material from the country rocks that comprise the walls of the conduit; but more frequently are a mixture of both' (Rittmann, 1964, pp. 121–2).

The relatively small quantity of ash emitted was a characteristic feature of the eruptions; moreover, the ashes of the Northern and Southern Breakthroughs differ not only chemically but morphologically.

Froth. Basalt froth occurs in hair-like tangles and very fine, weightless vitreous capillaries rather like 'Pele's hair'. Average length is 8–9 cm.

Plagioclase lapilli. These are characteristic only of the Southern Breakthrough. The plates vary from several millimetres to 4–5 cm in length and often form various aggregates, which are sometimes spherical and ellipsoid (Fig. 3). The plagioclase lapilli are denser than the scoriae, and can be easily crumbled out of them.

Lavoclasts. These are what is formed when the cooling lava fragments in the course of the eruption. They occur at the Northern and (more widely) Southern Breakthroughs where the lava flows, as they flood onto already cooled lava fields, pull them apart, creating various levées and accumulations.

Quantity and distribution of the pyroclastics

Most of the pyroclastic material was formed from the activity of the Northern Cones. In order to assess the volume and mass of it erupted in the region of the Great Tolbachik Fissure Eruption (GTFE), we dug several lines of exploration pits both during the eruption of the Northern Cones in August 1975, and again in August–September 1976 for the purpose of refining our estimates of the volume of pyroclastic material from the Northern Breakthrough. Seventy-one exploration pits were dug in all. The profiles from these were used to draw isopachs of the ash in the area of the Northern Breakthrough and six sectors of different thickness were distinguished (Fig. 4).

Sector 1. Area $S_1 = 233$ km² (a calculation of the areas using different methods gave similar results). Average thickness of ash layer $d_1 = \frac{1}{2}(0.1 \pm 0.3) = 0.2$ m $= 0.0002$ km. Volume $V_1 = S_1 \times d_1 = 233 \times 0.0002 = 0.0466$ km³.

Sector 2. Area $S_2 = 59$ km². Average layer thickness $d_2 = \frac{1}{2}(0.3 + 0.5) = 0.4$ m $= 0.0004$ km. Volume $V_2 = S_2 \times d_2 = 59 \times 0.0004 = 0.0236$ km³.

Fig. 3. Plagioclase lapilli.

Sector 3. Area $S_3 = 50$ km². Average thickness $d_3 = \frac{1}{2}(0.5+1.0) =$ 0.75 m = 0.00075 km. Volume $V_3 = S_3 \times d_3 = 50 \times 0.00075 = 0.0375$ km³.

Sector 4. Area $S_4 = 26$ km². Average thickness of ash $d_4 = \frac{1}{2}(1.0+1.5) =$ 1.25 m = 0.00125 km. Volume $V_4 = S_4 \times d_4 = 26 \times 0.00125 = 0.0325$ km³.

Sector 5. Area $S_5 = 18$ km². Average thickness of ash $d_5 = \frac{1}{2}(1.5+2.0) =$ 1.75 m = 0.00175 km. $V_5 = S_5 \times d_5 = 18 \times 0.00175 = 0.0315$ km³.

Sector 6. Area $S_6 = 12$ km². Average thickness of ash $d_6 = \frac{1}{2}(2+8) =$ 5 m = 0.005 km. Volume $V_6 = S_6 \times d_6 = 12 \times 0.005 = 0.06$ km³.

Fig. 4. Diagram of distribution of ash thickness at the Northern Breakthrough.
1, Cones I, II, III; 2, exploration pits; 3, isopachs for ash; 4, lava flows.

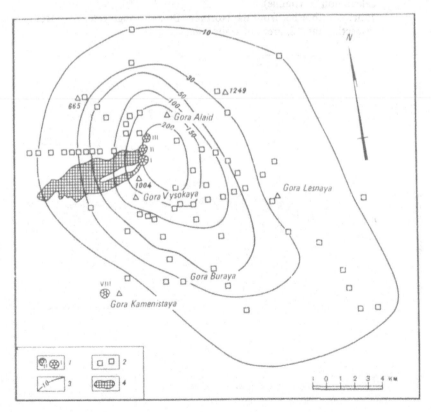

The total area of all six sectors covered with ash to a depth of 10 cm and more (up to 8 m close to the cones in the east and south-east), is $S_I = 398$ km². The total volume of ash that fell on this area is $V_I = 0.2317$ km³.

Several other areas of ash have also been distinguished, beyond the bounds of S_I (Fig. 5).

The study of the distribution of ash thicknesses using profiles (Fig. 6) established that as one moves away from the centre of the eruption the thickness diminishes by a factor of 10 every 10 km. For this reason the area S_{II} can be calculated from the following parameters: at a distance of 10 km from the 10 cm isopach the thickness will be 1 cm and the average thickness for this area $d_{II} = \frac{1}{2}(10+1) = 5.5$ cm $= 0.000055$ km; $S_{II} = 786$ km². $V_{II} = S_{II} \times d_{II} = 786 \times 0.000055 = 0.04323$ km³; S_{III} is calculated using ash samples collected at a distance of up to 100 km from the

Fig. 5. Sketch map showing situation of ash-covered areas (for calculation of volume).
1, areas with ashes from the Northern Breakthrough calculated from inspection pits; 2, areas of white ash.

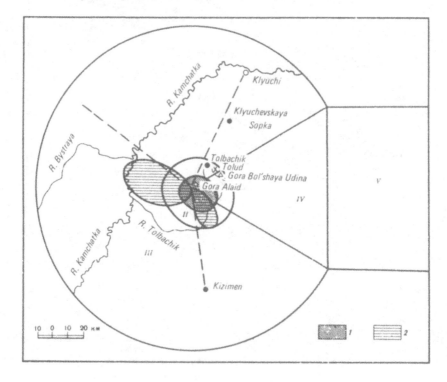

Northern Breakthrough during the eruption: the ash sample from the upper reaches of the Kavavli river ($154 \ g/m^2$) was collected on 2 August 1975 over a period of six hours; that from Klyuchi ($125 \ g/m^2$) on 19 July 1975 over a period of five hours; that from the region of Kizimen volcano ($156 \ g/m^2$) on 17 August 1975 over a period of six hours. In other words, all the samples were collected in the course of six hours and contain about the same amount of material. The average sample amounts to $\frac{1}{3}(154+125+156) = 145 \ g/m^2$ over six hours, and over a 24-hour period $(145 \times 4) = 580 \ g/m^2$, which gives an ash thickness of 0.58 mm (assuming that $1000 \ g/m^2$ gives a thickness of 1 mm). Thus the average thickness $d_{III} = \frac{1}{2}(10 \ mm + 0.58 \ mm = 5.29 \ mm = 0.00000529 \ km$. $S_{III} = \pi R^2 - (S_I + S_{III} + S_{IV}) = 26065 \ km^2$, $V_{III} = S_{III} \times d_{III} = 0.13788 \ km^3$.

The next area, S_{IV}, was calculated using the fact that the main axis of the ash-fall in each discrete segment of time lay only in one direction. Depending on the direction of the wind, this ash-fall axis shifted within a definable sector. The apex of this sector was the Breakthrough, whilst the base was 100 km away, since we have weight samples collected in this

Fig. 6. Decrease in ash thickness (in inspection pits). Western section, compiled 22 August 1975.
Ash beds: 1, dark grey; 2, brown; 3, dirty brown; 4, dark grey; 5, white; 6, dark grey, black.

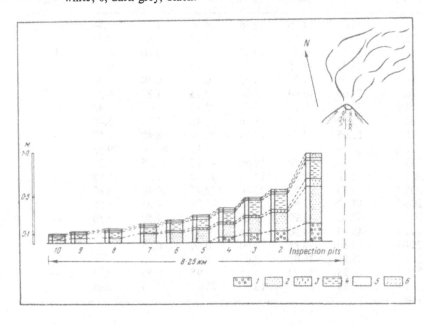

direction. The width of this base was taken as 100 km. This is the area of most intense ash-fall at a distance from the breakthrough of up to 100 km where the thickness of the ash is taken to be 1 mm = 0.1 cm.

The average thickness $d_{IV} = \frac{1}{2}(10 \text{ cm} + 0.1 \text{ cm}) = 5.05 \text{ cm} = 0.0000505$ km.

The area is defined as an equilateral triangle with sides of 100 km, minus part of area S_I and $S_{IV} = 4171 \text{ km}^2$, $V_{IV} = 4171 \times 0.0000505 = 0.21063$ km^3.

Finally, the area S_V is that of the trail (from satellite data), which was up to 1000 km long and about 100 km wide. The thickness of the ash is taken as 1 mm to 0.001 mm, i.e. practically 0. The average density $d_V = \frac{1}{2}(1 \text{ mm} + 0) = 0.5 \text{ mm} = 0.0000005 \text{ km}$, $S_V = 100 \times 1000 = 100000 \text{ km}$, $V_V = 100000 \times 0.0000005 = 0.05 \text{ km}^3$.

The total volume of ash thrown out by the Northern Breakthrough

Table 2. *Granulometric composition of the ash of the Northern Breakthrough*

Sample numbered according to exploration pit	Fraction sizes (mm), amount of ash (%)						
	> 2.0	2.0–1.0	1.0–0.5	0.5–0.25	0.25–0.1	0.1–0.01	< 0.01
14/1	42.05	54.50	12.50	9.55	1.95	—	—
15/1	40.50	39.25	7.90	6.40	1.40	4.55	—
16/1	34.40	35.20	10.85	15.95	0.60	0.55	2.45
17/1	76.75	14.65	3.10	3.50	1.25	0.75	—
18/1	56.05	37.9	4.80	0.75	0.50	—	—
18/1a	79.00	13.90	2.00	2.75	1.85	0.50	—
19/1	79.00	10.20	5.05	2.30	—	3.45	—
20/1	22.50	33.50	16.70	18.40	5.50	3.40	—
21/1	17.65	22.55	17.50	29.30	8.40	4.60	—
202/1	—	—	—	3.15	67.50	29.35	—
203	—	10.80	19.60	43.15	17.25	9.20	—
204	2.10	20.55	14.05	42.70	15.25	.5.35	—
205/1	4.50	27.15	22.40	29.75	10.40	5.80	—
206/1	13.10	33.60	12.35	31.25	4.65	5.05	—
207	37.45	42.90	9.10	6.55	1.50	2.50	—
208/1	45.15	34.30	9.85	9.35	0.55	0.80	—
209/1	57.10	34.85	7.60	0.40	0.05	—	—
210/1	53.85	38.95	7.0	0.20	—	—	—
211/1	57.95	31.70	5.15	3.95	0.50	0.75	—
213/1	16.85	31.3	15.7	25.95	8.35	1.85	—

(Cones I, II and III) during the entire eruption from 6 July to 15 September, i.e. over a period of 72 days, constitutes $V_\Sigma = 0.67344$ km³. Further, one can calculate the volume of pyroclastic material ejected by the Northern Breakthrough by using data from the granulometric analysis of the ashes (Table 2). In Table 3 these samples have been grouped into intervals, depending on how far they were collected from the centre of the eruption.

From Table 3 it is evident that as distance from the cone increases, the quantity of coarse pyroclastic material (> 2.0 mm) decreases sharply, whilst that of silty-psammitic and silty-pelitic material increases. Thus at a distance of 15 km the > 2.0 mm fraction disappears, and at 20 km the 2–0.5 mm fraction disappears. This all takes place on area S_I, where the volume of the ash $V_I = 0.2317$ km³ (calculated from exploration pits). In this area the < 0.01 mm fraction (Table 2) is completely absent and there is hardly any of the 0.1–0.01 mm fraction – i.e. the ash of these fractions was carried beyond the limits of area S_I. This means that at a certain distance from the cones the fraction size goes over entirely to 0.1–0.01 mm (line six in Table 3) and we may assume that here the same volume of ash $V_{VI} = V_I = 0.2317$ km³ ≈ 0.230 km³ falls. Continuing this pattern, which is also clearly visible in the graph (Fig. 7), we can assume that the same volume $V_{VII} = 0.230$ km³ (line seven in Table 3) will be carried to a distance where the size of the fractions will be 0.1–0.01 mm and down to 0. The total volume of the ash is then $V_I + V_{VI} + V_{VIII} = 0.2317 + 0.230 + 0.230 = 0.6917$ km³, which accords with the volume of ash calculated by us from the individual areas ($V_\Sigma = 0.67344$ km³). Then the true

Table 3. *Distribution of ash fractions in relation to distance*

Distance from Cone II (km)	Fraction sizes (mm) (%)				Notes
	> 2.0	2–0.5	0.5–0.1	0.1–0.01 and < 0.01	
0	100	—	—	—	Cones I, II, III
5	75	18	6	1	Over area where
10	20	40	30	10	vol $= 0.2317$ km³
15	0	50	40	10	(from exploration
20	0	0	70	30	pits)
			0	100	$V_6 = V_1$
				0	$V_7 = V_6 = V_1$

volume of the ash will be the average one $\frac{1}{2}(0.6917+0.67344) = 0.68257$ km³ = 0.6826 km³. The volume of the three cones of the Northern Breakthrough, I, II and III (Fedotov *et al.*, 1976) amounts to 0.33 km³, while the overall volume of the pyroclastic material ejected by the Northern Breakthrough of the GTFE will be 1.0126 km³.

In order to determine the weight of the pyroclastic material ejected, the specific gravity of the ash, cinder and lava was calculated. The average specific gravity of the ash is 1.11 t/m³ (from five calculations). The average specific gravity of the cinder is 0.77 t/m³ (from five calculations). The overall volume of the ash (without cones) is 0.6826 km³, of which 80% is of the fine fraction with a specific gravity of 1.11 t/m³ and 20% cinder with a specific gravity of 0.77 t/m³; therefore ash volume = 0.54608 km³ = 546080000 m³, cinder volume = 0.13652 km³ = 136520000 m³. The weight of the ash P_{ash} = 546080000 m³ × 1.1 t/m³ = 600688000 t =

Fig. 7. Granulometric analysis of black ash sampled at various distances to the east of Cone II.
Ash sampled, sample number (distance in kilometres); 19 (2.5); 17/1 (5.0); 16/1 (6.0); 15/1 (6.5); 14/1 (7.5); 20/1 (8.5); 21/1 (9.0); 206/1 (11.5); 205/1 (13.5); 204 (16); 203 (17); 202 (18).

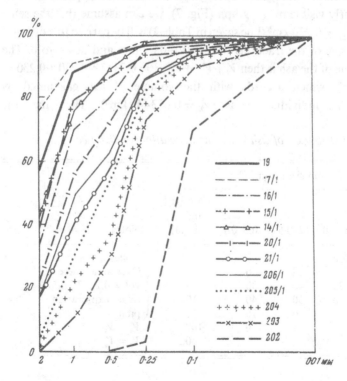

0.6×10^9 t. The weight of the cinder $P_{cinder} = 136\,520\,000$ m$^3 \times 0.77$ t/m$^3 = 105\,120\,400$ t $= 0.105 \times 10^9$ t. The total weight of the pyroclastic material (ash and cinder) $P = 600\,688\,000 + 105\,120\,400$ t $= 705\,808\,400$ t $= 0.706 \times 10^9$ t. The average specific gravity of the lava $= 2.0$ t/m^3 (from 5 determinations). The specific gravity of the material that forms the cones is calculated from the average specific gravities of the lava, cinder and ash and equals 1.29 t/m^3. The weight of the material that forms the cinder cones $P = 327\,000\,000 \times 1.29 = 419\,830\,000$ t $= 0.42 \times 10^9$ t. Thus the basic figures (by weight) are: weight of cinder and ash $P = 0.706 \times 10^9$ t; weight of cone material $P = 0.42 \times 10^9$ t; total weight of pyroclastics $P = 1.126 \times 10^9$ t.

Mineralogical and petrochemical characteristics of the pyroclastics

The pyroclastic material is fairly varied chemically, petrographically, mineralogically and in terms of its textures and structures.

The ashes of the Northern Breakthrough, which are chemically and mineralogically close to its lavas, differ from those of the Southern Breakthrough in SiO_2 content (49–50% for the Northern Breakthrough and 51% for the Southern Breakthrough), and in having high proportions of Mg (almost twice as much) and Ca (3–4% more); the ashes of the Southern Breakthrough are more alkaline (by 1.0–1.5%) (Table 4). Analyses were made of ash samples from the various stages of the eruption, ashes collected at various distances from the volcano, samples of fresh ash collected as it fell and samples from the exploration pits which were already covered by ash from subsequent ejections. The ashes from the exploration pits also differ from those gathered as they fell, in having more SiO_2 (49–50%) and less Fe, Mg and Ca. Aeolian differentiation is hardly noticeable in the chemical composition. The pale-grey powdery ash that fell in the night of 8/9 August is chemically and mineralogically completely different from the rest. Chemically it differs from the black ashes of the Northern Breakthrough in containing more silica (3–4% and up to 5% more), more H_2O^- and H_2O^+, and less Mg (3–4% less) and Ca (2–3% less). Granulometrically, too, the black ashes (see Table 2) differ radically from the pale ones (see Table 1). At a distance of up to 20 km the black juvenile ashes completely lack the pelitic fraction (see Fig. 7); there is a conspicuous decrease in the psephitic fraction from 77% near the Breakthrough to 3–4% at 20 km from it; and the pale ash contains up to 30% of the silty–pelitic fraction and 3–7% of the psephitic (Fig. 8).

Table 4. *Chemical composition of ash from the Great Tolbachik Fissure Eruption*

Component	203	1010	1014	1017	III-3	III-1	III-2
SiO_2	49.54	49.94	49.92	49.90	49.72	50.24	50.32
TiO_2	1.02	0.96	0.97	1.05	1.02	1.05	1.05
Al_2O_3	13.22	12.99	12.50	14.56	13.06	13.45	13.70
Fe_2O_3	3.71	2.93	3.65	5.19	3.81	4.80	3.68
FeO	6.09	6.84	6.06	4.59	5.92	5.69	5.62
MnO	0.16	0.16	0.16	0.16	0.16	0.17	0.17
MgO	10.02	10.70	11.39	8.04	10.02	9.46	9.65
CaO	12.48	12.35	12.41	12.48	12.41	11.42	11.34
Na_2O	2.39	2.28	2.18	2.70	2.60	2.56	2.48
K_2O	1.02	0.93	0.93	1.20	0.98	1.20	1.20
H_2O^-	0.18	0.16	0.12	0.14	0.16	0.04	0.06
H_2O^+	H/trace	H/trace	H/trace	0.26	H/trace	H/trace	0.21
P_2O_5	0.20	0.21	0.20	0.19	0.19	0.28	0.28
Total	100.45	100.45	100.49	100.46	100.05	100.36	99.76

Component	III-5	I	II	III	III-6	1018	252	253
SiO_2	50.20	52.04	52.84	53.06	54.10	51.02	51.00	51.26
TiO_2	1.00	1.00	1.02	0.95	0.96	1.76	1.76	1.76
Al_2O_3	13.65	14.28	14.53	14.36	15.86	15.92	15.48	15.67
Fe_2O_3	3.60	4.06	4.37	4.95	4.01	3.09	4.97	4.75
FeO	5.78	5.00	4.62	4.45	4.14	4.47	6.21	6.67
MnO	0.17	0.12	0.15	0.14	0.17	0.18	0.18	0.16
MgO	9.55	6.68	5.86	6.79	5.87	4.86	5.57	5.54
CaO	11.48	9.66	9.07	8.65	7.55	8.09	7.75	7.87
Na_2O	2.48	3.21	3.21	2.70	2.91	3.86	3.14	3.14
K_2O	1.20	1.20	1.20	1.20	1.42	2.40	2.45	2.34
H_2O^-	0.06	1.22	1.36	1.22	0.72	0.10	0.30	0.16
H_2O^+	0.22	0.52	0.96	0.68	1.50	0.47	0.10	0.09
P_2O_5	0.28	0.31	0.33	0.29	0.26	0.55	0.61	0.50
Total	99.67	99.40	99.54	99.36	99.47	99.77	99.52	99.91

Analysts: G. Novoseletskaya, T. Osetrova, N. Gusakova.
    203, ash from Cone I, sampled 24 July, 4.5 km south of Breakthrough; 1010, ash from Cone II, sampled 31 August, 9.0 km to east; 1014, ash from Cone I, sampled 26 July near Kizimen volcano, 60 km from Breakthrough; 1017, ash from Cone I, sampled 2 August in the headwaters of the Kavavli river, 80 km from Breakthrough; III-3, III-1, III-2, III-5, section of ash from Cone I sampled 19 August from a pit 2.5 km west of Breakthrough; I, II, III, white ash (A, P. Khrenov collection) sampled 9 August; III-6, white ash sampled from pit, 19 August; 1018, ash from Southern Breakthrough, sampled 6 December (Petrov collection); 252, 253, ash from southern Breakthrough, sampled 28 and 30 December.

An examination of heavy concentrates of the white and black ash under a binocular microscope yielded some interesting results. The ash of Cones I and II (black) consists mainly of opaque, dark glass with a negligible quantity of brown-green glass, while the ash from the eruptions of the last days of August (white) is mainly composed of melted, acute-angled fragments of green distended glass. As well as the glass, one finds single crystal clasts of plagioclase, pyroxenes and olivine in the ashes. Mineralogically, the pale ash differs markedly from those under and overlying it: it has almost no juvenile material such as volcanic glass (in all fractions), the magnetic fraction is represented by magnetite and fragments of effusive rocks, and in the electromagnetic fraction fragments of effusive rocks constitute 90%, pyroxenes 3–5%, plagioclase 1–2%, olivine 1–2%, altered fragments 1% and sulphides single %. In the non-magnetic fraction the principal minerals are quartz, plagioclase, pyroxene and single instances of pyrite.

Thus, in terms of composition and shape (poorly rounded, rounded) these fragments are mainly resurgent material. The ash samples from the exploration pits contain the same minerals and an insignificant quantity

Fig. 8. Granulometric analysis of white ash.
Ash sampled (sample number, date, distance): 216, 9 August; Sh-6, 19 August, pit 2.5 km from the cone; Sh-7/3, 21 August, pit 6 km from the cone.

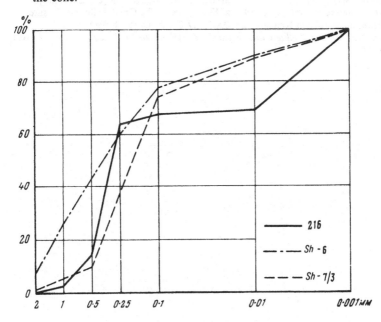

of volcanic glass fragments, the presence of which may have two causes: (1) this is glass from under or overlying ash layers – ash sample III 7/3 was taken from a layer 10 mm thick and, naturally, as it was being removed it may have been contaminated by material from other layers; (2) it could also be juvenile material ejected simultaneously with the resurgent material; (3) during the storm accompanying the ejection of the pale ash, light resurgent ash must certainly have become mixed with black juvenile ash.

The ejection of the resurgent material may be explained by the following. Before the pale ash was emitted, considerable pauses of up to nine hours had been occurring in the activity of Cone I, during which geodynamic data recorded tilts in the Earth's surface. Consequently, during the pauses the sides of the conduit may have collapsed as a result of tectonic movement, and their rocks been abraded partly as they collapsed and partly during the tremendous gas blast which accompanied the eruption of the white ash.

Thus the suggestion made earlier on the basis of chemical analysis alone, that the white ash was transitional between the eruption of the Northern Breakthrough and that of the Southern, is not quite correct, especially as after the ejection of the white ash, during the formation of Cones II, III and IV, only juvenile material was ejected, consisting of volcanic glass of a brown-green colour.

References

Fedotov, S. A., Khrenov, A. P. & Chirkov, A. M. 1976. Bol'shoye treshchinnoye Tolbachinskoye izverzheniye 1975 g. Kamchatka. (The Great Tolbachik Fissure Eruption of 1975 in Kamchatka.) *Dokl. AN SSSR*, **228**(5), 1193–5.
Rittmann, A. 1964. *Vulkany i ikh deyatel'nost'*. (*Volcanoes and their activity*.) M., Mir, 400 pp.

Features of pyroclastics of the Northern Breakthrough of the Great Tolbachik Fissure Eruption and the origin of its pale-grey ash

Ye. F. Maleyev and Yu. V. Vande-Kirkov

The eruption of the new volcanoes along the Great Fissure Breakthrough of Ploskiy Tolbachik volcano, which began on 6 July 1975 (Fedotov *et al.*, 1976b) produced a variety of pyroclastic material. A large quantity of ash, cinders and bombs was ejected onto the surface independently of or simultaneously with the effusion of lava. Furthermore, judging from the cumulative curve of the mass of lava material from the Northern Breakthrough (Fig. 1), the molten rock flowed to the surface at a practically constant rate averaging $0.02–0.025 \times 10^9$ t/24 h. From the relationship between the explosive products and the lavas it is possible to divide the eruption of the Northern Breakthrough into four periods.

1. Explosive; from 6 to 27 July, explosions of Cone I; explosive index $E = 100\%$ (weight).

2. Effusive/Explosive; from 27 July to 12 August, explosions of Cone I, 1st and 2nd lava flows; $E = 75–90\%$.

3. Explosive/Effusive; from 12 to 26 August, explosions of Cones II and V, 3rd and 11th lava flows; $E = 20–40\%$.

4. Effusive/Explosive; from 26 August to 15 September, explosions of Cone II, 5th, 6th, 12th and 15th lava flows; $E = 80\%$.

During the initial stages of eruption, of Plinian and Strombolian–Plinian types, Cone I ejected a continuous stream of pyroclastic material to a height of 8–11 km and dispersed fine ash for a distance of up to 500 km. The discharge of incandescent fragmentary juvenile material looked like the blast from a jet engine. Then eruptions of Strombolian–Vulcanian type began to take place. The height of the eruptive cloud gradually came down to 3–5 km, and at the end of July and beginning of August to 2–3 km. In the afternoon of 8 August Strombolian eruption ceased and at 18.50 hours a Vulcanian eruption began, with the ejection of fine ash. At the beginning

57

of the Vulcanian stage pale-grey, then black and subsequently dark-grey ash was thrown out. This concluded the eruption of Cone ɪ. The subsequent vents – Cones ɪɪ and ɪɪɪ – which formed in August, were characterised by lesser eruptions with an eruptive cloud extending mainly to a height of 2–3 km, although the domes grew at approximately the same rate. One of the authors witnessed the formation of the vent on the site of Cone ɪɪ 400 m north of Cone ɪ. The eruption began with a jet of white gas, then extraneous black material was thrown up to a height of 100–200 m, followed by juvenile material.

Strombolian and Hawaiian types of eruption are characteristic of the Southern Breakthrough, with pyroclastics being ejected to a height initially of several hundred metres and subsequently of a few tens of metres.

The force of the explosions influenced the process of aerial differentiation

Fig. 1. Cumulative curve and discharge of eruptive material from the Northern Breakthrough of the Great Tolbachik Fissure Eruption. The arrowed signs mark the time of appearance in the eruption of light-grey ash (square+arrow), blue gases (small circle+arrow); *A, B, C, D* and *E* represent lava not forming lava flows; 1–15 denote the period of effusion of individual lava flows; ɪ, ɪɪ, ɪɪɪ, ɪv, v denote periods of activity of individual cinder cones.
The cumulative curve has been constructed from instrument observations of the increments in time of the volume of the individual cinder cones (Fedotov *et al.*, 1976a), from calculation of the total volume of pyroclastics from the Northern Breakthrough (Fedotov *et al.*, 1976b), and calculations of volume of individual lava flows.

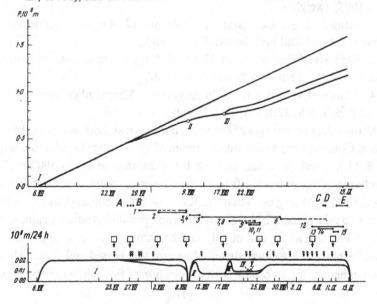

of the material. During the Strombolian eruptions hardly any pelitic and silty material was laid down near the cones, whilst 100–200 km from the Northern Breakthrough at the settlement of Klyuchi, in the lower reaches of River Kamchatka (Kamaki) and on Lake Azhabach'ye, primarily silty material was found.

Whilst Cone ɪ was being formed, coarsely-fragmented material 2–3 m and more in diameter was ejected in an incandescent state to a height of 1–2 km. Falling onto the cone, it smashed to pieces and rolled to the foot. Occasionally, during low, lateral explosions, the incandescent fragmented material would become concentrated in particular parts of the cone, and then form glowing avalanches descending from the top of the cone to the foot. The bright-red and orange colour of this material permits the assumption that agglutinates were formed. The fragments and blocks composing the cone were most commonly 0.2–1.0 m in size. The blocks were predominantly angular, occasionally rounded. The viscosity of the lavas meant that sculptured structures of 'bread crust' type were formed, the lavas were of low porosity, hummocky and sometimes had other material fused to them. Shaped bombs were virtually absent. The coarsely-fragmented material was mainly composed of non-porous and massive lava. The outside of the bombs and blocks was observed to be more coarsely porous, whilst their inner part was composed of dense, finely-porous or less porous lava with pores of an irregular, usually flat shape.

The fragmentary material of Cone ɪɪ consisted of smaller fragments of greater plasticity. During the eruption bubbles of lava probably formed and burst in the crater, leading to the formation and ejection of a considerable quantity of thin lava flakes, of which those 5–20 cm in diameter and 1–3 mm or more thick were the most common. Occasionally they would be as much as 30 cm in diameter and 1.5–2.5 mm thick. The edges of the flakes are usually torn, their surface deformed and the porosity slight. Among the pyroclastic products one can observe a large number of typical basalt 'shaped' bombs (spherical and spindle-shaped), as well as complexly rounded lumps of basalt lava.

The pyroclastic material of Cone ɪɪɪ consists of a large quantity of highly-porous scoriae, fragments of which have torn, uneven surfaces and acutely-angular edges. The pores are usually spherical, and the density of the scoriae is 0.6–0.9 g/cm^3. 'Shaped' bombs were also found side by side with the scoriae.

The pyroclastics of the Southern Breakthrough were even more porous. Some cinder fragments floated down after ejection like leaves from the trees. Individual crystals of plagioclase more than 2 cm in diameter were also ejected.

Around the volcanoes of the Northern Breakthrough cinder gravel and coarse ash fell within a radius of 3–4 km, as a result of which a composite layer of tephra formed around the cones. The thickness of the pyroclastics 1–2 km from the centre of the eruption was as much as 12 m reducing to 5 and 3 m; at 3–4 km it was 0.5–10 m, and at 10 km a few decimetres.

As an example one may cite the succession of ash and gravel that formed during the eruption of Cone I over the period 31 July to 9 August 1975 3.5 km to the south-west of the cone. In its upper part lies an horizon of fine ash (top downwards):

1. Black volcanic sand with an admixture of silty-pelitic material compacted by falling rain. Thickness 2 cm (specimen 21).

2. Dark-grey fine-grained volcanic sand with an admixture of silty-pelitic material. The layer contains two seams of pale-grey silty-pelitic ash 2–3 mm thick. Total thickness 7 cm (specimen 21a).

3. Black fine-grained sand with a large quantity of silty-pelitic material. Thickness 5 cm (specimen 21b).

4. Pale-grey volcanic sand with an admixture of pelitic material. Thickness 4 cm (specimen 21c).

Total thickness of the fine ash 5–20 cm.

Beneath the fine ash rests cinder. In it one may distinguish three layers (top downwards):

1. Ash with gravel material. Thickness 45 cm (specimen 31/1).

2. Reddish-black ash consisting of three seams: a lower, finely psammitic; a middle, coarsely psammitic with an admixture of gravel material; and an upper, psammitic. Thickness 7 cm (specimen 31/2).

3. Ash with an admixture of gravel material forming a layered unit in which the constituent layers are distinguished by the varying coarseness of their material. Thickness 26 cm (specimen 31/3).

The total thickness of the coarse pyroclastic material (9 August, 3.5 km south-west of Cone I) was 78 cm.

Ash and fragments of gravel material measuring up to 2 cm were present in the black–grey groundmass. The fragments were rounded and angular–indented with a lumpy, lacerated surface. The fragmentary material less than 2 mm was more angular.

Analyses of the granulometric composition of the ash (see Table 1) have shown that before the appearance of the pale-grey ash (in the period 31 July to 8 August) and after its appearance in the closing stages of

Table 1. *Granulometric composition of ash from the Northern Breakthrough of the Great Tolbachik Fissure Eruption, 1975*

Name of material	Distance from eruption centre (km)	No. of sample	2.0 wt (g)	2.0 %	2–1 wt (g)	2–1 %	1–0.5 wt (g)	1–0.5 %	0.5–0.25 wt (g)	0.5–0.25 %	0.25–0.1 wt (g)	0.25–0.1 %	0.1–0.01 wt (g)	0.1–0.01 %	0.01 wt (g)	0.01 %
Cinder, Cone I, 1–8 Aug. 1975, basal layer		31/3	7.48	37.40	8.62	43.10	2.10	10.50	1.27	6.35	0.26	1.30	0.27	1.35	—	—
Cinder, Cone I, 1–8 Aug. 1975, middle layer	3.0	31/2	9.50	47.50	7.97	39.85	1.73	8.65	0.80	4.00	—	—	—	—	—	—
Cinder, Cone I, 1–8 Aug. 1975, upper layer		31/1	9.71	39.55	7.91	39.55	1.82	8.10	0.54	2.70	0.22	1.10	—	—	—	—
Ash, Cone I, 8–9 Aug. 1975, undifferentiated		31/4	6.30	31.50	8.89	44.45	1.90	9.50	0.40	2.00	0.35	1.75	0.37	1.85	1.79	8.95
Ash, coarse with grit, underlying 21c, Cone I		21d	10.86	54.30	7.27	36.35	1.20	6.00	0.54	2.70	0.13	0.65	—	—	—	—
Ash, light-grey, Cone I, 8 Aug. 1975	3.5	21c	—	—	0.31	1.55	1.38	6.90	11.65	58.25	1.52	7.60	0.68	3.40	4.46	22.30
Ash, black, Cone I, 8 Aug. 1975		21b	—	—	0.89	4.45	1.20	6.00	4.15	20.75	4.55	22.75	5.34	26.70	3.87	19.35
Ash, grey–black, Cone I, 9 Aug. 1975, lower part of layer		21a	—	—	0.51	2.55	1.10	5.50	6.90	34.50	5.65	28.25	3.56	17.80	2.28	11.40
Ash, grey, Cone I, 9 Aug. 1975, upper part of layer		21	0.54	2.70	1.69	8.45	1.20	6.00	7.08	35.40	4.26	21.30	3.06	15.30	2.17	10.85
Ash, light-grey (an analogue of 21c), Cone I, 8 Aug. 1975, almost 9 cm thick	0.8–1.0	23a	4.25	21.25	7.80	39.00	2.31	11.55	0.13	0.65	0.17	0.85	0.52	2.60	4.82	24.10
Ash, last portion from Cone I, 9 Aug. 1975	2	23	9.90	49.50	6.10	30.50	2.25	11.25	1.43	7.15	0.32	1.60	—	—	—	—
Ash, undifferentiated, Cone I, taken from tent 8 Aug. 1975		20	—	—	0.80	4.00	0.73	3.65	5.10	25.50	6.10	30.50	4.49	22.45	2.78	13.90
Ash, light-grey, Cone I, 8–9 Aug. 1975, 9 cm thick	1.5	28	5.97	29.85	8.31	41.55	2.58	12.90	0.68	3.40	0.17	0.85	0.50	2.50	1.79	8.95
Ash, coarse, Cone I, 8–9 Aug. 1975	2.0	32	4.80	24.00	9.63	48.15	3.58	17.90	1.68	8.40	0.17	0.85	0.14	0.70	—	—
Ash, Cone I, 8–9 Aug. 1975		33	—	—	0.76	3.80	0.78	3.90	2.56	12.80	5.29	26.45	6.70	33.50	3.91	19.55
Ash, fine, Cone I, taken from bushes 8–9 Aug. 1975	1.2	35	—	—	1.30	6.50	0.63	3.15	2.05	10.25	4.40	22.00	8.30	41.50	3.32	16.60
Cinder, Cone II	0.8	34	12.21	61.05	7.37	36.85	0.35	1.75	0.07	0.35	—	—	—	—	—	—
Ash, Cone II	3	30	—	—	0.98	4.90	2.17	10.85	12.55	62.75	3.88	19.40	0.25	1.25	0.17	0.85

Fig. 2. Ashes from the Northern Breakthrough of the Great Tolbachik Fissure Eruption. Numbers by the histograms correspond to sample numbers.

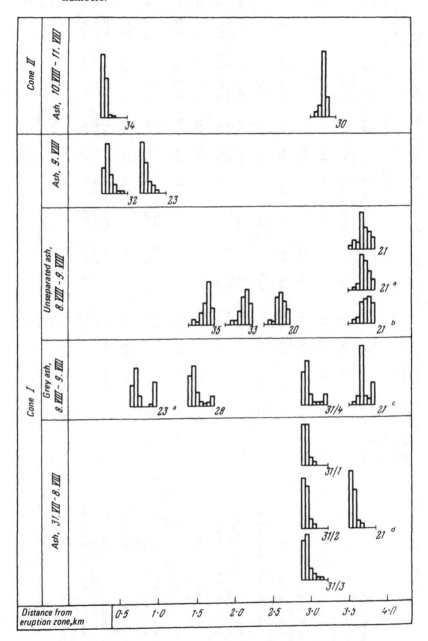

Cone I's activity (9 August), the material exploded from Cone I as black ash had a more or less homogeneous granulometric composition at a distance of up to 3.5–4.0 km, with an average degree of sorting (Fig. 2, specimens 31/1–31/3, 21d). The dimensions of the material are commensurate with psephitic – coarsely-psammitic varieties. The finely-psammitic, silty fraction comprises only a very few %, while the pelitic fraction is completely absent. Histograms of the ash of the first instalments of Cone II also demonstrate medium–good sorting of the ash material independent of its size (Fig. 2 specimens 30, 34).

In clean, sterile samples the pale-grey ash that fell on 8–9 August differed sharply in granulometric composition from the black ashes, i.e. from the entire mass of ash thrown out by Cone I. There was considerably more pelitic, silty, and finely-psammitic material in it. Moreover, the histograms each have two clear peaks, which indicates the presence of material from different sources. Somewhat less revealing in this respect are the histograms of samples of the unseparated ash and of the ash formed during the concluding part of Cone I's activity (8–9 August); but even these reveal the extraordinarily bad sorting of the material, confirming that it came from different sources.

An examination of ash of different fractions in thin sections and special preparations has shown that the black ashes of the Northern Breakthrough, like the lavas of the Northern Breakthrough, are composed of olivine–pyroxene basalts or fragments of them, represented by glass from the groundmass and crystals of olivine and pyroxene, in some cases of plagioclase. Lithic clasts comprise up to 80% of the total volume of ash and are rounded or ragged in appearance. The olivine–pyroxene basalts are characterised by seriate–porphyritic textures, and are compact or porous. As a rule the groundmass has a hyaline, hyalopilitic and more rarely an intersertal texture. The glass of the groundmass is usually saturated with microlites, micronolites of plagioclase, pyroxene and olivine, and with ore dust. The glass exists either as a light-brown transparent or a dark-brown, almost black, and opaque variety.

The pale-grey ash, apart from its olivine–pyroxene basalts, consists of 50–60% angular clasts of a plagioclase rock, and its fragments – groundmass glass in various states of devitrification, and plagioclase crystals. Fragments of this rock are distinguished by their porphyritic plagioclase crystals in the form of phenocrysts and subphenocrysts, with a markedly subordinate amount of pyroxene and olivine phenocrysts. The groundmass has a hyalopilitic, intersertal texture composed of laths of plagioclase, olivine and pyroxene microlites, and ore material. The glass of the groundmass is oxidised to various degrees; generally opaque brown, dark-brown, and more rarely transparent light-brown varieties are present. It is curious that the plagioclase rock easily divides along the glass–crystal contact and therefore the fragments are characteristically angular. The quantity of plagioclase rock fragments in the large psammite fractions reaches 30–40%, in the fine psammitic 60–65% (Table 2), and in the silty and pelitic fractions, judging from visual estimates, not less than 60–80%.

Granular material washed out of the pale-grey ashes shows up to 15–16% prismatic and angular single grains of amphibole and angular, angular–rounded grains of quartz (Table 3).

The pale-grey ashes differ radically in their chemical composition from the black ashes (Table 4). The black ashes, like most of the lava and tephra of the Northern Breakthrough, have the composition of high-magnesium basalts (Volynets *et al.*, 1976), with increased calcium (CaO) content, moderate alkalis ($Na_2O + K_2O$), and low aluminium (Al_2O_3). Thus the chemical composition of the black ash reflects the true chemical composition

Table 2. *Ratio of juvenile and admixed material in individual fractions of the grey ash from Cones* I *and* II *of the Northern Breakthrough*

| | Fraction size (mm) (%) | | | | | | | |
| | 2–1 | | 1.0–0.5 | | 0.5–0.25 | | 0.25–0.1 | |
No. of sample	juvenile (%)	extraneous (%)	juvenile (%)	extraneous (%)	juvenile (%)	extraneous (%)	juvenile (%)	extraneous (%)
21c	70	30	—	—	64	36	41	59
23a	66	34	—	—	—	—	—	—
33/1	60	40	—	—	—	—	36	64
21	74	26	—	—	68	32	—	—
21a	—	—	—	—	74	26	—	—
31/4	92	8	—	—	—	—	—	—
20	—	—	—	—	78	22	—	—
35	—	—	—	—	—	—	75	25
Cone II, X*			96	4	76	24	72	28

Determinations carried out on MIU-1 equipment; rocks numbered in the first column are described in Tables 1 and 4.
* A. P. Khrenov's specimen.

Table 3. *Mineralogical analysis of the pyroclastic products of the Northern Breakthrough of the new Tolbachik volcanoes*

Mineral concentrate	Total weight (g)	N/el. fraction	Electro-magnetic fraction	Magnetic fraction	Light fraction	Minerals			
						Apatite	Leucoxene	Pyrite	Marcasite
31/4–75	3.32	Grains	1.10	1.12	1.10	Single grains	—	Grains	Grains
11–75	2.50	Single grains	0.85	0.04	1.61	—	—	Single grains	—
23a–75	5.28	Grains	0.86	2.25	2.17	—	Single grains	Grains	Grains
21a–75	9.45	Grains	2.49	1.19	5.77	—	—	Grains	Single grains
30–75	8.80	—	2.0	0.6	6.2	—	—	—	—
Ts–75	7.57	Grains	0.50	1.04	6.03	—	—	Single grains	Single grains
31–75	4.68	Grains	1.28	0.35	3.05	—	—	Single grains	—

Table 3 *continued*

Mineral concentrate	Minerals								Pyroclastic formations
	Amphiboles	Ilmenite	Olivine	Pyroxenes	Epidote	Magnetite	Quartz	Feldspars	
31/4–75	Grains	Single grains	Single grains	38.3	—	10.1	15.0	Single grains	36.5
11–75	—	—	—	54.3	—	Grains	Single grains	Single grains	45.6
23a–75	Single grains	—	Single grains	12.3	Single grains	0.2	4.0	Single grains	83.6
21a–75	Single grains	Single grains	Single grains	19.7	—	1.3	16.3	Single grains	62.6
30–75	—	—	—	78.4	—	0.1	Single grains	—	21.4
Ts–75	Single grains	Single grains	Single grains	11.5	Single grains	Single grains	—	—	88.4
31–75	—	—	Single grains	33.7	—	—	—	—	66.2

Table 4. Chemical analyses of the pyroclastic products from the Northern Breakthrough of the new Tolbachik volcanoes

Name of material	No. of sample	SiO_2	TiO_2	Al_2O_3	Fe_2O_3	FeO	MnO	MgO	CaO	Na_2O	K_2O	H_2O^-	H_2O^+	P_2O_5	Total	Analyst
Cinder, 28 April 1975, summit crater, Ploskiy Tolbachik, Cone I	1	51.02	2.26	14.64	3.96	8.30	0.18	4.63	8.35	3.47	2.29	0.22	None	0.43	99.75	G. P. Novoseletskaya
Cinder 1-8 Aug., Cone I, lower part of layer	31/3	50.02	1.10	13.50	4.43	5.99	0.17	9.26	12.04	2.39	0.64	0.12	H/trace	0.40	100.06	L. A. Kartasheva
Cinder 1-8 Aug., Cone I, middle layer	31/2	49.74	1.10	13.70	4.34	5.76	0.17	9.18	12.16	2.49	0.84	0.12	H/trace	0.39	99.99	L. A. Kartasheva
Cinder 1-8 Aug., Cone I, upper layer	31/1	49.70	1.10	13.70	4.42	5.96	0.18	9.30	12.16	2.49	0.84	H/trace	H/trace	0.38	100.23	L. A. Kartasheva
Lava, south bocca, Cone I, last of flow No. 1	925–YuK	49.50	1.01	13.18	3.18	6.85	0.18	9.98	12.34	2.18	0.93	0.29	0.00	0.25	99.87	L. A. Kartasheva
Lava, north bocca, Cone I	27	49.74	1.00	13.72	6.04	4.76	0.33	9.62	11.44	2.28	0.88	None	0.18	0.18	100.17	L. G. Baytsayeva
Ash, light-grey, 8 Aug. 1975, Cone I	21c	53.16	0.98	15.52	4.38	4.31	0.29	7.07	8.23	2.96	1.20	0.80	0.70	0.12	99.72	L. G. Baytsayeva
Ash, black, 8 Aug. 1975	21b	53.08	0.98	16.17	3.90	4.31	0.29	6.96	7.89	3.21	1.20	0.94	0.63	0.12	99.66	L. G. Baytsayeva
Ash, grey-black, 9 Aug. 1975, lower part of layer	21a	52.42	0.98	15.88	4.12	5.02	0.31	7.68	7.71	3.21	1.20	0.72	0.32	0.16	99.73	L. G. Baytsayeva
Ash, grey, 9 Aug. 1975, upper part of layer	21	52.02	0.98	15.89	4.07	5.11	0.30	7.47	8.08	3.34	1.20	0.86	0.31	0.11	99.76	L. G. Batysayeva
Ash 8–9 Aug. 1975, total layer	31/4	53.32	1.00	18.63	2.38	4.35	0.15	4.85	8.85	3.47	1.26	0.70	0.97	0.26	100.19	L. A. Kartasheva
Coarse ash, 9 Aug. 1975, Cone II	30	50.24	1.10	13.25	3.16	6.66	0.17	9.83	12.04	2.49	0.84	0.08	H/trace	0.35	100.22	L. A. Kartasheva
Cinder and ash, 9 Aug. 1975, Cone II	25	49.42	1.00	14.21	6.44	4.31	0.24	9.86	11.28	2.28	0.88	None	0.30	0.12	100.34	L. G. Baytsayeva

Sample	No.														Total	Analyst
Bomb, 10 Aug. 1975, Cone II	29	49.58	1.00	14.22	5.99	4.36	0.30	9.86	11.28	2.60	0.84	0.10	0.02	0.12	100.27	L. G. Baytsayeva
Lava, Flow No. 7, 20 Aug. 1975	919-YuK	49.36	0.95	13.30	3.79	6.33	0.15	9.75	12.40	2.34	1.04	0.12	0.12	0.28	99.95	L. A. Kartasheva
Lava, Flow No. 8, 23 Aug. 1975	918-YuK	50.04	0.83	13.58	7.02	2.89	0.18	10.56	11.06	2.49	1.02	—	0.08	0.16	99.91	A. M. Okrugina
Lava, Flow No. 6, 1 Sept. 1975	926-YuK	49.52	1.02	13.18	2.87	7.08	0.17	9.99	12.22	2.39	0.96	0.27	0.00	0.21	99.98	L. A. Kartasheva
Lava, Southern Break-through, 24 Aug. 1975	930-YuK	50.76	1.31	16.77	3.30	6.75	0.16	6.60	8.76	3.47	1.85	—	0.19	0.52	100.44	A. M. Okrugina
Bomb, Southern Break-through, 27 April 1975	9	51.18	1.56	16.98	3.48	6.67	0.15	4.81	9.10	3.47	1.91	0.36	None	0.49	100.16	T. V. Dolgova
Lava, Southern Break-through, 5 Nov. 1975	929-YuK	51.46	1.55	16.99	4.46	5.82	0.16	4.91	8.65	3.72	2.02	—	0.11	0.57	100.42	A. M. Okrugina

Sample 1 (6074) collected by A. I. Farberov; Sample 9 (3872) by V. A. Yermakov; Sample 930-YuK by Ye. K. Markhinin; Sample 920-YuK by A. L. Ivanov; Samples 925-YuK, 919-YuK, 918-YuK, 926-YuK, by Y. V. Vande-Kirkov; remainder by Ye. F. Maleyev.

of juvenile products of the Northern Breakthrough eruption. The pale-grey ash is characterised by high aluminium content (Al_2O_3), average alkali ($Na_2O + K_2O$) and magnesium (MgO), moderate titanium (TiO_2) and reduced calcium (CaO) and iron ($FeO + Fe_2O_3$). Moreover, the pale-grey ash has a somewhat increased quantity of silica (SiO_2). Chemically, the pale-grey ash is fairly close to the aluminous subalkaline basalts (Volynets *et al.*, 1976), which in the form of megaplagiophyre lavas make up the entire bulk of the volcanic products of the Southern Breakthrough. But the amount of silica in the pale-grey ash is somewhat higher (Fig. 3) and that of iron and calcium noticeably lower, than in the olivine–pyroxene basalts, let alone the megaplagiophyric ones. On the other hand, the quantities of aluminium, magnesium and alkalis in the pale-grey ash are comparable with those in the megaplagiophyre basalts. Thus, even if the presence of

Fig. 3. Variation in chemical composition of the products of the Great Tolbachik Fissure Eruption.
1, olivine-pyroxene basalts of the Northern Breakthrough; 2, light-grey ash of the Northern Breakthrough; 3, megaplagiophyre basalts of the Southern Breakthrough; 4, admixture material in light-grey ash of the Northern Breakthrough.

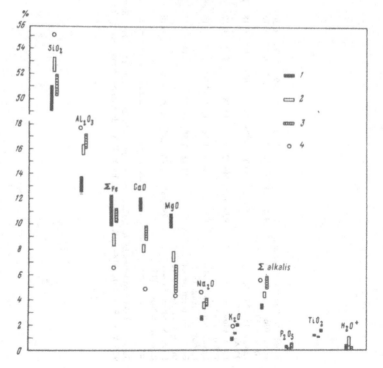

excess silica and the deficiency of iron and calcium can be explained, one still cannot ignore the considerable, if not total, predominance in the pale-grey ash of material corresponding chemically to the megaplagiophyre basalts of the Southern Breakthrough. This clearly does not agree with the factual data: the pale-grey ash contains not less than 40–50% juvenile material characteristic of the products of the Northern Breakthrough. What kind of lava material, then, could be mixed with the juvenile black ash to produce the pale-grey ash? The composition of this admixed material has been calculated by us as follows, on the basis of the known chemical composition of the pale-grey ash and the juvenile material of the pale-grey ash, taking into account the quantitative ratios in the pale-grey ash of juvenile material and admixed material (including free quartz revealed in concentrates): SiO_2, 56–58%; TiO_2, 1.0%; Al_2O_3, 17–18%; ΣFe, 6.0%; MnO, 0.3%; MgO, 4–5%; CaO, 5.0%; Na_2O, 4.0%; K_2O, 1.6% (see Fig. 3). It is interesting that andesite basalts with such ratios of rock-forming elements occur at the base of the Klyuchi group of volcanoes in the form of megaplagiophyre lavas (Piyp, 1956; *Petrokhimiya... (Petrochemistry...*), 1966). Furthermore, analogous complexes of megaplagiophyre lavas form part of the Pliocene basement of the Klyuchi group (see Shantser, this volume).

Detailed petrographic studies have shown that the fragments of plagioclase rock may indeed correspond in composition to andesite basalts. Thus, judging from refractive indices (Fig. 4), the groundmass glass of these fragments is considerably more acidic than that of the groundmass of juvenile products of the Southern and, particularly, the Northern Breakthroughs. It contains 55–61% SiO_2, which according to I. I. Gushchenko (1965), is characteristic of natural glasses of ashes from the early stages of eruption of the Tolbachik volcano.

The plagioclase of the admixed lava material has a wider field of composition than the plagioclase of juvenile products of the Northern and, particularly, the Southern Breakthrough (Fig. 5). Moreover, the qualitative characteristics of the plagioclase, i.e. the dependence of the structural state of the plagioclase on its composition, have features in common with andesite basalt and andesitic rock varieties.*

On the basis of the above facts, then, it is perhaps possible to speak of the presence in the pale-grey ash of extraneous material which differs from the juvenile products of the eurption and was brought up to the surface

* The degree of order in the plagioclase is taken from Yu. V. Vande-Kirkov's graphs (1974).

from deep horizons together with the juvenile material during the course of Vulcanian eruption.

Thus the pale-grey ash is finely-fragmented explosive material, of mainly finely-psammitic, silty and pelitic dimensions, consisting of juvenile material from the Northern Breakthrough probably admixed with extraneous material that is non-cognate with this eruption.

The pale-grey ash appeared repeatedly and regularly during the eruption of the Northern Breakthrough (see Fig. 1): it was ejected at the end of the activity of the explosion cones before the formation of new vents, as well as before the renewal of the explosive activity of Cone II and before the effusion onto the surface of new flows and certain lavas. Since this pattern

Fig. 4. Refractive index of natural glass in the products of the Great Tolbachik Fissure Eruption and Ploskiy Tolbachik volcano.
1, olivine–pyroxene basalt of the Northern Breakthrough; 2, olivine–pyroxene basalt of the black ash of the Northern Breakthrough; 3, megaplagiophyre basalts of the Southern Breakthrough; 4, admixture material in light-grey ash of the Northern Breakthrough.
A, ash from the early stages of the Ploskiy Tolbachik eruption; B, ash from the late stages of the Ploskiy Tolbachik eruption; C, ash from the young cones. (A, B, C are after Gushchenko (1965).)

Fig. 5. Composition fields of plagioclase from the lavas and ash of the Great Tolbachik Fissure Eruption.
1, olivine–pyroxene basalts of the Northern Breakthrough; 2, megaplagiophyre basalts of the Southern Breakthrough; 3, admixture material in light-grey ash of the Northern Breakthrough.

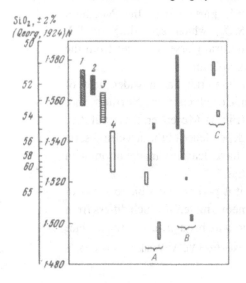 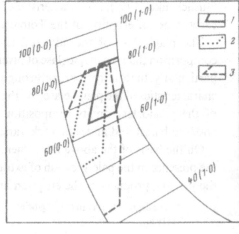

is broken at the end of the activity of Cone II, i.e. at the end of the activity of the entire Northern Breakthrough, the eruption of the Southern Breakthrough should probably be considered independently of the eruption of the former.

References

Fedotov, S. A., Enman, V. B., Magus'kin, M. A., Levin, V. Ye. & Zharinov, N. A. 1976a. Vnedreniye bazal'tov i obrazovaniye pitayushchikh treshchin Bol'shogo Tolbachinskogo izverzheniya 1975 g. po geologicheskim dannym. (The injection of basalts and the formation of feeder-fissures during the Great Tolbachik Eruption of 1975, from geological data.) *Dokl. AN SSSR*, **229**(1), 170–3.
Fedotov, S. A., Khrenov, A. P. & Chirkov, A. M. 1976b. Bol'shoye treshchinnoye Tolbachinskoye izverzheniye 1975 g., Kamchatka. (The Great Tolbachik Fissure Eruption of 1975, Kamchatka.) *Dokl. AN SSSR*, **228**(5), 1193–6.
Gushchenko, I. I. 1965. *Peply Severnoy Kamchatki i usloviya ikh obrazovaniya. (The ashes of Northern Kamchatka and the circumstances of their formation.)* M., Nauka, 144 pp.
Petrokhimiya kaynozoyskoy Kurilo–Kamchatskoy vulkanicheskoy provintsii. (The petrochemistry of the Cenozoic Kuril–Kamchatka volcanic province.) 1966. M., Nauka, 279 pp.
Piyp, V. I. 1956. *Klyuchevskaya sopka i yeye izverzheniya v 1944–1945 gg. i v proshlom. (Klyuchevskaya Sopka and its eruptions of 1944–45 and in the past.)* M., Izd-vo AN SSSR, 312 pp.
Vande-Kirkov, Yu. V. 1974. Novyy variant diagrammy dlya opredeleniya srednikh-osnovnykh plagioklazov na Fedorovskom stolike. (A new variation of a diagram for classifying intermediate and basic plagioclases on the Fedorov universal stage.) *Byul. vulkanol. stantsiy*, No. 50, 130–2.
Volynets, O. N., Yermakov, V. A., Kirsanov, I. T. & Dubik, Yu. M. 1976. Petrokhimicheskiye tipy chetvertichnykh bazal'tov Kamchatki i ikh geologicheskoye polozheniye. (Petrochemical types of Quaternary basalts of Kamchatka and their geological position.) *Byul. vulkanol. stantsiy*, No. 52, 115–26.

Basement xenoliths in the eruption products of the New Tolbachik Volcanoes and the problem of the formation of magma conduits in the upper crust

A. Ye. Shantser

During the period when Cones I and II of the Northern Breakthrough of the New Tolbachik Volcanoes were active, large quantities of xenoliths of sedimentary, volcanogenic–sedimentary and magmatic rocks of the basement were discovered in the products of eruption. Most frequently the xenoliths are associated with ejections of pyroclastics; extremely rarely are they found in flows of basalt blocks. Usually the xenoliths occur as the nuclei of rounded volcanic bombs. They consist of fragments (the vast majority angular) of both sedimentary and magmatic rocks, and are practically unaltered. In places the surface of the xenolith shows some heat effects: the rock has been somewhat compacted and in the course of oxidation it has changed colour from light-grey, grey and greenish-grey (unaltered varieties) to bright-red (the oxidised ones). Where the lava skin has been broken and a fragment of rock has come into direct contact with a gas stream, the heat effects are more considerable.

Unfortunately, in the course of studying the eruption not even an approximate quantitative and qualitative assessment of the varieties of xenoliths could be made, or the order established in which the varieties had made their appearance during the eruption. Let us simply note that on the chart of the eruption drawn up by Yu. V. Vande-Kirkov on the basis of a synthesis of all monitored observations, the greatest numbers of xenoliths in eruption products usually appeared after the ejection of grey (resurgent) ash (see Maleyev & Vande-Kirkov, this volume), the formation of which may generally be linked with underground explosions and the breaking up of the basement. Very broadly speaking, it has also been observed that during the activity of Cone II the xenoliths brought to the surface were mostly sedimentary at first and then volcanic.

Altogether, the xenoliths may be divided into three groups. The first

72

consists of volcanomict sandstones and siltstones with a predominance of littoral–marine varieties, and one often finds in the sandstones well-preserved shells and shell-fragments of marine molluscs (Fig. 1). According to palaeontologist V. N. Gladikova (Kamchatka Territorial Geological Survey) the fauna in these xenoliths is characteristic of the Upper Miocene/Pliocene beds of Kamchatka. The second group consists of sandstones and siltstones with a predominance of continental varieties, and remains of plant stems and carbonised detritus often occur in the rock fragments. Fragments of volcanogenic rock are extremely rare. The third group is predominantly basalt tuffs (usually psephitic and psammitic varieties), and more rarely xenoliths of olivine–pyroxene and mega-plagiophyre basalts, whilst sandstones and siltstones occur in lesser quantities. The xenoliths were compared with the rocks of a Pliocene sequence occurring in a rise immediately to the east. There was a quite distinct similarity between the first group of xenoliths and the lower part of the section, the second group and its middle part, and the third group and the upper parts of the section (upwards through the Neogene section there is a steady change in the facies from littoral–marine to continental terrigenous and volcanogenic).

Fig. 1. Xenolith of 'volcanomict' sandstone with shell of a marine mollusc in a volcanic bomb erupted from Cone II of the New Tolbachik Volcanoes of 1975.

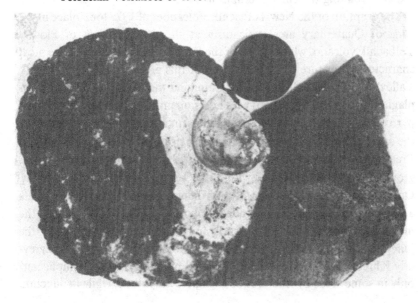

The presence of xenoliths in the eruption products leads one to wonder in the first instance how and under what conditions the magma is capable of capturing fragments of the basement. If one assumes the simplest model of conduit – a vertical dyke injected along a small fissure in the upper part of the crust – then it is on the whole improbable that this would pick up a xenolith: the rapid onset of increased viscosity (compared with normal fluids) of the magmatic mass, and the considerable pressures on the walls of the fissures, suggest that the country rocks would not so much be captured, as squeezed out in directions away from the contacts of the dyke, that the molten material could not mingle with anything if it were rising so fast to the surface, etc. Thus we are faced with a problem: how can the capture of fragments of country rocks by the basalt magma (with its relatively low viscosity compared with that of acidic and intermediate magmas) take place, and in particular how can it occur successively at different stratigraphic levels? Evidently, much depends here on the geometry of the conduit and the peculiarities of its development. Unfortunately, observations of only the dynamics of the eruption and the nature of its products do not supply us with an answer. For this purpose it is interesting to look to broader geological material in an attempt to find the closest analogies with the current Tolbachik eruption in extant successions revealed by erosion to considerable depth. In order to get to terms with this, it is necessary to give a short outline of the geological structure of the area framing the current eruption.

The eruption of the New Tolbachik Volcanoes of 1975 took place in the zone of Quaternary areal volcanism at the southern foot of Ploskiy Tolbachik volcano, which is one of the Klyuchi volcanic group. The most characteristic feature of the volcanism of this area (called the Tolbachik Valley) is the mass effusion of high alumina meso-, mega- and giganto-plagiophyre lavas (Piyp, 1956) with lesser quantities of magnesian olivine-pyroxene varieties. On the whole, the present eruption is characterised by the same petrographic varieties – megaplagiophyre lavas in the Southern Breakthrough and olivine–pyroxene basalts in the Northern. The Klyuchi group of volcanoes is situated within the Central Kamchatka Depression. On the latitude of the southern tip of the group the Depression presents a wide, monoclinal graben bounded on both the west and the east by the monoclinal horsts of the Kozyrevskiy and Tumrok ranges. Moreover, the junction of the Central Kamchatka Depression with the horst of Kozyrevskiy Khrebet is a smooth one (monoclinal plunging), and is complicated only in some areas by fractures of negligible amplitude, while its junction with the Khrebet Tumrok horst is abrupt and along a series of step faults. The horst uplift of Khrebet Tumrok borders directly on the southern zone of lava fields of the Klyuchi group. Within it the pre-Quaternary basement is well-exposed in the form of Cretaceous, Palaeogene and Upper Neogene rocks that make up a number of structures characteristic of this region (Fig. 2). In the Neogene section (Shchapina Formation), as has been pointed out above, laterally and in time a steady succession of facies can be observed, from littoral–marine to continental terrigenous and volcanogenic (Shantser *et al.*, 1966). Characteristic features of the volcanism of this period are effusions, considerable in area and thickness, of plagiophyre (up to mega- and giganto-plagiophyre) lavas of basalt composition, with subsequent massive eruptions of acidic (dacitic) welded pyroclastics. Palaeovolcanogenic reconstructions enable us to speak of the presence in the Late Neogene of extensive basalt zones occurring in conjunction with large lava volcanoes (evidently of the shield type). The formation of the later ignimbrite sheets is closely associated with that of large calderas and volcano-tectonic ring structures.

A study of deep erosion levels (1.5–1.8 km) in the Khrebet Tumrok horst with exceptionally fine, often practically 100% exposure, has enabled geologists to identify at various stratigraphic levels a number of layered bodies that correspond in composition to the plagiophyre basalts of the upper parts of the Neogene section. These bodies are joined to the lavas of the section by both inclined and vertical intersecting dykes. The most deeply exposed bodies of this kind are noted in the tuffogenic part of the Cretaceous section. Further up through the section similar layered bodies constantly occur, their intrusion controlled both by lithological and tectonic factors. In relatively uniform beds the intrusion of the sills is confined to packets of poorly-cemented or fairly plastic rocks (sandstones, siltstones, mudstones, tuffs). Thus bodies of this kind are found in flyschoid packets of the Palaeogene, as well as among the relatively unconsolidated sandstones, tuffites and tuffs of the Shchapina Formation. Of considerable interest are the interformational layered basalts intruded along the boundary division of radically different sets of rocks. A particularly large body of this type is exposed on the boundary between Cretaceous siliceous–volcanogenic beds and the terrigenous and terrigenous–volcanogenic Neogene (Fig. 3). In this instance, as a result of neotectonic dislocations connected with the creation of the monoclinal Tumrok horst, it takes the form of a thick (50–100 m in its widest parts), gently-inclined dyke, whose underside consists of Cretaceous dislocated effusives, tuffs and cherts with a very small chilled margin, while its hanging wall is made up of littoral–marine sandstones, conglomerates and siltstones in which the thermally affected contact extends for 40–50 m. Along the watershed ridge dividing the basin of the River Andrianovka from the basins of the Levaya

Fig. 2. Tectonic sketch-map of the northern part of Khrebet Tumrok. 1, Cretaceous–Palaeogene structural stage; 2, Neogene structural stage; 3, Quaternary structural stage; 4, anticlinal axes; 5, synclinal axes; 6, strike of fold; 7, main fractures; 8, geological boundaries; 9, marker boundaries in Cretaceous; 10, dykes of megaplagiophyre basalts; 11, stratigraphical elements.

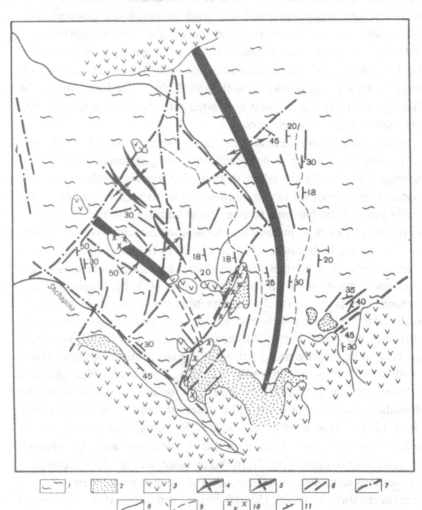

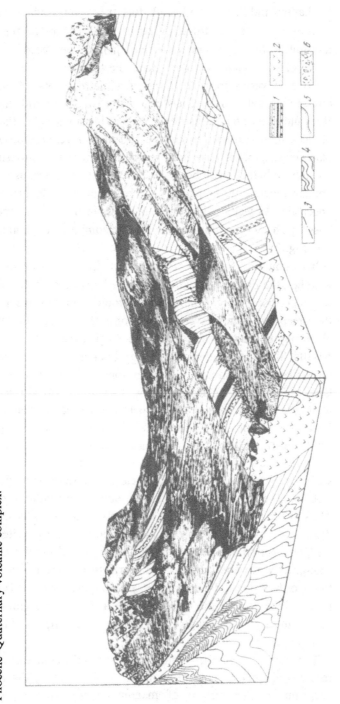

Fig. 3. Block diagram of the contact of Cretaceous and Neogene beds with an intraformational intrusion of a body of megaplagiophyre basalts.
1, Neogene Shchapina Formation; 2, megaplagiophyre basalts of a concordant (sometimes discordant) dyke, stocks and the top of the Shchapina Formation; 3, faults; 4, Cretaceous–Palaeogene basement; 5, unconformity; 6, Tumrok Upper Pliocene–Quaternary volcanic complex.

Shchapina and Pravyy Tolbachik, this dyke can be followed as practically continuous outcrops for a distance of 10–12 km. In the topographic incisions of the upper Levaya Shchapina it can be traced to a depth of the order of 80–1000 m. On the whole, the thickness of the layered bodies discovered varies from fractions of a metre to 50–100 m, and of the intersecting dykes from a few centimetres to 3–4 m and rarely 10–15 m. It should also be noted that in the deepest incisions within the effusive part of the Cretaceous section layered and more rarely discordant bodies of plagioclase gabbros have been recorded, similar in composition to anorthosites. Whether these bodies are peculiar to Cretaceous magmatism or are connected with the plagiophyre basalt cycle of the Neogene, remains uncertain. As well as the system of dykes in this region, small bodies of boss type have been found which are also composed of plagiophyre basalts (see Fig. 3).

In examining sections of other areas of Pliocene volcanism (Valaginskiy Khrebet, the Eastern Volcanic Zone, Sredinnyy Khrebet), the products of which consist wholly or partly of plagiophyre basalts, we also observe in the steep exposures layered intrusions of plagiophyre (up to megaplagiophyre) basalts and pyroxene–plagioclase and pyroxene–olivine basalts. Sills of the latter are not very thick and do not extend far along the strike. As has already been mentioned, the effusions of plagiophyre lavas in Khrebet Tumrok are followed – without any visible stratigraphic interval – by eruptions of acid pyroclastics. In the Pliocene sections of the Eastern Volcanic Zone and in the central part of Sredinnyy Khrebet a similar picture can be observed: basalt flows spread over large areas give place to sheets of dacitic ignimbrites with erupted material on a similar scale. The feeder channels of the acidic eruptions, as in the case of the basalts, take the form of a system of concordant and discordant dykes. Thus in the Cretaceous section of Khrebet Tumrok sills and discordant dykes are observed of hornblende granodiorite porphyrites, and in the Neogene section of the Eastern Volcanic Zone (within the raised block of Valaginskiy Khrebet) multi-stage sills with numerous dykes connecting them, of both granodiorite and diorite composition (Fig. 4). In the upper parts of the section the dykes acquire a dacitic and ignimbrite–dacitic appearance, which strongly suggests that their origin is linked with the ignimbrite sheets.

The enormous number of facts collected in the course of studying both recent and ancient volcanism indicates that during the injection and eruption to the surface of magmatic rocks varying quantities of material from the succession making up the country rock are captured in the form of xenoliths. Moreover, the largest number of xenoliths tends to be captured by acidic magma, the smallest (or practically none) by basic magma. On the surface we usually find masses of xenoliths of the country rocks in acid agglomerate flows, ignimbrites and dacitic flows, but far fewer of them in the blocky, fairly viscous basalt flows of generally olivine–pyroxene composition and the basalt pyroclastics contemporary with them; while in the flows of plagiophyre and megaplagiophyre lavas formed of liquid, very mobile gas-saturated magma, xenoliths are practically absent. Thus xenolith capture is a direct function of the viscosity of the magma, and coincidentally of the speed at which it rises to the surface – which is in turn linked to the nature of the conduit, its configuration, mechanics and the time it takes to form.

The analysis of the geological data enables us to state that the conduits for both basic and acidic volcanism are of a single shape in the rocks of a volcanic area. Within Eastern Kamchatka one can trace systems of conduits (various combinations of dykes) by direct observation to a depth of 2–3 km in deep incisions in the Tumrok and Valaginskiy ranges. Such observations permit construction of a fairly reliable model of the successive development of a magma conduit and suggest the manner in which

Fig. 4. Multi-stage sills of diorite–porphyrites and granodiorite–porphyries in the eastern part of Valaginskiy Khrebet.

xenoliths were captured at different levels of a geological section of the upper crust.

Seismological data indicate that during the eruption of the Northern Group of the New Tolbachik Volcanoes the magma rose from fairly great depths very rapidly – at about 100–150 m/h (Fedotov *et al.*, 1976), through zones that had been weakened possibly both tectonically and, evidently, by the advancing magma itself (opening up fissures, etc.). However, in such an unstable zone, where tension evidently alternates with compression, the optimum conduit is undoubtedly formed during a period of eruption. It is most probably external factors like these which cause the brief intervals in volcanic activity that are frequently observed during the eruption, as well as the irregular pulsations, intensification or indeed abatement of the eruption, extinction of one cone and birth of the next. Both tectonic and lithologic obstacles may arise in the path of ascending magma. These could include the closing up of fissures, or the screening by some dense strata of the increasing stream of ascending magma. In such circumstances, in areas like these the magma will, at the same time as advancing upwards, tend to pour out along the bedding planes, forming sills. These large sheets, which take a considerable time to form, may be regarded as small intermediate magma chambers in which partial differentiation of the melt could occur. Thus, by destroying the tectonic and lithologic obstacles before it, the magma will form (from bottom to top) a ramified conduit-cum-chamber consisting of a combination of concordant and discordant injections, such as we actually encounter when studying the appropriate sections.

In places of constraint, the laminar flow characteristic of melts of this kind is evidently disrupted and various turbulences may form, which will facilitate on the one hand the proliferation of the conduit upwards and sideways (along a bed), and on the other hand the detachment and capture of a xenolith from an overlying bed (evidently a process like the collapse of a roof), and the xenolith may then be carried right up to the surface. Vortices of this type (elastic turbulence) were revealed during an experiment with a viscous–elastic liquid – polyacrylamine (PAM) – in the Institute of Thermal Physics of SOAN (Kutateladze *et al.*, 1971).

Viscous–elastic liquids with a very low Reynolds number are closer in their physical parameters to magmatic melts than the usual Newtonian liquids. The experiment showed that in liquids of the PAM type turbulence arises before an obstacle, and not after it as in the case of Newtonian liquids. It may be assumed that elastic turbulence also comes about in magma when it encounters certain barriers on its way to the surface. Thus xenoliths are most likely to be captured when the conduit is evolving in places where various barriers occur in front of the ascending magma column, i.e. where sills may potentially be formed. It is possible that in such places the accumulation of volcanic gases leads to underground explosions which shatter the basement rocks, whose fragments are then carried to the surface by the gas stream. In the upper part of the conduit, as the gas escapes violently from the magma, the xenoliths (being bodies of different composition and different aggregate state) are fairly easily separated from the melt and ejected together with the pyroclastic material, leaving their lava skin more or less intact. Moreover, since the xenoliths are conveyed to the surface extremely quickly, they remain practically unaltered. Some of them, of course, may remain in the lava flows that are being discharged, and be partially reworked. By analogy with the extremely similar volcanism in the Pliocene (the sections of the volcanic areas described above), one may assume that a similar magma conduit with concordant intrusions formed during the current eruption as well. It is quite probable too that the formation of sills occurs successively upwards through the section and that the majority of xenoliths are ejected in the same sequence. The bedded parts of the section, where a sill is formed, are simply sites where very large numbers of xenoliths may be captured; chance captures of fragments of the country rocks by the molten lava may, of course, occur at any time and on any stratigraphic level. Bearing in mind that the ejecta of the Tolbachik volcanoes contain only xenoliths of the Neogene part of the section (with the exception of some fragments of basalts reminiscent of Cretaceous effusives), it may be assumed that the possible formation of sills and the brief delays in the rise of the magma to the surface occurred at relatively shallow depths. From the outcrops of rocks of the pre-Quaternary basement in the basin of the River Tolbachik one may at present estimate the thickness of the Quaternary lavas of the Tolbachik volcanic area as 400–500 m, and the maximum thicknesses of the Neogene as 1200–1300 m. Thus the concordant bodies were most probably formed at depths of from 500 to 1800–2000 m.

The products of the eruption of the Northern Cones and the Southern Cone contain fairly numerous inclusions of white and light-grey pumiceous rocks. Recently a group of scientists from the Institute of Volcanology (Flerov *et al.*, 1978) carried out a study of the pumice and pumiceous inclusions of the basaltic Quaternary volcanoes of Kamchatka and the Kurils, with particular reference to the pumiceous inclusions of Ploskiy

Tolbachik. The authors have come to the conclusion that the pumiceous inclusions are pyrometamorphic products of fragments of magmatic, volcanoclastic and siliceous basement rocks, and that such selective melting may therefore occur at not very great depths. In the lavas of the Southern Breakthrough (personal communication by F. Sh. Kutyyev) very small, isolated olivine–pyroxene segregations have been found. Whether these are xenoliths is not yet clear.

In conclusion, we should like to emphasise that the interpretation of the data presented above in no way claims to be exhaustive. There is no doubt that further study of both the geology and the immediate material of the eruption may produce quite unexpected alternatives. However, one of the basic objects of this paper is to show the possibilities that are opened up by a study of the volcanic process proper and its products in combination with data on similar phenomena in the recent geological past.

The author offers his sincere thanks to the director of the Institute of Volcanology, S. A. Fedotov, and colleagues of the Institute, O. N. Volynets, G. B. Flerov and Yu. B. Slezin, who took part in the preliminary discussion of the data.

References

Fedotov, S. A., Gorel'chik, V. I. & Stepanov, V. V. 1976. Seysmologicheskiye dannyye o magmaticheskikh ochagakh, mekhanizme i razvitii bazal'tovogo, treshchinnogo Tolbachinskogo izverzheniya v 1975 na Kamchatke. (Seismological data on the magma chambers, mechanics and development of the Tolbachik fissure basalt eruption of 1975 in Kamchatka.) *Dokl. AN SSSR*, **228**(6), 1407–10.

Flerov, G. B., Volynets, O. N., Khrenov, A. P. & Petrova, V. V. 1978. Pemzovyye i pemzovidnyye vklyucheniya v bazal'takh chetvertichnykh vulkanov Kamchatki i Kuril (pervichnaya priroda, sostav i protsessy pirometamorfizma). (Pumice and pumiceous inclusions in the basalts of the Quaternary volcanoes of Kamchatka and the Kurils (their initial nature, composition and processes of pyrometamorphism).) In: *Vklyucheniya v vulkanicheskikh porodakh Kurilo–Kamchatskoy ostrovnoy dugi. (Inclusions in the volcanic rocks of the Kuril–Kamchatka island arc.)* M., Nauka.

Kutateladze, S. A., Popov, V. I. & Khabakhpasheva, Ye. M. 1971. K voprosu ob obtekanii tsilindra laminarnym potokom vyazko–uprugoy zhidkosti. (Flow round a cylinder by a laminar body of viscous–elastic liquid.) *Dokl. AN SSSR*, **197**(3), 545–6.

Piyp, B. I. 1956. Klyuchevskaya sopka i yeye izverzheniya v 1944–1945 gg. i v proshlom. *(Klyuchevskaya Sopka and its eruptions of 1944–45 and in the past.)* Trudy Labor. vulkanol., Issue 11, 308 pp.

Shantser, A. Ye., Chelebayeva, A. I. & Geptner, A. R. 1966. Stratigrafiya i korrelyatsiya neogenovykh otlozheniy khrebta Tumrok i nekotorykh drugikh rayonov Kamchatki. (The stratigraphy and correlation of the Neogene deposits of Khrebet Tumrok and some other regions of Kamchatka.) In: *Stratigrafiya vulkanogennykh formatsiy Kamchatki. (The stratigraphy of the volcanogenic formations of Kamchatka.)* M., Nauka, pp. 86–98.

Age divisions of the Holocene volcanic formations of the Tolbachik Valley

O. A. Braytseva, I. V. Melekestsev and
V. V. Ponomareva

The Tolbachik Valley is an extensive lava plain formed of effusions from the numerous cinder cones of the regional (Piyp, 1956) or linear (Sirin, 1968) zone of the same name. Piyp (1956) and Melekestsev (Melekestsev *et al.*, 1970) identified the lavas of the Klyuchi volcanic field as post-glacial (Holocene), while Yermakov & Vazheyevskaya (1973) have identified two groups of volcanites – Lower Holocene and Middle to Upper Holocene – on the basis of different lava compositions and states of preservation of the lava complexes. The main object of the investigations described in this paper, was to achieve a detailed age division of the Valley's volcanic formations using tephrochronological methods and to determine the stratigraphic position of virtually every major cone and its flows. From this point of view, all other geological studies, even the most detailed, have limited possibilities, since they enable us to establish the time sequence and interrelations of volcanic deposits only where they are in direct contact. For spatially separated volcanic formations, determining their relative age without resort to tephrochronology raises further problems, while for absolute dating it appears to be the only practical key.

The method of dating volcanic formations by tephrostratigraphic studies is based on the fact that the younger the volcanites, the thinner the pyroclastic layer covering them must become and the fewer the horizons of tephra and buried soils. The task consists of obtaining tephra sections on the lava flows or in the craters of cinder cones and correlating them with standard sections of pyroclastic deposits of the area, thus enabling one to determine the position of each datable item in the overall history of volcanic activity and so achieve a relative age division of the volcanites. If datings of the individual horizons of tephra and buried soils are available, it is possible to determine chronological limits in absolute figures

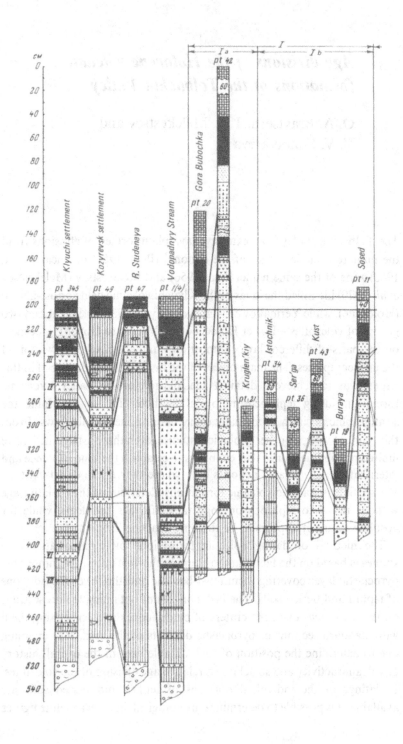

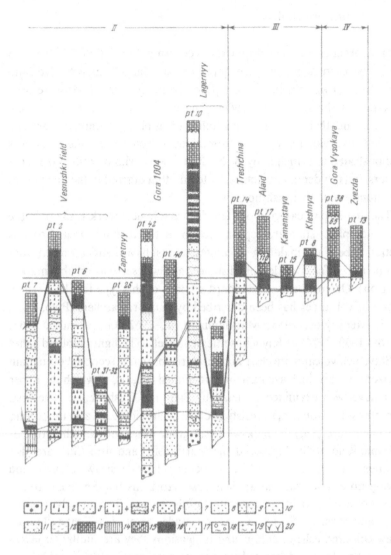

Fig. 1. Correlation of tephra sections in lava flows of the Tolbachik
Valley with standard sections in the pyroclastic mantle of the Klyuchi
group of volcanoes.
1, volcanic bombs and cinder cone agglomerates; 2, scoriaceous lapilli;
3, volcanic grit; 4, volcanic grit and lapilli; 5, stratified tephra:
intercalated horizons of lapilli, volcanic grit and sand; 6, pumice grit;
7, volcanic sand; 8, volcanic sand with grit; 9, fine white and
straw-coloured ashes from Shiveluch volcano: a, uniform, b, with
grey, medium-grained sand at the base; 10, fine yellow and cream
ashes from Shiveluch; 11, coarse-grained grey–yellow volcanic sand;
12, fine pink and grey volcanic sands; 13, tephra from the 1975–76
eruption; 14, sandy loams; 15, sandy loams, saturated: a, with
volcanic grit, b, with volcanic sand; 16, buried soils; 17, basalts; 18,
moraines; 19, fluvioglacial deposits; 20, plant remains. I-VIII, main
marker horizons of Klyuchi ash.

for the formation of the volcanic rocks covering them. Studies of this kind were first carried out in Kamchatka at the Malyy Semyachik volcano (Braytseva *et al.*, 1978), where tephrochronological methods were used to establish the sequence in which the volcano's lava complexes were formed, and their age. The possibility of applying tephrostratigraphic methods to divide the volcanic formations of Tolbachik Valley itself was demonstrated in principle by A. N. Sirin (1968), who described from the craters of its cinder cones sections of tephra that differed in the thicknesses and numbers of their ash horizons.

The first step in carrying out tephrochronological work is to study the most complete, standard tephra section in the research area. For this section to be of use in dating volcanites, it must have distinct ash beds that can be traced for considerable distances and will serve as reliable marker horizons. One of the main standard sections of this type for the Klyuchi group of volcanoes has been described by many researchers (Piyp, 1948, 1956; Menyaylov, 1955; Markhinin *et al.*, 1962; Gushchenko, 1965; Dikov, 1969, 1974) at Klyuchi settlement itself. The light-coloured ashes at Shiveluch volcano are clear marker horizons in this section. Seven main horizons of Shiveluch ash have been identified (Fig. 1), of which the upper four have been examined in detail by Piyp and Gushchenko. The first, uppermost horizon (Sh_1) consists of chalk-white silt-grade ash, the second (Sh_2) and third (Sh_3) of chalk-white silt-grade ashes with an admixture of volcanic sand at the base, and the fourth (Sh_4) and fifth (Sh_5) are also composed of fine ash but differ in being yellow or straw-coloured and having no coarse material at their base. Horizons Sh_1–Sh_5 constitute a packet of ash layers in the upper part of the section. The underlying sixth (Sh_6) and seventh (Sh_7) horizons are made up of silty light-yellow or straw-coloured ash; in colour and dimensions they are similar to layers Sh_4 and Sh_5, but are separated from them in the section by a considerable time interval and belong rather to the lower part of the column. Apart from the seven main horizons identified, the Klyuchi section contains several other beds of Shiveluch ash. However, these are rather thin and often wedge out along the strike, whereas our research has shown that the horizons described occur over the entire area around Klyuchevskaya Sopka and the middle course of the River Kamchatka and serve as basic markers for correlating the successions in the pyroclastic mantling. The same intercalations of Shiveluch ash were described by Dikov (1969, 1974) on the shore of Ushkovskoye ozero near the settlement of Kozyrevsk.*

* Dikov's numbering of the horizons is slightly different since he identified a layer ɪvₐ, which we have termed the fifth (Sh_5).

To answer conclusively the question of the age of the horizons of Shiveluch ash one needs enough datings. Present notions are highly contradictory, however (see Table 1). Gushchenko (1965) determined the time of formation of the first ash horizon (agreeing with Piyp) as ≈ 120 years ago (ash from the eruption of 1854), the second as ≈ 500, the third as ≈ 1000 and the fourth as ≈ 1800 years ago. Dikov (1974) has published radiocarbon datings of sections along the River Kamchatka which make the two upper ash horizons considerably older. For a conclusive answer one also needs radiocarbon datings of the standard section itself at the Klyuchi settlement; at the moment, for the three upper horizons of Shiveluch ash we are taking as approximations those figures given in Fig. 3. As far as the older ash horizons are concerned, the date of about 2500 years obtained by Dikov for the culture layer beneath ash Sh_5 (according to our numbering) and figures of the order of 6700–7000 years (4th–5th millenia BC) for the upper material culture beneath layer Sh_6, appear realistic. The radiocarbon datings in the interval from 10 000 to 13 000 years given by Dikov (1969) for the floor of the pyroclastic cover are in good agreement with the fact that in the area of Ushkovskoye ozero and Klyuchi it rests on fluvioglacial deposits of the final phase of the Upper Pleistocene glaciation, which ended 10 000 to 11 000 years ago.

By directly tracing along the River Kamchatka the ash horizons from the standard section at Klyuchi, it was possible to correlate them with the section at Kozyrevsk and, further across the valley of the River Studenaya, with sections of the upper reaches of the Vodopadnyy stream, i.e. directly with sections from the area of the Tolbachik Valley. Tracing the ash horizons was facilitated by the fact that these three sections occur within a single north–south belt corresponding to a particular sector of the ash fall zone. In the Studenaya and Vodopadnyy valleys the sections were taken on fairly old elements of the relief – an Upper Pleistocene moraine and fluvioglacial plains, in order to obtain a complete Holocene column as at Klyuchi. In the Studenaya river valley all seven ash horizons identified at Klyuchi were recorded; furthermore, at the bottom of the section there is an eighth horizon (Sh_8), of fine, straw-coloured ash, resting immediately on the moraine and marking an eruption of Shiveluch apparently at the very beginning of the Holocene. In the Vodopadnyy valley the seventh and eighth horizons of Shiveluch ash are absent, although this is of no general significance and is explained by the fact that the fluvioglacial plain on which the section was taken at points 1 and 4 is younger than the moraine. Far more important is the absence in the upper part of the section of one

Table 1. *Characteristics of Shiveluch ash horizons within the Tolbachik Valley*

| Numbering of sections at Klyuchi settlement | Absolute age | | | Thickness (cm) | Colour and granularity of the Shiveluch ashes in the Tolbachik Valley |
	According to B. I. Piyp	According to I. I. Gushchenko	According to N. N. Dikov (years ago)		
Sh₁	1854	1854	675±80		Absent in the Tolbachik Valley sections
Sh₂	1780	15th cent.	1145±80	2–4	Straw-coloured/white ash, in some instances with fine-grained sand at the base
Sh₃	16th cent.	10th cent.	—	7–10	Fine straw-coloured/white ash with medium-grained sand at the base
Sh₄	15th cent.	2nd cent.	—	5–7	Fine straw–yellow ash
Sh₅	—	—	2440±80	1–1.5	Fine straw–yellow ash
Sh₆	—	—	6000±7000	3	Fine straw–yellow ash
Sh₇	—	}	10300±360	—	Absent in Tolbachik Valley; fine straw-coloured ashes in the Studenaya valley
Sh₈	—				

of the horizons of chalky-white ash. Evidently the first Shiveluch horizon, which in the Studenaya valley already occurs in the form of thin lenses, pinches out in the Tolbachik Valley. The remaining Shiveluch ash horizons are very clearly expressed in the section, differing in colour, thickness and coarseness. Their basic characteristics are given in Table 1.

We have numbered the ash horizons of the Tolbachik Valley after the Klyuchi section, although the reliability of the correlations has still to be confirmed by radiocarbon datings obtained independently in these remote areas. Thus at present such a correlation may be regarded as a working outline. It is important to stress that the upper horizon of Shiveluch ash in the Tolbachik Valley cannot be younger than 350 years since at point 10 near the Lagernyy cone we felled a larch that had an age of 342 years. It had grown on soil that had formed on the cinder of Gora Vysokaya, and this cinder is underlain by the first layer of Shiveluch ash (Sh_2) in the Tolbachik Valley.

The presence of clear marker beds of Shiveluch ash in the Tolbachik Valley tephra sections, has created favourable pre-conditions for the tephrostratigraphic division of the volcanites. From aerial photographs Melekestsev drew up a scheme of the cinder cones and lava flows of the area which was the basis for our work on tephrostratigraphic dating. On lava flows at the points indicated on the scheme (over 40 points), exploration shafts 1–5 m deep revealed a pyroclastic cover overlying effusives. The considerable depth of the exploration shafts in a number of cases was caused by the substantial thickness of the 1975 tephra, which meant that it was more difficult to work in the axial zone, near the site of the breakthrough, and in places actually impossible. A study of tephra for dating purposes in the craters of the cinder cones can yield no practical results, as the permafrost is too close to the surface for shafts to be sunk as far as the country rocks. Moreover, in the craters exposed by erosion the tephra has been washed away, while in the closed ones it is overlain by thick scree. The volcanites were therefore dated according to the tephra covering the flows and the result checked against the stratigraphic position in the overall section of tephra from cinders of the cones themselves, in places where there was no doubt which cones they came from.

The tephra sections on the lava flows from the cones are given in Fig. 1. Comparison with the marker horizons of the Shiveluch ashes reveals several age groups of volcanic formations.

Group I includes volcanites that are covered by an horizon of ash Sh_4 or Sh_5 and were formed in the time interval between deposits of these two

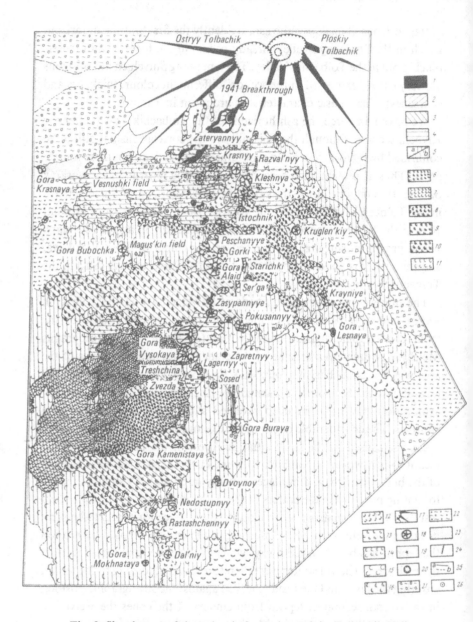

Fig. 2. Sketch map of the volcanic formations of the Tolbachik Valley. Age groups of volcanic formations: 1, fifth; 2, fourth; 3, third; 4, second; 5, first: *a*, first subgroup, *b*, second subgroup.
Volcanic formations: 6–16 lava flows: from 6, Zvezda cone; 7, Gora Vysokaya; 8, Kleshnya cones; 9, Gora Kamenistaya; 10, Gora Alaid; 11, Treshchina (the Fissure); 12, Lagernyy cone; 13, inferred flows from Gora 1004 and Zasypannyye cones; 14, Vesnushka field;

horizons and horizon Sh_6 (Fig. 2). Within this group one can distinguish an older sub-group of volcanic forms that originated before horizon Sh_5 was laid down. Belonging to this sub-group are the cones and flows of Gora Bubochka and the Kruglen'kiy cone.* The younger group of cones and lava flows was formed between the deposition of ash horizons Sh_4 and Sh_5 and includes Gora Buraya, and the Ser'ga, Istochnik, Sosed and other cones. As a whole, the volcanic formations shown by vertical hatching on the sketch map constitute a single group of the oldest cones of the Tolbachik Valley. The volcanites of this complex cover considerable areas and are only partially overlaid by younger formations in the axial part of the zone. Within this valley we have discovered no effusives older than ash Sh_6.

Age group II consists of the cinder cones and lava flows that are covered by ash horizon Sh_3, i.e. that were formed in the period between the deposition of horizons Sh_3 and Sh_4. These include the volcanic formations of the Vesnushka field, which is a collection of lava flows and cones of the northernmost part of the Valley that are similar in appearance and state of preservation. At the moment, we have placed the whole of this volcanite field in age group II, although further detailed studies of its numerous component cones may lead to some of them being reallocated to other age groups. Lagnernyy and Zapretnyy cones, and Gora 1004, also belong to age group II. The stratigraphic position of the volcanites of Gora 1004 was determined according to where its thick layers of cinders are covered by ash horizons Sh_3 (point 42), whilst for Lagernyy cone mutually confirmative data were obtained from the position both of its lava flow (point 12) and the cinders of the cone itself (point 10), which are directly overlaid by ash Sh_3.

Age group III contains cones and flows that are overlaid by ash horizon Sh_2, i.e. that were formed in the time interval between the depositions of layers Sh_3 and Sh_2. The oldest lava flow in this group is the one that is attributed to the Fissure cones. The Gora Alaid, Peschanyye Gorki,

* The names of the cones are given for convenience of description and interpretation of the data.

Caption to Fig. 2 (*cont.*)

15, cones of the first age group; 16, of undetermined age. 17, Ostryy and Ploskiy Tolbachik volcanoes; 18, cinder cones; 19, small flow centres and bocci; 20, craters.
Other conventional signs: 21, moraines; 22, fluvioglacial plains; 23, proluvial plains; 24, volcanic tectonic scarps as a relief feature; 25, boundaries: *a*, established, *b*, inferred; 26, observation points.

Kleshnya and Kamenistaya cones are younger and everywhere their lava flows are directly overlaid by layer Sh_2. It is significant that the thick cinder beds of Gora Alaid and Peschanyye Gorki (1–2 m), which have been identified on an E–W line not far from the cones themselves (points 20, 28, 18, 25), occupy the same stratigraphic position as the flows of these cones, between ash horizons Sh_2 and Sh_3. However, we can assume that the youngest cone in this group is Kleshnya, the lava flows of which skirt Gora Alaid and Peshchanyye Gorki.

Age group IV contains cones and flows that were formed after the deposition of ash Sh_2, namely Gora Vysokaya and the Zvezda cone. All that rests on their lava flows is soil covered by the pyroclasts of 1975. Tephra from Gora Vysokaya can be very clearly observed in its immediate vicinity (points 10, 11, 12, 26, 14, 42) and everywhere occupies the same stratigraphic position as the lava flows – it rests on horizon Sh_2 and is itself covered by present-day soil. Possibly the eruption of Tolbachik in 1740 recorded by S. P. Krasheninnikov, during which lava burnt the woods at the foot of the volcano (*Katalog deystvuyushchikh vulkanov... (Catalogue of the active volcanoes...*), 1957), is connected precisely with the formation of the Zvezda cone. The maximum age of the larches on the lava flows of Zvezda is 190 years. Some of these trees were felled by us at the edge of the flow close to older remains of the woods, so they were the first colonisers of the Zvezda lavas.

We have placed the eruptions known in historical times – those of 1941 and the latest, 1975–76 – in age group V.

The five age groups of volcanic formations of the Tolbachik Valley are shown schematically in Fig. 3.

As the oldest formations within the part of the Tolbachik Valley studied have proven to be volcanites resting beneath ash horizon Sh_5 (and without Sh_6 above them), one can assume that lava flows over this area of the Valley commenced 2500–3000 years ago. Thus they can all be placed in the Late Holocene, within the framework of which the five age groups of volcanites above have been identified. The Ploskiy Tolbachik structure as a whole formed considerably earlier (in the Upper Pleistocene to beginning of Holocene); the Tolbachik Valley's areal activity can evidently be compared in time with the creation of the volcano's complexly-structured caldera.

In the first stage volcanic activity was dispersed all over the Valley. Then volcanic manifestations were 'gathered together' in the axial part of the zone, which is where the cones of age groups II–V arose. Evidently, the reason Yermakov and Vazheyevskaya concluded that areal activity was

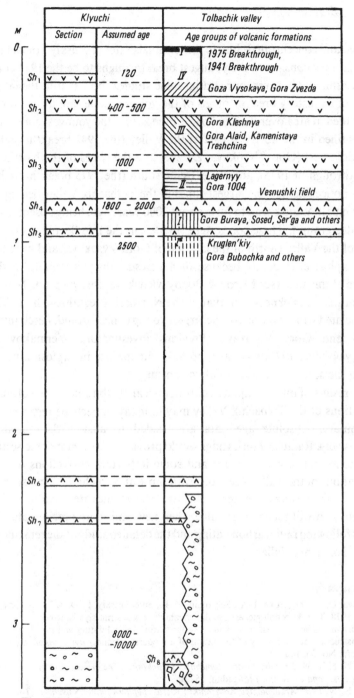

Fig. 3. Diagram showing the relationships of Shiveluch marker ash horizons and the age groups of the volcanic formations in the Tolbachik Valley. For legend see Figs. 1 and 2.

reduced and concentrated close to the terminal crater of Ploskiy Tolbachik, was that they considered the youngest breakthroughs to be the 1941 cone and Kleshnya. However, our stratigraphic studies show that in the latter stages of its history breakthroughs of the volcanic cones are fairly evenly distributed in area over the axial part of the zone – the youngest cones can be identified in the northern part of the Valley (the 1941 breakthrough), its middle part (Gora Vysokaya, Zvezda, and the cones of the Northern Breakthrough of 1975), and the extreme south (the 1975 breakthrough).

It is noticeable that the largest cones of the Tolbachik Valley belong to age groups II–V and began to form about 2000 years ago, being absent from the older group I. These cones were situated in the middle and northern parts of the Valley (north of the latitude of Gora Vysokaya) and were the most explosive, evidently because of magnesian magma coming to the surface at this time (see Flerov & Bogoyavlenskaya, this volume). Tephra comparable in thickness with that of the Northern Breakthrough of 1975 is associated with eruptions of the largest young cones – Alaid, Peschanyye Gorki and Gora Vysokaya. Previous investigators (Yermakov & Vazheyevskaya, 1974) have also pointed out the rise in explosion index during the second phase of areal volcanism.

The results of our attempt at a chronological attribution of the volcanic formations of the Tolbachik Valley may in many respects be regarded as preliminary. Absolute age data are needed to answer the question conclusively. Radiocarbon datings would provide us with more precise age boundaries between the groups and some important corrections to our correlations of the Valley's sections with those at Klyuchi, Kozyrevsk, etc. However, the results of our tephrostratigraphic studies are valuable at this moment and will retain their value when new and more precise data are added following radiocarbon dating and the detailed study of the remainder of the Tolbachik Valley.

References

Braytseva, O. A., Yegorova, I. A., Selyangin, O. B., Sulerzhitskiy, L. D. & Nesmachnyy, I. A. 1978. Tefrokhronologicheskoye datirovaniye i rekonstruktsiya istorii formirovaniya sovremennogo vulkana. (Tephrochronological dating and the reconstruction of the history of formation of a modern volcano.) *Byul. vulkanol. stantsiy*, No. 55, 100–15.
Dikov, N. N. 1969. *Drevniye kostry Kamchatki i Chukotki.* (*The ancient fires of Kamchatka and Chukotka.*) Magadan, 252 pp.
Dikov, N. N. 1974. *Arkheologiya Kamchatki i vozrast peplov vulkana Shiveluch.* (*The archaeology of Kamchatka and the age of the ashes of Shiveluch volcano.*) Trudy SVKNII, Issue 59, pp. 132–5.

Gushchenko, I. I. 1965. *Peply Severnoy Kamchatki i usloviya ikh obrazovaniya.* (*The ashes of Northern Kamchatka and the conditions of their formation.*) M., Nauka, 144 pp.

Katalog deystvuyushchikh vulkanov Kamchatki. (*Catalogue of the active volcanoes of Kamchatka.*) 1957. *Byul. vulkanol. stantsiy*, No. 25. 180 pp.

Markhinin, Ye. K., Pugach, V. B. & Markhinina, S. N. 1962. *Ob yestestvennoy namagnichennosti peplovykh sloyev rayona Klyuchevskoy gruppy vulkanov.* (*On the natural magnetism of the ash layers in the area of the Klyuchi volcano group.*) *Byul. vulkanol. stantsiy*, No. 33.

Melekestsev, I. V., Krayevaya, T. S. & Braytseva, O. A. 1970. *Rel'yef i otlozheniya molodykh vulkanicheskikh rayonov Kamchatki.* (*The relief and deposits of the young volcanic areas of Kamchatka.*) M., Nauka, 104 pp.

Menyaylov, A. A. 1955. *Vulkan Shiveluch – yego geologicheskoye stroyeniye, sostav i izverzheniya.* (*The Shiveluch volcano – its geological structure, composition and eruptions.*) Trudy Labor. AN SSSR, Issue 9, 263 pp.

Piyp, B. I. 1948. Novoye eruptivnoye sostoyaniye vulkana Shiveluch s kontsa 1944 goda po may 1945 goda i nekotoryye zamechaniya o geologicheskoy strukture etogo vulkana i yego proshlykh izverzheniyakh. (The new eruptive state of the Shiveluch volcano from the end of 1944 to May 1945 and some remarks on its geological structure and past eruptions.) *Byul. vulkanol. stantsii na Kamchatke*, No. 14, 38–51.

Piyp, B. I. 1956. *Klyuchevskaya sopka i yeye izverzheniya v 1944–45 gg. i v proshlom.* (*Klyuchevskaya Sopka and its eruptions in 1944–45 and the past.*) Trudy Labor. vulkanol. AN SSSR. Issue 11, 310 pp.

Sirin, A. N. 1968. *O sootnoshenii tsentral'nogo i areal'nogo vulkanizma.* (*The relationship between central and areal volcanism.*) M., Nauka, 196 pp.

Yermakov, V. A. & Vazheyevskaya, A. A. 1973. Vulkany Ostryy i Ploskiy Tolbachik. (The Ostryy and Ploskiy Tolbachik volcanoes.) *Byul. vulkanol. stantsiy*, No. 49, 43–54.

Geology and petrochemistry of the Tolbachik regional zone of cinder cones

G. B. Flerov and G. Ye. Bogoyavlenskaya

Study of the volcanism of the Tolbachik regional zone of cinder cones (TRZCC), as it is called following Piyp's nomenclature (1956), and of its spatial relationship with the stratovolcanoes Ostryy and Ploskiy Tolbachik, has been the concern of a number of investigators. Their results have been published in works by Piyp (1946, 1956), Sirin & Timerbayeva (1971), Vazheyevskaya (1976), Yermakov (1971), Kirsanov & Ponomarev (1974), and Yermakov & Vazheyevskaya (1973).

According to these investigators and the authors of the present paper, the TRZCC was imposed on the Tolbachik volcanoes and so is of younger (Holocene) age. The formation of the actual structure of Ploskiy Tolbachik is dated as Upper Pleistocene to the beginning of the Holocene (Melekestsev *et al.*, 1970). The manifestation of the volcanism shows it to be of areal type. This volcanism occurs over a wide area to the south-west of Ploskiy Tolbachik and to a lesser extent over its north-eastern slope, and was apparently connected with the initial phases of the formation of the great crater caldera of Ploskiy Tolbachik, subsequently compensated by the products of the eruption of the central crater. Piyp (1956) had earlier associated the post-caldera activity of Ploskiy Tolbachik with the areal volcanism, and the authors of this paper adhere to this point of view.

We find the most detailed account of the geology of this region in a paper by Yermakov & Vazheyevskaya (1973) which, together with a scheme of age relationships, provides information concerning the geological structure and composition of the rocks of all the volcanoes of the Tolbachik group, including the Ostryy Tolbachik stratovolcanoes, whose activity in historical time preceded the areal volcanism. According to these authors, the time of formation of the entire Tolbachik group embraces the period from Middle to Upper Pleistocene (Q_2–Q_3^1) to the present, the areal volcanism itself being ascribed to Holocene time.

96

The present article is concerned with the material composition of the rocks of the TRZCC and uses some field and analytical data kindly provided by O. N. Volynets, A. A. Abdurakhmanov, A. A. Vazheyevskaya and V. A. Yermakov, to whom the authors express their gratitude.

On the basis of detailed geological, petrographical and petrochemical mapping of cinder cones and lava flows, we have compiled a sketch map locating the products of Holocene eruptions in the cinder cone region to the south of Ploskiy Tolbachik volcano. This is a diagrammatic representation of the spatial distribution of the petrochemical varieties of basalts in relation to the ages of the volcanic complexes (Fig. 1). The principles involved in distinguishing these types and their compositional features will be discussed later. This classification has been adopted by analogy with the similar products of the 1975–76 eruption (Volynets *et al.*, 1976b). The separation of magmatic complexes according to age has been performed from tephrogeochronological data (Braytseva *et al.*, this volume). Braytseva and her co-authors have established at least five large volcanic age groups within the Upper Holocene in our cinder cone area: I, 2000–2500 years; II, 1000–2000 years; III, 400–1000 years; IV, 120–400 years; V, present.

The region of areal volcanism that we have studied (see Fig. 1) encompasses the terrain from the foot of Ploskiy Tolbachik southwards to the River Tolbachik, and from the Tolud river in the east to the Gora Krasnaya–River Ozernaya line in the west, as well as the area further to the north-east of Ploskiy Tolbachik as far as the River Studenaya. The volcanic formations to the west of the Krasnaya–Ozernaya line have been associated with the activity of Ploskiy Tolbachik by Sirin (1968) and Melekestsev (1970). The areal dimensions of these effusions from Ploskiy Tolbachik are listed in Table 1: the total area of eruption products over the entire zone is about 908 km^2.

As can be seen on the geological sketch map (Fig. 1), the largest areas of effusions in the terrain south of Ploskiy Tolbachik pertain to the early stage of formation of the Tolbachik areal zone (group I). The forms the volcanism takes (extensive lava fields with relatively low cinder cones, lava domes, fissure effusions) are evidence that the areal volcanism of this stage bears an essentially effusive character. The lavas erupted at this stage are represented by mega- and mesoplagiophyric alumina type basalts that are also extensively developed in older beds (Pleistocene, Pliocene) at the base of volcanoes of the Klyuchi group (Piyp, 1956; Yermakov, 1971; and others) and form, in particular, the Ploskiye Sopki (Sirin, 1968). The extent of the areal volcanism is reduced with time and its manifestations become more and more localised in the axial part of the zone (see Fig. 1; Braytseva

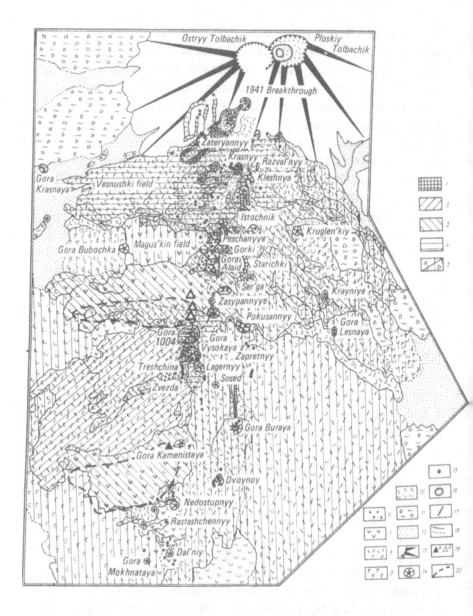

Fig. 1. Geological and petrochemical sketch map of the volcanic
formations of the Southern zone of the Tolbachik regional zone of
cinder cones.
1–5 age groups of volcanic formations: 1, fifth (present); 2, fourth
(120–400 years); 3, third (400–1000 years); 4, second (1000–2000
years); 5, first (2000–2500 years); *a*, first subgroup, *b*, second
subgroup.

et al., this volume; groups II–V). The eruption centres and their products (groups III–V) forming the cinder cones occur in a band that stretches north-eastwards for 45 km with a width of 3 km. The volcanism changes in character. Although effusives continue to predominate in the total volume of volcanic material, there is a marked increase in the role of explosive eruptions: large cinder cones are formed (Peschanyye Gorki, Gora Alaid, the Krasnyy cone, Gora 1004, Gora Vysokaya, the 1941 breakthrough, the 1975 Northern Breakthrough). The number of eruption centres is reduced (groups III–V) and the volcanism acquires a more autonomous character. The eruptive centres are now concentrated along an approximately north–south line forming an uninterrupted chain of cones of different geological ages. At the same time the composition of the magmatic material reaching the surface undergoes a gradual change: along with the typical aluminous subalkaline plagiophyric basalts that inherit the chemical composition of the lavas of the preceding geological phases, there appear chemically different olivine–pyroxene basalts that play, however, a subordinate role in the overall balance of the material of the entire areal zone. Magnesian intermediate-type basalts are also widely developed. Olivine–pyroxene basalts mainly form the large cinder cones mentioned above.

A similar evolution of geological structure associated with volcanism (the gathering of eruption centres on to an axial line), and accompanied by alteration of its character (from effusive to effusive–explosive) and of the type of deep-seated magmatic material (the appearance of two contrasting basalt varieties), is in our opinion evidence that the volcanic activity of the Tolbachik regional zone becomes in time aligned along a fairly narrow linear structure of a disjunctive character, similar in the first approximation to a rift structure. The movement of magmatic material is

Caption to Fig. 1 (*cont.*)

6–9 petrochemical types of basalts: 6, magnesian proper, type I; 7, magnesian of intermediate type, type II; 8, aluminous subalkaline, type III; 9, aluminous of intermediate type, type IV.
10, lava flows; 11, moraines; 12, fluvial deposits; 13, cones of the stratovolcanoes Ostryy and Ploskiy Tolbachik; 14–15, cinder cones and lava domes; 14 shown to scale, 15 not to scale; 16, craters; 17, volcanic tectonic scarps, expressed in the relief; 18, boundaries: *a*, established, *b*, inferred; 19, positions of cinder cones in the activity zone of the Great Tolbachik Fissure Eruption from, *a*, the Southern Breakthrough, *b*, the Northern Breakthrough; 20, Diagrammatic outline of the lava fields of the Northern and Southern Breakthroughs.

Table 1. *Volume of magmatic material from the eruptions, and areas made up of the volcanic products of the Tolbachik regional zone of cinder cones*

Regions, age groups	Magnesian basalts proper (type I)			Magnesian basalts of the intermediate type (Type II)			Alumina basalts (Types III–IV)			All basalts	
	Area (km^2)	Volume (km^3)	(%)	Area (km^2)	Volume (km^3)	(%)	Area (km^2)	Volume (km^3)	(%)	Area (km^2)	Volume (km^3)
Southern zone											
Age groups:											
I	1	0.1	3	6	0.3	9	800	40.0	100	800	40.0
II	5	1.5	48	61	1.2	38	157	2.8	88	164	3.2
III	0.5 (?)	0.1 (?)	6	0.6	0.06	4	48	0.4	13	114	3.10
IV	16.4	1.43	60	—	0.02*	1	78	1.45	90	79	1.60
V	—	—	—	—	—	—	30	0.9	39	46	2.35
Volcanic groups of undetermined age	—	0.03	3	—	0.8	71	—	0.3	26	40	1.13
Rocks of complexes II–V	—	3.16	28	—	2.38	21	—	5.8	51	443	11.34
The entire Southern zone	—	3.16	6	—	2.38	4	—	45.85	90	800	51.39
Caldera of the volcano											
Ploskiy Tolbachik	—	—	—	—	—	—	8	2	100	8	2
Northern zone	—	—	—	—	—	—	100	5	100	100	5
The entire areal zone	—	3.16	5	—	2.38	4	—	52.8	91	908	58.4

The areas according to age groups have been calculated from reconstructions of the volcanic complexes.
* Volume of intermediate varieties from the 1975 eruption.

thus controlled by deep-seated fractures of sub-meridional strike which act as magma conduits; a view that has in fact been expressed by previous workers (Piyp, 1956; Fedotov, 1976). The existence of a fault is confirmed by seismic (GSZ) data and gravimetry surveys, according to which the cinder cones are situated in the zone of a steeply dipping deep-seated fault about 5 km wide that is fixed on the surface by a series of parallel fractures (Zubin *et al.*, 1976; Balesta *et al.*, this volume). At the same time, with regard to the structural control of volcanism in the areal zone, it is characteristic that the region (see Fig. 1) in which cinder cones composed of olivine–pyroxene magnesian basalts proper occur, is strictly limited: the eruption centres (cinder cones) of this composition are situated only to the north of the latitude of Gora Vysokaya. We have not observed any volcanic phenomena involving the magnesian basalts proper south of this line. Interpretation of these observations allows us to postulate the existence here of some concealed structural barrier, trending north-west or approximately east and west, that acts as a deep-seated fault.

All the rocks of the TRZCC are basaltic in composition and include melanocratic, leucocratic and transitional varieties that can be distinguished according to the ratio of dark minerals (olivine and clinopyroxene) to plagioclase in the phenocrysts and groundmass. The petrographic characteristics are given in Table 2. A whole range of varieties can be distinguished according to the qualitative mineralogical composition of the phenocrysts, the extreme varieties being olivine–pyroxene subaphyric and aphyric basalts (typical representatives of melanocratic basalts) and mega–mesoplagiophyric (leucocratic) basalts. These varieties are analogous in petrography to the basalts of the 1975–76 Northern and Southern Breakthroughs respectively (Volynets *et al.*, 1976). The first variety is subordinate in volume among the rocks of the areal zone and is characterised by the constant presence of olivine and pyroxene phenocrysts (2–3%) and subphenocrysts. Plagioclase occurs only rarely as single small crystals. The mega- and mesoplagiophyre basalts are widely developed and are distinguished by a characteristic coarsely porphyritic texture formed by large megacrystals or phenocrysts of plagioclase (0.5–3 cm) comprising up to 15–20% of the volume of the rock. Olivine occurs in single crystals among the phenocrysts but is always present among the subphenocrysts and microlites, in amounts subordinate to plagioclase. Clinopyroxene is rare. The mega- and mesoplagiophyric lavas have been distinguished as a petrographic type following Piyp (1956), who proposed this name on the basis of textural features and the size of the phenocrysts. A whole series

Table 2. *Petrographic varieties of basalts from the areal zone of Ploskiy Tolbachik*

Petrographic variety of basalt	Petrographic characteristics of the basalts					Content of phenocrysts and subphenocrysts (%)
	Mineral associations of phenocrysts and subphenocrysts	Macro- and microtextures	Groundmass		Texture	
			Composition			
Olivine–pyroxene, pyroxene–olivine	Ol + Px, Ol	Oligophyric, aphric, seriate–porphyritic	Plagioclase–olivine–pyroxene; olivine–pyroxene–plagioclase		Intersertal, hyalopilitic, microlitic, doleritic	3–15
Plagioclase–olivine–pyroxene, pyroxene–olivine–plagioclase	Ol + Px + Pl	Oligophyric, coarsely porphyritic, weakly megaplagiophyric, aphyric	Pyroxene–olivine–plagioclase		Hyalopilitic, intersertal, vitrophyric, microlitic	5–25
Pyroxene–olivine–plagioclase	Px + Ol + Pl	Mega- and mesoplagiophyric	Olivine–plagioclase		Microlitic, intersertal	15–30
Plagioclase–olivine, olivine	Ol ± Pl ± Px	Aphyric, oligophyric	Olivine–plagioclase, plagioclase essentially dominant		Microlitic	3–5
Olivine–plagioclase	Ol + Pl ± Px	Mega- and mesoplagiophyric, aphyric (rarely)	Olivine–plagioclase, plagioclase essentially dominant		Intersertal, hyalopilitic, rarely vitrophyric	5–20

The petrographic varieties of basalts are distinguished by the ratios of phenocrysts and subphenocrysts; the second column indicates of transitional varieties exists between the basalts described, with every possible variation of quantitative relationship between the mineral phases of the phenocrysts and microlites, and their texture.

Petrochemical characteristics of the basalts of the areal zone are given on the basis of 119 analyses, of which 35 have already been published (Yermakov & Vazheyevskaya, 1973; Kirsanov & Ponomarev, 1974). Rocks of many cinder cones and their associated lava flows have been chemically analysed (within all the age groups described). Wide variations in chemical and petrographic composition are noted in these areal zone rocks: they can be arranged in general as a single series from high magnesian moderately alkaline basalts (type I) to subalkaline high-alumina (type III) with intermediate varieties (types II, IV) between them (Table 3).* The table of average compositions shows that while silica variation is insignificant (49.8–50.8%), considerable variation can be seen in the content of MgO (4.8–10.1%), Al_2O_3 (13.9–17.4%), CaO (8.9–10.7%), as well as of total alkalis (3.5–5.3%) and K_2O (1–1.8%). The 1975–76 Tolbachik eruption provided us with concrete evidence for the possibility of a single autonomous eruption of basalt types that are chemically contrasting but united by intermediate varieties that comprise, however, minimal amounts (3–7%) of the volume of erupted material (Volynets *et al.*, 1976b). Since these two extreme types differ above all in MgO and Al_2O_3 content, in classifying the areal basalts we can use the ratio of these oxides as the main correlation criteria (Figs. 2 and 3). Basalts whose chemical composition is represented by points plotted on an MgO–Al_2O_3 diagram (see Fig. 2) are designated further according to petrography and sometimes texture. In spite of the transitional character of the chemical composition, two separate composition fields, A and B, can be distinguished, separated by a minimal but marked break between values of MgO/Al_2O_3 = 0.42–0.45. Field A encloses basalts of magnesian type, field B of aluminous type. According to studies of the types of Quaternary basalts in Kamchatka (Volynets *et al.*, 1976a), the magnesian basalts (field A) correspond to the magnesian varieties of the calcalkaline and partly the subalkaline basalts, the aluminous basalts to the subalkaline aluminous (tephrite basalts group) and partly to the aluminous calcalkaline series. Since in the 1975–76 eruption we saw with our own eyes the effusion of basalts of these intermediate compositions from each of the breakthroughs (Volynets *et al.*, 1976b), it is worthwhile to distinguish their analogues

---

* The classification of the main types of these basalts is based on that proposed by Volynets *et al.* (1976a).

Table 3. *Average chemical composition of basalts in the areal zone of Ploskiy Tolbachik, the Tolbachik regional zone of cinder cones and the Ostryy and Ploskiy Tolbachik volcanoes*

Types of basalt; number of determinations (in brackets)	SiO$_2$	TiO$_2$	Al$_2$O$_3$	Fe$_2$O$_3$	FeO	MnO	MgO	CaO	Na$_2$O	K$_2$O	P$_2$O$_5$
A. Magnesian:											
I, magnesian proper (15);	49.80	1.20	14.29	4.63	6.05	0.19	9.43	10.28	2.58	1.17	0.32
Ia, Northern Breakthrough (1975) type (6);	49.68	1.12	13.90	4.29	5.92	0.20	10.12	10.73	2.43	1.05	0.32*
Ib, basalts without Ia subgroup (9);	49.88	1.25	14.55	4.86	5.62	0.18	8.97	9.99	2.68	1.24	0.32
II, magnesian of intermediate type (17)	50.54	1.37	14.86	3.63	7.28	0.18	7.64	9.56	2.84	1.48	0.41
B. Aluminous:											
III, aluminous subalkaline of Southern Breakthrough (1975) (49);	50.83	1.52	17.40	3.68**	6.53**	0.17	4.81	8.88	3.44	1.84	0.47**
IV, aluminous of intermediate type (17);	50.48	1.46	16.16	3.82	6.89	0.19	6.10	9.22	3.07	1.70	0.39
Basalts from the volcanoes Ostryy and Ploskiy Tolbachik (17)****	52.08	1.30	17.80	3.40	6.31	0.21	5.38	8.70	3.07	1.12	0.24***

Number of determination: * , 5; ** , 47; *** , 6; ****, from Yermakov.

within the main types of basalts in the entire areal zone as well (Fig. 2, Table 3). Among the magnesian basalts (A) can be distinguished: (1) a group of magnesian basalts proper (I) with olivine–pyroxene varieties only, and within them the subgroup Ia that corresponds in its range of MgO and Al_2O_3 values to the basalts that are the predominant types of the Northern Breakthrough of 1975; (2) a group of magnesian basalts of intermediate type (II). Corresponding groups are distinguished among the aluminous basalts (B): (1) high-alumina subalkaline basalts (III), the most widespread and most similar in composition to the predominant rock types of the Southern Breakthrough of the 1975–76 eruption; (2) aluminous basalts of intermediate composition (IV).

Approximations of the volume of basaltic material erupted over the

Fig. 2. MgO–Al_2O_3 ratios in basalts of the Tolbachik regional zone of cinder cones.
1, olivine–pyroxene subaphyric and aphyric basalts; 2, olivine–pyroxene, plagioclase, olivine–plagioclase basalts of subaphyric, aphyric texture, sometimes with rare megacrystals of plagioclase; 3, olivine–pyroxene–plagioclase basalts of mega- and mesoplagiophyre aspect; 4, mega- and mesoplagiophyre basalts (olivine–plagioclase); 5, average composition of basalts of the 1975–76 eruption (Volynets *et al.* 1976b): predominant type of Northern Breakthrough (1), intermediate type of Northern Breakthrough (2), predominant type of Southern Breakthrough (3), intermediate type of Southern Breakthrough (4); 6, composition fields of basalts from the Tolbachik areal zone: *A*, magnesian basalts, *B*, aluminous basalts; 7, boundaries of composition fields of basalts; 8, evolution lines of basalt compositions.
I–IV, basalt types given in Table 3.

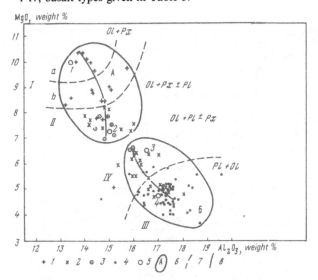

106   *Flerov and Bogoyavlenskaya*

areal zone (as at this stage of investigation) by analogy with the eruptions of the Northern and Southern Breakthroughs of 1975–76 and from reconstructions of buried flows, gave the relationships reflected in Table 1. For all age groups: types (III) and (IV) together comprise about 91% (52.8 km³), types (I) and (II) are close with 5% (3.2 km³) and 4% (2.4 km³) respectively; type (IV) is of insignificant development. For the period from the beginning of age group II (the first appearance of magnesian basalts) to the present and for the Southern zone only, the volume relationships of the different basalt types are now: types (III)+(IV) – 51% (5.8 km³), types (I) and (II) – 28% (3.2 km³) and 21% (2.4 km³) respectively.

In Fig. 2 it is possible to see a definite dependence of the petrography of the rocks in these groups on chemical composition. Olivine is the predominant mineral in rocks of every group. At the same time there is a direct correlation between the MgO/Al₂O₃ and Px/Pl ratios as the rock chemistry changes towards the high-alumina subalkaline varieties; thus pyroxene could be an indicator of the chemical composition of the rock.

Fig. 3. Magnesia–alkali ratios in basalts of the Tolbachik regional zone of cinder cones.
1, magnesian basalts proper (type I); 2, magnesian basalts of intermediate type (type II); 3, high-alumina subalkaline basalts (type III); 4, alumina basalts of intermediate type (type IV); 5–6, see 5 and 6 for Fig. 2. *a*, boundary of field of basalts of subalkaline and moderate alkalinity (Volynets *et al.*, 1976).

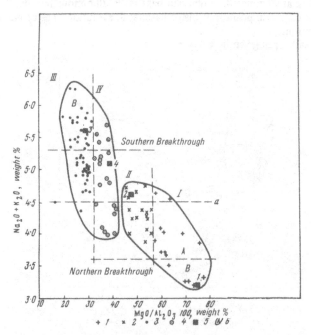

Differences are observed also in the degree of crystallinity of the rocks as shown by their phenocrysts. For the magnesian proper olivine–pyroxene basalts of group (I), independently of which facies they belong to, subaphyric and aphyric textures are typical, while the aluminous subalkaline (tephrite-basalts) are predominantly represented by megaplagiophyric lavas: their aphyric and subaphyric varieties (with single megacrystals of plagioclase) are usually found in the lava intercalations of cinder cones and in bombs. The subaphyric varieties (with or without plagioclase mega-crystals) and to a lesser degree the coarsely porphyritic olivine–pyroxene–plagioclase rocks of megaplagiophyric aspect, are characteristic of the intermediate groups (II and IV) under the same distribution control of textural varieties by facies (predominance of porphyritic textures in flows). The megaplagiophyric rocks form lava flows occasionally paragenetically related to large cinder cones made up of olivine–pyroxene subaphyric varieties. They form little extrusions within the actual cinder cones and are also found as a dyke at the foot of Gora Alaid (see Fig. 5).

The textural inhomogeneities of these volcanic rocks in their various manifestations had already been noted by a number of students of the volcanic complexes from the prehistoric phase of activity (Vazheyevskaya, 1976; Sirin, 1968; and others); the present eruptions are also being observed in relation to the dynamics of eruption activity. In particular, during the eruption of the Olimpiyskiy lateral breakthrough in Alaid in 1972, the first juvenile material consisted of basalts of subaphyric aspect; as the flow became greater the crystallinity of the rocks increased (Avdeyko *et al.*, 1974; Droznin & Filosofova, 1976). When all this regular behaviour is analysed, the differences in the crystallinity (in porphyritic texture particularly) can be considered due to different rates of cooling in a melt that was crystallising under different temperature gradients. Higher temperature gradients exist of course under conditions of explosion (when there is little restraint on the melt in the crater), and of eruption caused by a higher speed of supply of melt to the surface and its higher temperature. At the same time temperature gradients are lowered when the discharge of magmatic material into flows, extrusion domes and minor outlets is gentle. For the same reason, the composition and amount of phenocrysts in the rocks of an eruption reflect the cooling of the crystallising melt along definite paths of a liquidus–solidus system. Accordingly we have an opportunity to observe a textural rock series from aphyric to densely porphyritic, from seriate-porphyritic to porphyritic.

The plot in Fig. 3 shows distinctly that our aluminous and magnesian basalts exhibit different trends in the behaviour of the alkalis. The

magnesian basalts are characterised by relatively uniform increase, from
3.2 to 4.75% when MgO/Al_2O_3 is falling from 76 to 45, whereas the
aluminous basalts feature wide fluctuations in alkali content (from 4 to
6.3%) for lesser variation of $MgO/Al_2O_3 = (45–30)$. This inversion in the
behaviour of the alkali content is fixed between 4 and 4.75% in the
intermediate varieties (II) and (IV) of the two main types of basalts,
magnesian and aluminous, marked as A and B in Fig. 3.

A similarly stepped but not linear evolution of the chemical composition
of the two basalt types can also be seen on the MgO/Al_2O_3 plot (see Fig.
2) in the field of intermediate compositions of magnesian type basalts. The
general linear dependence of MgO and Al_2O_3 values is disrupted for a
number of rocks while the general tendency to gradual decrease of the
relative MgO content is preserved: alterations in Al_2O_3 content are
insignificant. At the same time an increase in total alkalinity is seen for
this rock group (see Fig. 3, type (II)).

Fig. 4. $MgO–Al_2O_3$ (a) and $Na_2O+K_2O–MgO/Al_2O_3$ (b) ratios in
basalts of age groups in the Tolbachik regional zone of cinder cones
and stratovolcanoes Ostryy and Ploskiy Tolbachik.
Basalts of the Tolbachik regional zone of cinder cones: 1–5, see Fig. 3,
1–5 for these conventional signs; 6, basalts of stratovolcanoes Ostryy
and Ploskiy Tolbachik; 7, basalts of the post–caldera eruptions of the
summit crater of Ploskiy Tolbachik.
An unbroken line joins compositions of basalts of volcanic complexes
due to a single eruption: Gora 1004 (1), Gora Alaid – Peschanyye
Gorki (2), Gora Vysokaya (3), 1941 Breakthrough (4), 1975–76
eruption (5). Arrows indicate the trend of basalt compositions. The
dashed line is the boundary between basalts of different composition:
magnesian (A), aluminous (B). I–V are age groups.

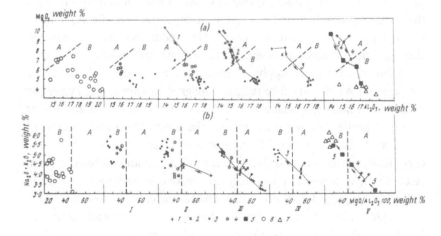

The presence of two different evolutionary series among the rocks of the areal zone corresponding to magnesian (*A*) and aluminous (*B*) basalts is evidence for the existence of two independent rock series (melts) whose chemical evolution is due above all to a specific initial magma for each series. The variations in alkalinity are without doubt greater during the evolution of the aluminous melts.

The inhomogeneity and contrast of the material composition of the basalts of the areal zone established by geological studies (see Fig. 1) are periodically repeated in its geological history (Fig. 4, I–IV). The composition of the rocks of the stratovolcanoes Ostryy and Ploskiy Tolbachik (Yermakov & Vazheyevskaya, 1973) plotted in Fig. 4, forms a wider field and corresponds to that of aluminous (calcalkaline) basalts of moderate alkalinity. These differ markedly from the corresponding type in the areal zone, first of all in lower alkali content, especially of K_2O, and to a lesser degree in increased alumina content (see Fig. 4 and Table 3). Thus direct inheritance of the composition of the basalts of the central structures by basalts of the areal zone has not been observed.

Although the eruption of magnesian basalts proper is not characteristic of the earlier complex (Fig. 4, I), in all the subsequent age cycles including the present one we can see manifestations of all the petrochemical types in their various relationships. These petrochemical varieties occur both within the scale of a single eruption and in eruptions closely associated in time and occasionally in space. Geological material that reveals the character of the spatial and temporal relationships of eruptions that produced the different types of basalts enables us to distinguish three types of eruption.

1. Autonomous eruptions of typically extreme petrochemical types of basalts (types (I) and (III)) – magnesian proper (olivine tholeiites) and subalkaline aluminous (tephrite basalts) that are characterised by independent eruption centres. Although basalts of intermediate varieties are indeed observed, they reflect only a definite time stage of the eruption and are clearly of subordinate significance in the volume of material erupted. Examples are the 1975–76 eruption (the Northern and Southern Breakthroughs), the 1941 breakthrough, the Krasnyy cone and many other areal eruptions of aluminous basalts. One of these constantly active volcanic centres that has been periodically erupting megaplagiophyric aluminous basalts right up to the present time is the summit crater of Ploskiy Tolbachik (Kirsanov & Ponomarev, 1974), whose post-caldera life is linked paragenetically with areal volcanism. Eruptions of this type of basalt

appeared both independently of and simultaneously with eruptions of magnesian basalts in the breakthroughs of 1941 and 1975.

2. Eruptions of intermediate-type basalts with a significant predominance of magnesian plagioclase–olivine–pyroxene basalts of type (II) – from independent eruptive centres unconnected with the centres described for type (I). In the nature of their eruption they resemble more the aluminous type. This type of eruption characterises the copious eruptions of Gora Kamenistaya and is found in various degrees over the entire areal zone. In the area we studied, these independent eruptions of intermediate type aluminous basalts are found only in association with the Kruglen'kiy cone eruption.

3. Eruptions of complex type that are characterised by close paragenetic and spatial relationships between the petrochemical types of basalts (magnesian proper, aluminous subalkaline, intermediate). The volcanic centres (cinder cones, effusive fissures, lava domes) that erupt these different types of basalts are spatially close and in a number of cases hereditary. In the areal zone to the south-west of Ploskiy Tolbachik (see Fig. 1) this type of eruption produced immense volcanic complexes that include the large cinder cones Peschanyye Gorki – Gora Alaid (both the result of a single eruption), Gora Vysokaya, and Gora 1004. Study of these large cinder cones has shown that they possess a complex polycentric structure and are characterised by a two-stage development (Fig. 5). The rocks of these cinder cones and their associated lava flows reflect a general regular petrochemical variation in composition from magnesian to high-alumina subalkaline basalts (see Fig. 4). The first phase of the eruption formed thick cinder and lava flows of basalts mostly buried under subsequent lava effusions of the magnesian proper kind (type (I)) in the case of Gora Alaid, Gora Vysokaya, Peschanyye Gorki and Gora 1004. It is now possible to establish that for Gora 1004 the eruption terminated with magnesian basalts of intermediate (II) type. The subsequent phase of the eruption in the Gora Alaid – Peschanyye Gorki complex and of Gora Vysokaya was marked by basalts of intermediate type (type II, less of type IV) and culminated in eruptions of megaplagiophyric aluminous basalts (type III). The basalts of this phase flood out from fissures and domes at the foot of the cones and also form Sommalian-Vesuvian type structures in the craters of phase I cones and minor extrusions, dykes and fissure flows on their slopes (see Fig. 5(a)). In their turn these small volcanoes, when situated within crater centres, can produce lava flows – the eastern cone of Peschanyye Gorki, for instance, and the south-eastern cone of Gora

Fig. 5. Geological and petrographical sketch map of the volcanic complexes of Gora Alaid – Peschanyye Gorki (a) and Gora Vysokaya (b).
1, olivine–pyroxene subaphyric and aphyric basalts; 2, olivine–pyroxene–plagioclase, olivine–plagioclase subaphyric, aphyric and megaplagiophyre textures; 3, olivine–pyroxene–plagioclase basalts of mega- and mesoplagiophyric aspect; 4, mega- and mesoplagiophyric (olivine–plagioclase) basalts; 5, dyke of megaplagiophyre basalts; 6–8, petrochemical types of basalts: 6, magnesian proper, 7, magnesian of intermediate type, 8, aluminous subalkaline; 9, areal fallout of bombs of aluminous subalkaline basalts; 10, cinder cones.

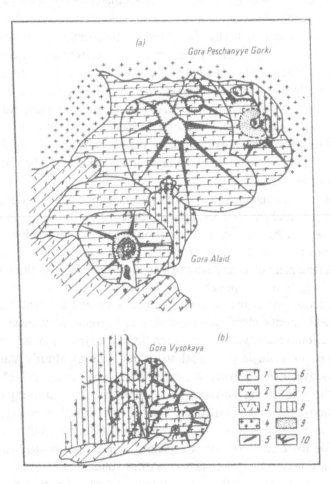

Vysokaya (Fig. 5(*a*), (*b*)). The thick eruptions of subalkaline aluminous basalts with which this phase culminates are especially characteristic of the Gora Vysokaya volcanic complex, where they form a cinder cone eccentrically situated on a cone of the previous phase and pour out a series of lava flows over a considerable area (see Fig. 5(*b*)).

The material that we have obtained on the geology, petrochemistry and volume of eruption products in the Tolbachik areal zone demonstrates that the differences in chemical composition of the basalts are fairly difficult to explain in terms of crystal differentiation (to which Sirin & Timerbayeva (1971) assigned a leading role) or the magmatic differentiation in the liquid state of a single basaltic magma in a volcanic channel or reservoir.* Such explanations are contradicted by: the tendency of the chemical composition of the basalts to change during the eruption process from magnesian to aluminous, culminating in aluminous subalkaline basalts, a process that is systematically repeated in volcanic complexes of different age groups; the paragenesis of contrasting petrochemical types of basalts, which likewise are repeated periodically in the geological history of the areal zone; and the subaphyric and aphyric aspect of the rocks, which is typical of magnesian type basalts and observed in a number of cases in bombs of aluminous basalts.

At the same time the separation of chemical composition fields of the rocks on the petrochemical diagrams, their different evolution within the various series and the presence of an inversion step in the composition field of transitions between them, rather testifies to the relative independence of two initial magmas of magnesian and aluminous types. Thus at this stage of the investigation it is possible to speak of the existence under the Tolbachik areal zone of two magma chambers, each at a different depth and each with specific chemical composition and evolutionary direction. This conclusion is once again confirmation that these types of magma can be generated in a single geological structural zone as already stated (Vazheyevskaya, 1976; Volynets *et al.*, 1976a; Volynets *et al.*, 1976b).

The authors, following Volynets *et al.* (1976b), are inclined to explain the formation of these many varieties of basalt united through intermediate types, by an hypothesis involving mixing of melts during uplift to the surface. The mechanism of this is explained in principle in the article 'Geochemical features...' (Volynets *et al.*, this volume).

It is possible to say something concerning the depth of occurrence of

---

* The role of differentiation processes in the genesis of these types of melts is treated in greater detail by Volynets *et al.* in this volume.

the basalt magma chambers from eruption dynamics. According to data for the 1975–76 eruption (Fedotov *et al.*, 1976), the dynamics of the Northern and Southern Breakthroughs were essentially different. The work of the Northern Breakthrough erupting high-magnesian basalts is evidence of the huge part played by the gaseous phase in its activity. Throughout the period of activity of these cones, a fiery column of gas and ash rising to 8–12 km persisted over the craters. As a result of explosive activity, three cones were formed, one of which attained a height of 330 m in less than a month (Cone I). The explosion index of the Northern Breakthrough was 90% for a total volume of magmatic material of 1.44 km^3 ejected in two months. The eruptive activity of the Southern Breakthrough, which produced high-alumina subalkaline basalts, was of a calm effusive character. Its explosion index did not exceed 10% for a total volume of magmatic material of about 0.8–0.9 km^3 ejected in fifteen months. This vent operated in pulses, pyroclastics were ejected to a height of the order of 200 m and the height of the single cone was 180 m by the end of the eruption. The discharge of magmatic material at the Northern Breakthrough was 130 m^3/s, at the Southern Breakthrough three times less (Fedotov, 1976).

These features of the eruption dynamics for the respective magmas are systematically characteristic of the volcanism over the entire Tolbachik areal zone. In effect, the most substantial cinder cones are made up of olivine–pyroxene magnesian basalts proper of Northern Breakthrough type. At the same time the aluminous basalts and magnesian basalts of intermediate type form small cones, while the cinder cones of phase II of the complex type of eruption (type III, Peschanyye Gorki – Gora Alaid) centrally emplaced onto phase I volcanic craters, are formed without any destruction of the previous structures and the areal scatter of the bombs does not exceed the crater limits (see Fig. 5(*b*)). One indication of the power with which the magnesian basalts were erupted, to which Fedotov duly drew attention, was the formation during the eruption, as an effect of the high-velocity gas jet, of very smooth, rounded and spherical bombs often with inclusions of extraneous material.

There is no doubt that the main energy factor and moving force of the magma was the gas contained in it. Thus when the dynamic characteristics described above for the different types of basalt are taken into account, it is possible to speak of the greater energy capacity of the magnesian proper type compared with the subalkaline high-alumina type, and to some degree (as a consequence) of the greater depth of occurrence of its

114   *Flerov and Bogoyavlenskaya*

chamber. Powerful eruptions of magnesian basalts of greater energy potential might trigger subsequent eruptions of less energetic aluminous subalkaline basalts that are apparently uplifted from relatively lesser depths.

On the basis of calculations of magma discharge from a dyke during a basaltic fissure eruption, Fedotov (1976) had already come to the same conclusion: he considered that magmas in large volcanoes (in his case, magnesian cinder cones) in fields of areal volcanism should rise from chambers at greater depths. That the chambers of the magnesian and aluminous subalkaline basalt magmas are situated at different depths is borne out by the seismic circumstances prevailing before the Northern and Southern Breakthroughs of the GTFE (Fedotov, 1976).

It is understandable from what has been said that the magnesian basalts proper are absent from the early age group of the areal eruption zone and that the systematic succession of eruptions of the two magma types are close in time and space.

The mechanism proposed for the advent of magma from chambers at different depths does not exclude – on the contrary, it increases – the probability of magma conduits being inherited, which would facilitate magma mixing.

In conclusion, the authors thank S. A. Fedotov and O. N. Volynets for discussion of their results and for editing the paper in preparation for publication.

References

Avdeyko, G. P. *et al.* 1974. Izverzheniye vulkana Alaid v 1972 g. (The 1972 eruption of the Alaid volcano.) *Byul. vulkanol. stantsiy*, No. 50, 64–80.
Droznin, V. A. & Filosofova, T. N. 1976. Termodinamicheskaya informativnost' kristallichnosti produktov izverzheniya. (Thermodynamic information obtainable from the crystallinity of eruption products.) *Dokl. AN SSSR*, **229**(2), 447–50.
Fedotov, S. A., Khrenov, A. P. & Chirkov, A. M. 1976. Bol'shoye treshchinnoye Tolbachinskoye izverzheniye, 1975, Kamchatka. (The Great Tolbachik Fissure Eruption, 1975, Kamchatka.) *Dokl. AN SSSR*, **228**(5), 1193–6.
Fedotov, S. A. 1976. O pod"yeme osnovnykh magm v zemnoy kore i mekhanizme treshchinnykh bazal'tovykh izverzheniy. (On the uplift of basic magmas in the crust and the mechanism of basaltic fissure eruptions.) *Izv. AN SSSR*, Ser. geol., No. 10, 5–23.
Kirsanov, I. T. & Ponomarev, G. P. 1974. Izverzheniya vulkana Ploskiy Tolbachik i nekotoryye osobennosti ikh produktov. (The eruptions of the Ploskiy Tolbachik volcano and some features of its products.) *Byul. vulkanol. stantsiy*, No. 50, 53–63.
Melekestsev, I. V., Krayevaya, T. S. & Braytseva, O. A. 1970. Rel'yef i otlozheniya molodykh vulkanicheskikh rayonov Kamchatki. (*Relief and deposits of the young volcanic regions of Kamchatka.*) M., Nauka, 104 pp.
Piyp, B. I. 1946a. Novyy pobochnyy krater vulkana Tolbachik. (A new lateral crater of Tolbachik volcano.) *Byul. vulkanol. stantsii na Kamchatke*, No. 13, 19–21.

Piyp, B. I. 1946b. Vulkan Ploskiy Tolbachik. (The Ploskiy Tolbachik volcano.) *Byul. vulkanol. stantsii na Kamchatke*, No. 12, 70–3.

Piyp, B. I. 1956. *Klyuchevskaya sopka, yeye izverzheniya v 1944–45 gg. i v proshlom. (Klyuchevskaya Sopka, its eruptions in 1944–45 and the past.)* Trudy Labor. vulkanol., No. 11, 310 pp.

Sirin, A. N. 1968. *O sootnoshenii tsentral'nogo i areal'nogo vulkanizma. (On the relationship between central and areal volcanism.)* M., Nauka, 196 pp.

Sirin, A. N. & Timerbayeva, K. M. 1971. O dvukh tipakh bazal'tov i sostave iskhodnoy magmy vulkanov Klyuchevskoy gruppy na Kamchatke. (On the two types of basalts comprising the initial magma of the Klyuchi group of volcanoes in Kamchatka.) In: *Vulkanizm i glubiny Zemli. (Volcanism and the depths of the Earth.)* M., Nauka, pp. 147–50.

Vazheyevskaya, A. A. 1976. Dva tipa bazal'tov na Kamchatke. (Two types of basalts in Kamchatka.) *Byul. vulkanol. stantsiy*, No. 52, 127–34.

Volynets, O. N. *et al.* 1976a. Petrokhimicheskiye tipy chetvertichnykh bazal'tov Kamchatki i ikh geologicheskoye polozheniye. (Petrochemical types of the Quaternary basalts of Kamchatka and their geological position.) *Byul. vulkanol. stantsiy*, No. 52, 115–26.

Volynets, O. N. *et al.* 1976b. Petrologiya vulkanicheskikh porod treshchinnogo Tolbachinskogo izverzheniya 1975 goda. (Petrology of the volcanic rocks of the Tolbachik fissure eruption of 1975.) *Dokl. AN SSSR*, **228**(6), 1413–22.

Yermakov, V. A. 1971. Megaplagiofirovyye lavy – analog anortozitovykh porod. (Megaplagiophyric lavas are the analogues of anorthositic rocks.) *Izv. AN SSSR*, Ser. geol., No. 10, 56–72.

Yermakov, V. A. & Vazheyevskaya, A. A. 1973. Vulkany Ostryy i Ploskiy Tolbachiki. (The Ostryy and Ploskiy Tolbachik volcanoes.) *Byul. vulkanol. stantsiy*, No. 49, 43–54.

Zubin, M. I., Smirnov, V. S. & Tarakanovsky, A. A. 1977. Plotnostnyye i geoelektricheskiye neodnorodnosti zemnoy kory i verkhney mantii v rayone Klyuchevskoy gruppy vulkanov. (Density and geoelectric inhomogeneities of the crust and upper mantle in the area of the Klyuchi group of volcanoes.) In: *Geodinamika vulkanizma i gidrotermal'nogo protsessa. (Geodynamics of volcanism and the hydrothermal process.)* Trudy IV Vses. vulkanol. soveshchaniya. M., Nauka, pp. 136–44.

Geochemical features of the rocks of the Great Tolbachik Fissure Eruption 1975–1976 in relation to petrogenesis

O. N. Volynets, G. B. Flerov, V. N. Andreyev,
E. I. Popolitov, V. A. Abramov, L. L. Petrov,
S. A. Shcheka and G. I. Selivanova

A study has been made of the distribution in the rocks produced by the eruption of the major (Si, Al, Mg, Ca, Fe, Ti, Na, K, P) and minor (Rb, Li, F, B, Be, Ag, Nb, Ta, Zr, Hf, Ba, Sr, Ni, Co, Cr, V, Cu, Zn, Pb, Sn, Ga, Mo, Mn, Sc, As, Ge, Y, La, Ce, Nd, Yb) elements. The degree of detail involved and the accuracy of the methods used varied. The most detailed data were obtained for the major elements and also for Rb, Li, F, B, Be, P, Ti, Ni, Co, Cr, V, Cu, Zn, Pb, Sn and Ga; data on the remaining elements are so far only preliminary. The quantities of the major elements were determined by the usual methods of complete silicate analysis of the rocks in the chemical laboratory of the Institute of Volcanology of the DVNTs AN SSSR. The Na, K, Pb and Li contents were determined in the Vinogradov Institute of Geochemistry of SO AN SSSR (using flame photometry), as well as F, B, Be and Ag (by the quantitative spectrum method using special techniques evolved for F, B and Be by L. L. Petrov, and for Ag by V. A. Legeydo), Ba, Sr, La, Ce, Nd, Y and Yb (using quantitative X-ray spectroscopy), Sr, Mg, Ca and Al (using atomic absorption on the Perkin-Elmer 403 instrument), and Nb, Ta, Zr and Hf (using the spectrochemical method). The Ni, Co, Cr, V, Ga, Pb, Zn, Cu and Sn contents were determined by quantitative spectroscopy in the physical chemistry research laboratory of the Far East Geological Institute of the DVNTs AN SSSR. Preliminary data on the Ni, Co, Cr, V, Ag, Cu, Zn, Pb, Mo, Sn, Sc, Ga, Ge, Y, Zr, P, Ti, Mn, B and Nb contents were obtained using refined semi-quantitative spectrum analysis in the laboratory of the Geochemical Expedition of the Central Geophysical Trust (in the town of Aleksandrov). The As and Sb contents were also determined there, using a special technique of quantitative spectrum analysis. At the Institute of Geochemistry of SO AN SSSR two determinations were made of the isotope composition of Sr from basalts of the eruption.

Mention must first be made of information published earlier (Volynets *et al.*, 1976a) on the petrochemical properties of the solid volcanic products of the eruption. As was noted there, all the rocks of the eruption correspond to basalts in their composition. And although in Yoder & Tilley's classification (1965) they are all placed in one group – that of the olivine–tholeiites, since they contain normative olivine and hypersthene – the basalts of the Northern and Southern Breakthroughs differ greatly in Al_2O_3, MgO, CaO, K_2O, Na_2O, P_2O_5 and TiO_2 content. The MgO and CaO content is higher in the basalts of the Northern Breakthrough, whilst that of the other components is higher in the Southern Breakthrough (Table 1, Fig. 1). The difference in silica and iron oxide content in the two Breakthroughs is very small (0.6–0.7%), both being slightly higher in rocks of the Southern Breakthrough. Because of the sharp decrease in Mg content in the rocks of the Southern Breakthrough compared with the Northern, the former are much more ferruginous than the latter (50–60 at. %, as against 30–40%). The basalts of both Breakthroughs have a similar degree of iron oxidisation (25–35 at. % for the vast majority of the rocks).

These very different basalts of the Northern and Southern Breakthroughs are united in the course of the eruption by the appearance of basalts of an intermediate composition (see Fig. 1). The latter occur mainly at the end of the Northern and beginning of the Southern Breakthroughs and over an extremely limited interval of time. The composition of the basalts of the Northern Breakthrough began to change only a week before it ceased activity (the Northern Breakthrough lasted a total of ten weeks). That of the basalts of the Southern Breakthrough (which lasted almost fifteen months) also changed radically only in the first week of the eruption, although over the next two months the composition gradually evened out. One must also remember that the volumes of the intermediate basalts are quite small, comprising only 5–10% of the volume of rocks from each Breakthrough.

According to the classification scheme offered earlier for Quaternary basalts of Kamchatka (Volynets *et al.*, 1976b), the basalts of the Northern Breakthrough are of the magnesian type with moderate alkalinity and the basalts of the Southern Breakthrough are of the aluminous subalkaline type. In terms of the ratio of MgO to Al_2O_3/SiO_2 the basalts of the Northern Breakthrough are similar to Hawaiian tholeiites, those of the

Table 1. *Average chemical composition of basalts from the Tolbachik eruption of 1975–76*

Oxides	Northern Breakthrough		Southern Breakthrough		
	Predominant lava types (7 July 75–10 Sept. 75)	Intermediate lavas from the end of the Northern Breakthrough (11–15 Sept. 75)	Intermediate lavas from the first days of the Southern Breakthrough (18–24 Sept. 75)	Intermediate lavas from the start of the eruption (25 Sept.–30 Nov. 75)	Predominant lava types (1 Dec. 75–9 Dec. 76)
SiO_2	49.76 / 49.18–50.91	50.02 / 49.46–50.74	50.30 / 49.58–50.76	50.90 / 50.20–51.46	50.69 / 49.72–51.88
TiO_2	1.02 / 0.83–1.10	1.30 / 1.10–1.59	1.51 / 1.31–1.78	1.70 / 1.45–1.85	1.66 / 1.30–1.95
Al_2O_3	13.48 / 12.10–14.22	15.32 / 13.83–16.35	16.62 / 16.02–17.07	16.83 / 16.12–17.30	17.10 / 15.91–18.60
Fe_2O_3	3.06 / 2.08–4.86	3.47 / 2.22–4.87	3.14 / 2.03–5.09	3.42 / 2.74–4.46	3.55 / 1.49–6.01
FeO	6.99 / 5.48–8.19	6.88 / 5.72–7.47	6.95 / 5.11–7.61	7.05 / 5.82–7.56	6.99 / 4.78–8.79
MnO	0.16 / trace–0.21	0.17 / 0.15–0.18	0.18 / 0.13–0.20	0.17 / 0.17–0.21	0.17 / 0.08–0.22
MgO	9.88 / 9.21–10.66	7.69 / 6.21–8.83	6.44 / 5.74–6.95	5.39 / 4.42–6.26	4.87 / 3.83–5.83
CaO	11.60 / 10.46–12.34	9.83 / 9.12–10.53	9.20 / 8.67–9.75	8.74 / 8.06–9.17	8.65 / 7.84–9.42
Na_2O	2.44 / 2.18–2.75	3.14 / 2.83–3.96	3.30 / 3.09–3.52	3.36 / 3.21–3.72	3.60 / 3.24–3.97
K_2O	1.03 / 0.84–1.44	1.62 / 1.36–1.89	1.83 / 1.75–2.02	1.99 / 1.85–2.19	2.10 / 1.88–2.40
P_2O_5	0.25 / 0.16–0.37	0.35 / 0.27–0.40	0.40 / 0.20–0.52	0.40 / 0.13–0.57	0.53 / 0.42–0.68
H_2O^+	0.13 / 0.00–0.68	0.09 / 0.00–0.43	0.02 / 0.00–0.19	0.06 / 0.00–0.20	0.06 / 0.00–0.39
H_2O^-	0.18 / 0.00–0.34	0.11 / 0.00–31	0.19 / 0.00–0.34	0.25 / 0.00–0.50	0.11 / 0.00–0.46
n	21	11	8	14	85

n is the number of analyses for calculation of the average analysis. The numerator gives the average content for basalts, the denominator the range of content in individual samples.

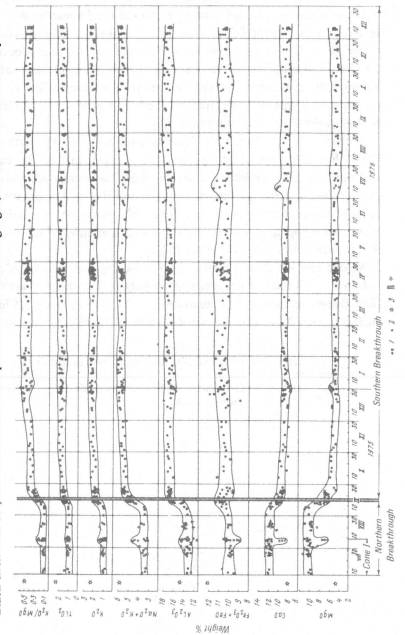

Fig. 1. Variation of chemical composition of rocks during eruption. 1, lavas, bombs; 2, ashes; 3, cinders from summit crater of Ploskiy Tolbachik; 4, pause between eruptions of the Northern and Southern Breakthroughs. Black circle with dot, data from atomic absorption method. Remaining signs, results of bulk silicate analysis.

120   *Volynets* et al.

Southern Breakthrough to Hawaiian alkaline basalts (Fig. 2), although in terms of alkali and titanium content, K/Na ratio and a number of other features the rocks of the Tolbachik eruption and the Hawaiian lavas differ considerably. The average composition of the intermediate basalts does not correspond to any of the calculated average types of basalts of Kamchatka, although it is extremely similar to the average composition of the corresponding type of basalts from the Tolbachik areal zone (see Flerov & Bogoyavlenskaya, this volume).

The amounts of rare and trace elements in the basalts of the North and South Breakthroughs differ essentially as the amounts of the main rock-forming elements differ (Tables 2 and 3). The average amounts of minor elements in the various types of basalts of the eruption are similar to those of the corresponding average petrochemical types of basalts of Kamchatka (Popolitov *et al.*, 1976b; Leonova & Kirsanov, 1974; Leonova & Ogorodov, 1975; Leonova *et al.*, 1974).

As one can see from Tables 2 and 3, the regional features of the volcanites of Kamchatka recorded previously by a number of authors (Leonova & Kirsanov, 1974; Leonova *et al.*, 1974; Popolitov *et al.*, 1976a), such as increased B content compared with the Clarke (5–10 times higher),

Fig. 2. The relation between MgO and Al₂O₃/SiO₂ in rocks of the eruption.
1–2, Northern Breakthrough; 3–4 Southern Breakthrough; 1, 3, lavas and bombs; 2, 4, cinder and ash.

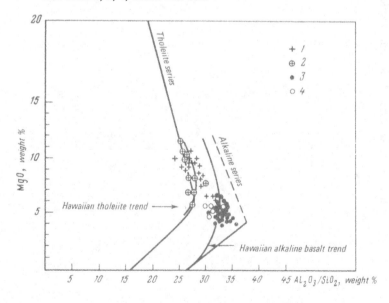

a lower Nb content (3–5 times less) and Ti content (1.3–2 times less), also occur in the rocks of this eruption. Moreover, the first quantitative As and Sb determinations carried out for the Kamchatka basalts also showed increased concentrations (2–5 times greater) of these elements relative to the Clarke (Turekian & Wedepohl, 1961).*

In comparison with the predominant type of Kamchatka basalts – aluminous basalts of moderate alkalinity (or, to use a different terminology, basalts of the calc-alkalic series) – the basalts of the Northern Breakthrough have higher Mg, Cr, Ni, Ag and B contents and less Cu, Be, Nb and Ta, with similar amounts of Ca, Fe, Ti, Sr, P, Mn, Co, La, Ce, Nd, Y and Yb and slightly less Al, Na, K, Rb, Li, Ba, F, Pb, Sn, Mo, Zn and V.

The basalts of the Southern Breakthrough differ from this dominant basalt in having higher Na, K, Rb, Ti, P, B, Be, Cu, Nb, La, Ce, Nd, Y and Yb contents, and slightly less Mg, Ca, Zn, V and Co, with similar amounts of the remaining elements.

In terms of the magnitude of the relative variation in the contents of major and minor elements (n) in the basalts of the Southern Breakthrough compared with those of the Northern, the following groups of elements can be distinguished: $n > 3$ Hf, Rb, Zr, La; $n = 2$–3 K, P, Pb, Ba, Nd, Be; $n = 1.5$–2.0 Ti, B, Mo, Nb, Y, Yb; $n = 1.25$–1.5 Na, Al, Li, Ga, F; $n = 1.1$–1.25 Cu, Mn, Sb, Ge; $n = 0.9$–1.1 Si, Fe, Sn, Sr, Ta; $n = 0.8$–0.9 Sc, V; $n = 0.7$–0.8 Ca, Zn, Co; $n = 0.5$–0.7 Mg; $n < 0.5$ Cr, Ni.

It follows from the above that the basalts of the Southern Breakthrough differ from those of the Northern in having higher concentrations of almost all the minor and trace elements studied (especially those whose geochemical history is to some extent related to that of K–Rb, P, Ba, Pb), and consequently they have a noticeably higher coefficient of accumulation (Shaw, 1969). Only a small group of elements characteristic of highly magnesian types of rocks (Cr, Ni, Co), as well as Sc, V and Zn, are concentrated (together with Mg and Ca) mainly in the basalts of the Northern Breakthrough.

The basalts of intermediate type occupy a regular intermediate position in relation to the basalts of the Northern and Southern Breakthroughs in terms of their minor and trace as well as major element contents. Moreover, during the period of the eruption in which they appeared (the end of the Northern Breakthrough and the beginning of the Southern),

---

* Data in the authors' possession on the Sb content in intermediate–acidic Quaternary lavas of Kamchatka, obtained independently by the neutron activation method, are also 5–10 times higher than the Clarke.

Table 2. *Content of minor elements in basalts of the Tolbachik eruption from results of refined semiquantitative analysis*

| Elements | Northern Breakthrough | | Southern Breakthrough | | | Holocene |
	Predominant mass of rocks 5 July–10 Sept. 75 (43)	Intermediate varieties 11–15 Sept. 75 (12)	Intermediate varieties 18–24 Sept. 75 (31)	Intermediate varieties 24 Sept.–30 Nov. 75 (27)	Predominant mass of rocks 1 Dec. 75–9 Dec. 76 (55)	Megaplagiophyre lavas of the Tolbachik zone of cinder cones (9)
1	2	3	4	5	6	7
Cu	160 / 50–300	160 / 120–300	230 / 100–400	210 / 100–400	340 / 150–500	200 / 100–400
Zn	230 / 80–800	170 / 120–250	210 / 100–400	200 / 120–300	270 / 200–600	180 / 120–250
Pb	2.5 / 2.0–3.0	4 / 2.0–6.0	5 / 2.0–10.0	4.5 / 4.0–20	7 / 4.0–15.0	7 / 3.0–15.0
Ni	55 / 20.0–100.0	40 / 30.0–80.0	40 / 20.0–80.0	35 / 20.0–50.0	35 / 20.0–60.0	30 / 20.0–40.0
Co	25 / 8.0–50.0	20 / 12.0–40.0	20 / 10.0–40.0	20 / 12.0–30.0	23 / 12.0–40.0	20 / 8.0–30.0
Cr	390 / 150–1000	300 / 200–400	270 / 120–500	210 / 150–400	160 / 100–300	130 / 60–250
V	230 / 80–400	220 / 150–300	280 / 120–500	260 / 200–400	260 / 200–600	250 / 200–500
Mo	0.5 / 0.3–0.8	0.6 / 0.4–1.0	0.7 / 0.3–2.0	0.9 / 0.6–1.5	1.0 / 0.6–1.5	1.0 / 0.3–1.5
Ag	—	0.08 / 0.08–0.08	0.07 / 0–0.1	0.05 / 0–0.4	0.08 / 0.08–0.6	0.08 / 0–0.12
Mn	1100 / 100–1500	950 / 600–1200	950 / 500–1500	950 / 600–1200	1300 / 1000–1500	950 / 600–1000
As	5 / 2.0–15	4 / 3.0–8.0	4 / 3.0–6.0	5 / 3.0–15.0	5 / 3.0–20.0	3 / 2.0–4.0
Sb	0.4 / 0.0–3.0	1.7 / 0.0–8.0	1.4 / 0.0–6.0	2.3 / 0.0–5.0	1.5 / 0.0–6.0	1.2 / 0.0–4.0
Sn	1.5 / 0.0–3.0	1.7 / 1.0–4.0	1.1 / 0.0–5.0	1.3 / 1.0–3.0	2.6 / 2.0–5.0	2.0 / 1.0–5.0
Sc	9 / 0.0–12.0	10 / 8.0–12.0	8 / 0–12.0	10 / 8.0–12.0	8 / 8.0–10.0	8 / 8.0–10.0
Ti	1700 / 500–8000	2100 / 500–3000	1900 / 1000–4000	1900 / 1500–5000	3900 / 40–6000	2800 / 2000–5000
Nb	0.02 / 0.0–8.0	2 / 0.0–8.0	0.5 / 0.0–8.0	1.2 / 0.0–8.0	2.2 / 0.0–8.0	1.0 / 0.0–8.0
Y	8 / 8.0–10.0	10 / 8.0–12.0	9 / 8.0–10.0	9 / 8.0–15.0	10 / 8.0–12.0	9.0 / 8.0–12.0

Table 2. (*cont.*)

Elements	Northern Breakthrough		Southern Breakthrough			Holocene
	Predominant mass of rocks 5 July– 10 Sept. 75 (43)	Intermediate varieties 11–15 Sept. 75 (12)	Intermediate varieties 18–24 Sept. 75 (31)	Intermediate varieties 24 Sept.– 30 Nov. 75 (27)	Predominant mass of rocks 1 Dec. 75– 9 Dec. 76 (55)	Megaplagio- phyre lavas of the Tolbachik zone of cinder cones (9)
1	2	3	4	5	6	7
Zr	70	75	70	70	80	75
	50–80	60–100	50–100	50–80	50–100	60–100
P	1400	2700	3700	3100	5400	3200
	600–2000	300–4000	1000–6000	2500–6000	2000–8000	2000–5000
Ge	1.0	0.9	1.1	0.8	1.3	0.8
	0.0–2.0	0.8–1.0	0.0–2.0	0.0–2.0	1.0–2.0	0.0–1.2
Ga	11	12	14	11	19	9
	6.0–15.0	8.0–20.0	8.0–25.0	6.0–20.0	12.0–30.0	6.0–20.0
B	25	30	30	25	60	35
	12.0–30.0	15.0–40.0	15.0–60.0	20.0–50.0	30.0–100.0	20.0–50.0

Numbers in brackets are the number of analyses made; content of elements in g/t. Numerator is average content of minor element; denominator is range from minimum to maximum content.

the amounts of these elements generally changed uniformly from those typical of magnesian basalts with moderate alkalinity to ones typical of subalkaline aluminous basalts (see Figs 1 and 3). Correspondingly, the dispersions of the amounts of the individual elements in the basalts of intermediate composition greatly exceed (by 1.5–3 times) those for the predominant mass of rocks of each Breakthrough.*

Appreciable differences can also be observed in the composition of basalt samples of the intermediate stage collected during a single day of the eruption. Thus, whereas in those from one day's eruption of the Northern Breakthrough the variations in Na content do not exceed 0.15–0.18%, K 0.08–0.10%, Rb 1–2 g/t, B 3–10 g/t, F 100–150 g/t, Be 0.3–0.5 g/t, and MgO 1.0–1.3%, in basalts of the last day's eruption of this Breakthrough the variations in Na content attain 0.3%, K 0.4%, Rb 7 g/t, B 20 g/t, F 420 g/t, Be 1 g/t, and MgO 2.6%. For the purposes of comparison we should point out that during the first two months of the eruption of the Northern

* This does not apply to F, B and H_2O^+, whose behaviour during the eruption is examined in a separate paper.

Table 3. *Content of minor and rare elements in basalts of the Tolbachik eruption*

	Northern Breakthrough		Southern Breakthrough		
Elements	Predominant lava types (7 July 75–10 Sept. 75)	Intermediate lavas from the end of the eruption (11–15 Sept. 75)	Intermediate lavas from the first days of the eruption (18–24 Sept. 75)	Intermediate lavas from the start of the eruption (25 Sept. 75–30 Nov. 75)	Predominant lava types (1 Dec. 75–9 Dec. 75)
1	2	3	4	5	6
Na	1.87 (68)	2.27 (15)	2.54 (18)	2.67 (19)	2.71 (136)
	1.62–2.18	2.04–2.58	2.35–2.80	2.38–2.76	2.48–3.40
K	0.75 (68)	1.31 (17)	1.49 (18)	1.60 (19)	1.75 (136)
	0.66–0.86	1.00–1.60	1.40–1.57	1.52–1.68	1.66–2.21
Rb	11.6 (56)	23.7 (7)	31.4 (11)	40.0 (15)	46.5 (94)
	10–14	20–30	27–35	31–50	38–60
Li	9.8 (56)	12.4 (7)	13.7 (11)	15.1 (14)	15.7 (81)
	4.5–23.0	8–16	10–19	11–21	10–48
F	285 (65)	350 (19)	350 (22)	310 (27)	405 (147)
	150–560	190–680	180–540	250–500	110–830
B	34 (72)	43 (19)	57 (22)	56 (27)	60 (167)
	19–48	20–57	20–85	45–66	40–99
Be	0.67 (75)	1.1 (19)	1.4 (23)	1.3 (27)	1.5 (167)
	0.45–1.30	0.70–1.90	1.0–2.2	0.90–1.70	0.80–2.60
Ba	330 (15)	440 (4)	535 (7)	535 (6)	870 (11)
	280–360	420–470	430–600	450–690	570–1100
Sr	315 (16)	425 (5)	410 (8)	470 (6)	325 (11)
	220–450	300–530	240–530	360–570	240–450
Ni	120 (28)	78 (10)	71 (11)	41 (10)	52 (50)
	55–182	40–125	36–130	26–55	24–100
Co	42 (28)	32 (10)	31 (11)	29 (10)	31 (50)
	30–57	24–42	25–45	23–35	20–46
Cr	265 (28)	195 (10)	145 (11)	120 (10)	102 (48)
	160–460	100–430	85–200	57–190	38–210
V	250 (28)	220 (10)	225 (11)	170 (10)	210 (50)
	150–320	150–310	158–300	93–240	93–320

Table 3. (*cont.*)

Elements	Northern Breakthrough		Southern Breakthrough		
	Predominant lava types (7 July 75–10 Sept 75)	Intermediate lavas from the end of the eruption (11–75 Sept. 75)	Intermediate lavas from the first days of the eruption (18–24 Sept. 75)	Intermediate lavas from the start of the eruption (25 Sept. 75–30 Nov. 75)	Predominant lava types (1 Dec. 75–9 Dec. 75)
1	2	3	4	5	6
Ga	12 (28) 7–20	11 (10) 5–19	12 (11) 5.5–22	12 (10) 6–19	16 (50) 7–28
Pb	2.7 (28) 0–5	3.2 (10) 0–6	4.4 (11) 2–7	3.5 (10) 2–6	5.9 (50) 0–12
Zn	75 (27) 42–91	52 (10) 28–110	58 (10) 36–79	45 (10) 36–63	59 (50) 30–100
Cu	145 (28) 100–220	140 (10) 110–190	170 (11) 110–240	155 (10) 110–190	175 (50) 110–290
Sn	3.3 (28) 1–5	2.2 (10) 1–4	3.1 (11) 1–9	2.4 (10) 1–4	3.3 (50) 1–5
Ag	0.15 (51) 0.08–0.25	0.13 (7) 0.11–0.16	0.15 (9) 0.12–0.19	—	—
F	0.11 (29) 0.07–0.16	0.16 (15) 0.12–0.19	0.18 (10) 0.13–0.23	0.18 (18) 0.14–0.25	0.24 (74) 0.17–0.34
Ti	0.61 (22) 0.50–0.66	0.78 (11) 0.66–0.95	0.90 (8) 0.79–1.07	1.02 (11) 0.87–1.11	1.0 (72) 0.64–1.38
Nb	2.7 (10) 1.0–5.8	2.9 (5) 1.0–4.3	4.8 (3) 2.3–7.0	3.2 (4) 2.2–5.5	4.7 (20) 2.6–7.0
Ta	1.1 (6) 0.4–1.9	0.5 (5) 0.3–0.7	1.2 (3) 1.0–1.6	0.7 (4) 0.3–1.4	1.0 (8) 0.5–1.5
Zr	70 (10) 10–143	145 (5) 70–188	239 (4) 125–362	175 (4) 165–200	231 (20) 125–330
Hf	1.0 (9) 0.2–2.0	3.6 (5) 2.0–4.2	3.3 (4) 2.8–4.6	4.3 (4) 3.4–6.0	5.5 (20) 3.3–8.0
La	$\approx$ 10 (3) ~ 10–10	—	—	—	30 (8) 25–33

Table 3. (*cont.*)

	Northern Breakthrough			Southern Breakthrough	
Elements	Predominant lava types (7 July 75–10 Sept. 75)	Intermediate lavas from the end of the eruption (11–15 Sept. 75)	Intermediate lavas from the first days of the eruption (18–24 Sept. 75)	Intermediate lavas from the start of the eruption (25 Sept. 75–30 Nov. 75)	Predominant lava types (1 Dec. 75–9 Dec. 75)
1	2	3	4	5	6
Ce	< 30 (3) / < 30– < 30	—	—	—	43 (8) / 20–50
Nd	≈ 10 (3) / 5–15	—	—	—	23 (8) / 20–30
Y	24 (3) / 23–25	—	—	—	43 (8) / 41–45
Yb	2.6 (3) / 2.2–2.9	—	—	—	4.6 (8) / 4.3–5.0
K/Na	0.40 / 0.34–0.48	0.58 / 0.42–0.66	0.59 / 0.54–0.63	0.60 / 0.56–0.70	0.65 / 0.55–0.71
K/Rb	645 / 540–740	550 / 435–615	475 / 435–505	400 / 435–500	375 / 315–475
Ba/Sr	1.05 / 0.73–1.83	1.04 / 0.9–1.24	1.30 / 0.8–1.41	1.14 / 0.8–1.57	2.68 / 1.42–3.80
Nb/Ta	2.5 / 1.5–4.5	5.8 / 3.3–8.6	4.0 / 4.4–5.0	4.6 / 3.9–7.7	4.7 / 3.2–7.3
Zr/Hf	70 / 13–116	40 / 34–50	72 / 37–129	41 / 33–48	42 / 30–51
Ni/Co	2.85 / 1.6–4.4	2.4 / 1.6–3.3	2.3 / 1.4–3.2	1.4 / 1.0–1.8	1.7 / 1.1–2.2
V/Ni	2.1 / 1.3–3.2	2.8 / 1.5–5.0	3.2 / 2.3–5.6	4.1 / 3.4–7.5	4.0 / 2.5–5.9
La/Yb	3.8 / 3.4–4.5	—	—	—	6.5 / 6.0–7.7

Na, K, F, Ti contents in weight %, other elements in g/t. Numerator, average content of element (in brackets the number of analyses used in calculating average); denominator, variation in content of element in individual samples from minimum to maximum.

Breakthrough the variations in the content of these elements were: Na 0.56%, K 0.2%, Rb, 4 g/t, B 30 g/t, F 410 g/t, Be 0.9 g/t and MgO 1.5%. Intermediate basalts of the first few days of the eruption of the Southern Breakthrough display a similar tendency. Thus basalts of the intermediate stage of the eruption, as distinct from the overwhelming mass of extremely homogeneous basalts of the Northern and Southern Breakthroughs, are characterised by an heterogeneous (non-equilibrium) composition. It should, however, be emphasised that the composition of the basalts from the last day of the eruption of the Northern Breakthrough is practically identical to the average composition of those from the first day of the eruption of the Southern Breakthrough.

As a result of the substantial difference in the composition of the basalts of the Northern and Southern Breakthroughs and the very small variations in the concentrations of their individual elements, the plots of the main mass of rocks of the Northern and Southern Breakthroughs give isolated clusters on the various petro-geochemical diagrams. However, these clusters are united by the presence of basalts of intermediate composition. Nevertheless, in the majority of cases the correlational links between the various elements (if all the rocks of the eruption are considered) are not linear. This can be clearly seen on the plots correlating $MgO-K_2O$, K-Na, K-Rb, etc. (Figs 4, 5 and 6). We would point out that a different type of relationship between the elements has been established for the differentiated series of Hawaiian lavas, where the quantities of all the elements vary linearly with the Mg content (Murata & Richter, 1966).

The order in which the different compositions of basalts come to the surface (first the magnesian basalts with moderate alkalinity, then the intermediate and finally the subalkaline aluminous) casts doubt on the possibility that all these different rocks were formed by the differentiation of magmas in a single magmatic column. Indeed, any processes of differentiation involving a gravitational separation of material (crystallisation, emanation, magmatic differentiation) should lead to the upper part of the column being enriched by a number of relatively light and mobile elements (in the first instance alkalis) and, consequently, could not ensure the sequence of changes in the rocks observed in the course of the eruption. The sharp difference in the average compositions of the rocks of the Northern and Southern Breakthroughs, the consistently homogeneous composition of the predominant mass of lavas in each Breakthrough, with a generally low coefficient of variation in the amounts of the individual elements and the extremely limited volume of rocks of the intermediate composition – all this suggests that the primary melts of the Northern and Southern Breakthroughs were independent. However, this does not dispose of the question of whether the different types of basalts could have been formed by processes of differentiation of a single primary magma in a relatively deep magma chamber (or chambers). This is a question that requires special discussion.

According to the most widely held opinion, the main cause of the variety in magmatic rocks is the crystallisation differentiation of magmatic melts. Let us examine how legitimately this hypothesis can be applied to the Tolbachik eruption. In accordance with the chemistry of the rocks, the phases immediately below the liquidus of the basalts of the Northern Breakthrough are olivine and clinopyroxene, and of the Southern Breakthrough mainly plagioclase. The groundmass of the basalts of the Northern Breakthrough, by comparison with their overall composition, is accordingly

Fig. 3. Variation of the Ti, P, Na, K, Rb and Li content and K/Rb ratios in rocks of the eruption.
1, lavas and bombs; 2, ashes; 3, pause between eruptions of the Northern and Southern Breakthroughs.

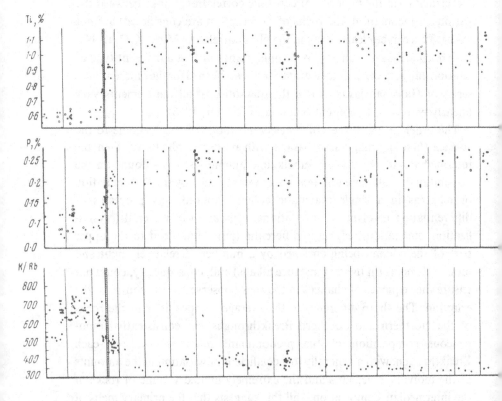

enriched with Al, alkalis, Ti, P, etc., and the groundmass of the basalts of the Southern Breakthrough is enriched with Mg, Fe, K, Ti, P, etc. This can also be clearly seen from the graphs in Figs. 4, 5 and 6. The compositions of the groundmass of rocks show the possible directions of shifts in the composition of the melts should crystallisation differentiation occur. It follows that if there is in principle a possibility that melts similar to the Southern Breakthrough basalts originated from crystallisation differentiation of magma corresponding to the basalts of the Northern Breakthrough, then the reverse is impossible. Theoretically, it is even easier to obtain melts of the Southern Breakthrough basalt type from melts with the composition of the intermediate basalt. In this case, the basalts of the Northern and Southern Breakthroughs could have formed a comp-lementary pair, in which the basalts of the Northern Breakthrough would

Fig. 3. (*cont.*)

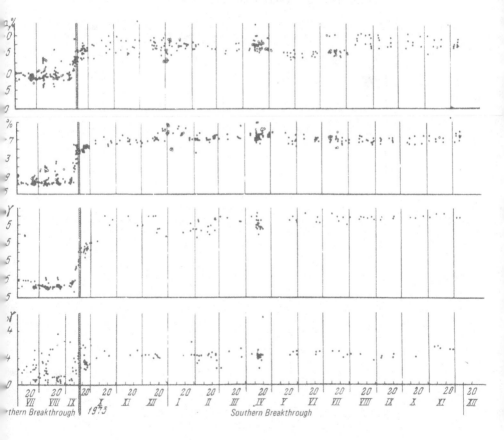

130   *Volynets* et al.

have been a cumulative series and the basalts of the Southern Breakthrough a residual melt.

The likelihood of all the types of rocks of the eruption originating in the crystallisation differentiation of a single primary melt corresponding to the basalts of the Northern Breakthrough or to intermediate basalts, can be assessed with the aid of some simple calculations, using the data presented in Tables 1–3, as well as data on the compositions of the

Fig. 4. Relations between contents of alkalis, MgO and Al$_2$O$_3$ in rocks of the eruption.

I, lavas and bombs; II, ash. Dots, rocks of the Northern Breakthrough; Crosses, rocks of the Southern Breakthrough. Arrows indicate direction and magnitude of the alteration in composition in paired values (bulk composition of the rock – composition of its ground mass). In constructing the arrows, as well as the authors' own material, use was made of data from Kirsanov & Ponomarev (1974) for the aluminous subalkaline basalts of the last eruptions of the summit crater of Ploskiy Tolbachik.

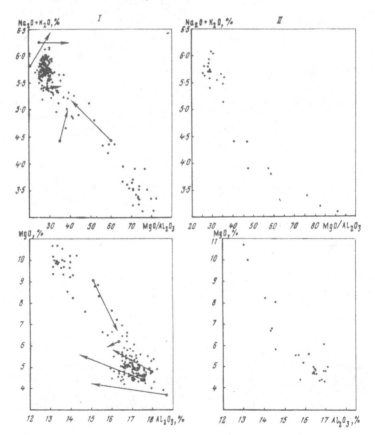

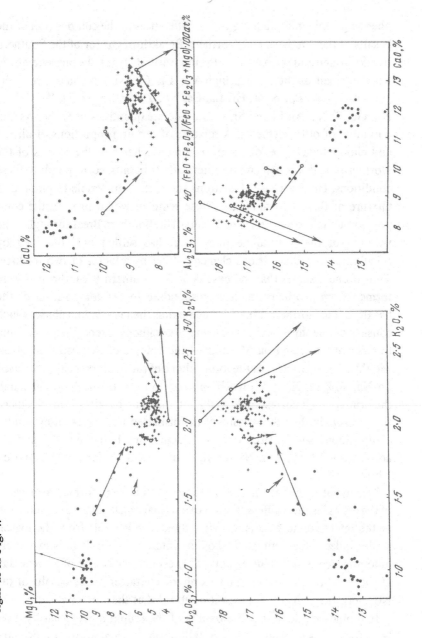

Fig. 5. Relations of contents of different pairs of petrogenic elements in rocks of the eruption. Conventional signs as in Fig. 4.

phenocryst minerals. Judging by the differences in the composition of the basalts of the Southern and Northern Breakthroughs (or of the Southern Breakthrough and the basalts of the intermediate type), the phase removed from the melt ought to have high Mg, Ca, Cr, Ni, Co and Zn contents and low Rb, K, P, Pb, Zr, Hf, La, Ce, Nd, Y, Yb, Ba, Sr, Ti, Be, B, Mo, Nb, Ga, F, Na, Al, Li and Sr, with moderate quantities of Si, Fe, As and a number of other elements. A mixture of certain proportions of olivine and clinopyroxene, i.e. of the near-liquidus phases of the basalts of the Northern Breakthrough and the intermediate basalts, corresponds to these conditions. However, as calculations show, it is impossible to pinpoint a mixture of these two minerals whose removal would provide the composition actually observed in the basalts of the Southern Breakthrough – the volumes of the 'crystalline phase that has settled out' obtained by calculation from the various elements differ by a factor of two to three. Thus if one assumes that the basalts of the Southern Breakthrough were formed from a primary melt corresponding to the composition of the basalts of the Northern Breakthrough minus their crystalline phase, which consists of one third olivine and two thirds clinopyroxene, then according to calculations based on Mg and Al the volume of the displaced phase should account for 25% of the total volume of the primary melt, 35% based on Na, 60% on K, 75% on Rb and 80% on Ca. If, on the other hand, the primary melt corresponds to the intermediate basalts of the Northern Breakthrough, then the volume of the displaced phase (assuming similar composition) should according to calculations based on Mg and Al account for 12–15%, on Na for 18%, on K and Ca for 25–30% and on Rb for 40–50%.

The hypothesis that the rocks of the eruption were formed as a result of the crystallisation differentiation of a primary melt with the composition of the intermediate basalts, is evidently also vulnerable from the logical angle. Indeed, the composition of the intermediate basalts is extremely inhomogeneous and changes noticeably even in the course of a single day of the eruption. Moreover the subaphyric character of the basalts of the Northern Breakthrough makes it doubtful that they are cumulates.

It should be added that the basalts of the eruption could hardly have been formed as a result of the crystallisation differentiation of a primary melt that corresponded to the composition of the plagiobasalts (or aluminous basalts of moderate alkalinity, in our terminology) – i.e. the most widespread type of Quaternary basalt on Kamchatka – as a number of authors assumed earlier for similar types of rocks (Yermakov, 1971;

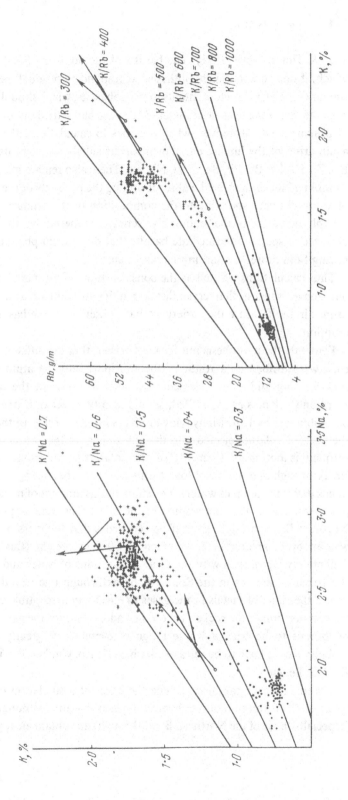

Fig. 6. Relations of contents of K and Na and K and Rb in rocks of the eruption. Conventional signs as in Fig. 4.

Sirin & Timerbayeva, 1971). The basalts of the Southern Breakthrough, which, if one accepts this hypothesis, correspond to residual melts, have similar Ca, Al, Cr, Ni, etc. contents to plagiobasalts, which should not have been the case if the near-liquidus Ca–Mg phase had settled out of this kind of primary melt. Moreover, when it comes to calculations, the problem again arises of the intense enrichment of the subalkaline basalts with K, Rb, Zr, Hf, La, Ba, Pb, Be, P, Ti, Nb, etc. There also remain many logical stumbling blocks, such as the absence among the products of the eruption of any rocks corresponding to the composition of the primary melt, the subaphyric nature of the 'cumulative series' (magnesian basalts) and the presence of specific intermediate basalts that differ from plagiobasalts in having both more Mg and more alkalis, etc.

Thus calculations, as well as the considerations of a general nature set out above, suggest that crystallisation differentiation of a single melt cannot in itself ensure the variety we have observed in the basalts of the eruption.

Evidently the hypothesis put forward earlier, that the different types of rock were formed as a result of the transport into the atmosphere of rock-forming and minor elements by volcanic gases in the course of eruption (Volynets *et al.*, 1976a), should also be rejected, if one uses the data obtained by I. A. Menyaylov *et al.* (1976). According to these data, the mass of volatiles ejected into the atmosphere at the Southern Breakthrough is not more than 1% of the total mass of products of the Breakthrough and the metal load forms not more than 0.1% of the mass of gas (0.9 is made up of water). The efflux of this amount of metal cannot possibly influence the composition of the rocks. The same applies to the Northern Breakthrough. Even if one concedes that the mass of volatiles was an order greater here, as much as 10% by weight (this figure is deliberately high), then with the same proportions of water and metal in the volcanic gases as in the Southern Breakthrough one would still not expect the efflux of metals in the gases to effect any perceptible change in the composition of the rocks. It should be added that for the gas transport hypothesis to be true, a whole range of elements of greatly differing mobility would have to be released, such as K, Rb, Cu, Na, P, Ti, Al, Ba, Zr, Nb, Be, etc.

It seems that we can also exclude the effect of assimilation of crustal rocks on the formation of the different types of basalts. Although basalts (especially those of the Northern Breakthrough) do contain an appreciable amount of inclusions of volcanogenic–sedimentary, siliceous rocks and occasional inclusions of granites and peridotites, there is no noticeable incorporation of xenoliths, even if completely melted, and the xenolith–basalt boundaries remain clear-cut and distinct in all cases. The absence of any noticeable traces of contamination of the basalts of the eruption by crustal rocks is also borne out by data on the Sr isotope composition (analyses carried out in the Institute of Geochemistry of SO AN SSSR by G. S. Plyusnin and G. P. Sandemirova): Northern Breakthrough, sample A_7B_1, lava from 1 August 1975, $^{87}Sr/^{86}Sr = 0.7041$, $^{86}Sr/^{88}Sr = 0.1194$ where the K content $= 0.76\%$, Rb $= 12 \times 10^{-4}\%$, Sr $= 300 \times 10^{-4}\%$; Southern Breakthrough, sample $A_{40}B_3$, a bomb from 22 September 1975, $^{87}Sr/^{86}Sr = 0.7043$, $^{86}Sr/^{88}Sr = 0.1196$ where K content $= 1.48\%$, Rb $= 31 \times 10^{-4}\%$, Sr $= 300 \times 10^{-4}\%$. Although the $^{87}Sr/^{86}Sr$ values are slightly higher than the average value for this ratio for the basalts of the Klyuchi volcano and the intermediate–acidic Quaternary rocks of Kamchatka, 0.7034 ± 0.002 (Khedzh & Gorshkov, 1977), they are generally close to the average value of 0.7035 ± 0.0005 (Khedzh & Gorshkov, 1977) for volcanites of the Pacific Ocean belt and oceanic islands. The main point, though, is that the value for this ratio is practically the same in the basalts of both the Northern and Southern Breakthroughs.

Thus analysis of the petro-geochemical properties of the rocks of the eruption reveals differences between the predominant types of basalts of the Northern and Southern Breakthroughs of a kind which cannot be explained by the hypothesis of crystallisation or emanation differentiation of a single primary melt, nor by the hypothesis that the latter was contaminated by basement rocks.

The fact that the basalts of intermediate composition are restricted to the period at the end of the activity of the Northern Breakthrough and the beginning of the Southern (i.e. to the period of transition from one type of basalt to another during the eruption); the rapid change in the composition of rocks during this intermediate stage of the eruption; the comprehensive and stable range of these intermediate-composition varieties, from magnesian moderately alkaline basalts to aluminous subalkaline basalts; the extremely limited volume of basalts of this intermediate type in relation to the overall volume of products of the eruption; and, finally, the conspicuous heterogeneity of the rocks in this period, even over a single day of the eruption – all this leads one to think that these rocks of intermediate composition were formed from the mixing of magmatic melts

corresponding to the basalts of the Northern and Southern Breakthroughs.*
Such an hypothesis is also supported by features of the mineralogical
composition of the intermediate-composition basalts, such as the presence
of generations of olivines that are similar in composition to those of both
the Northern and Southern Breakthroughs, and the fact that the lavas of
this stage contain individual megacrystals of plagioclase which are
characteristic of basalts of the Southern Breakthrough.

The data presented above and the critical analysis of the various
petrogenetic conceptions lead to two important conclusions: firstly, that
the melts which were the sources of basalts of the Northern and Southern
Breakthroughs were to some extent independent, and secondly that the
rocks of intermediate composition were formed by the mixing of these two
initial melts. These conclusions do not contradict the geophysical data
obtained at the eruption.

Thus according to seismological data (Fedotov *et al.*, 1976a; Fedotov,
1976), in the earthquake swarm that preceded the formation of the
Northern Breakthrough the epicentres of the deepest earthquakes were at
depths of 20–30 km, and as time passed they were observed to shift
upwards. But the earthquakes which preceded the formation of the
Southern Breakthrough were near the surface (depth of epicentres 0–5 km).
These observations enabled Fedotov and his co-authors (Fedotov, 1976;
Fedotov *et al.*, 1976a) to assume that the source of the basalts of the
Northern Breakthrough lay at sub-crustal depths, and that of the basalts
of the Southern Breakthrough within the upper part of the crust. The
presence of crustal subsurface magma chambers seems to be confirmed by
seismic and gravimetric data (Balesta *et al.*, 1976), which show that in the
region of the eruption within the upper zones of the crust there are parts
where the seismic waves are attenuated and there is a gravitational
anomaly, interpreted by these authors as shallow magma chambers.

Thus the interpretation of geophysical evidence gives grounds for
supposing that the basalts of the Northern and Southern Breakthroughs
came to the surface from magma chambers situated at different depths.
This accords well with petrological data indicating a certain independence
of the basalts of the two Breakthroughs. Moreover, the very presence of
chambers at different depths creates the pre-conditions for the mixing of
magmatic melts during the eruption, when the deep magmatic melt is rising

* Hypothetical mixing of magmatic melts was successfully used by T. L. Wright &
R. S. Fiske (1971) to account for the compositions of rocks from a series of
eruptions in the rift zone of Kilauea.

and on its way meets the crustal reservoir of magma (the intermediate chamber). The action of the deep melt could be the impulse that causes magma to erupt from the crustal reservoir, and the nature of the interaction of the melts is reflected in the compositional features of the rocks of the eruption's transitional stage (the end of the Northern and the beginning of the Southern Breakthrough).

If one assumes that the change in the composition of basalt samples collected chronologically during the transitional stage reflects the mixing of the two magmatic melts in the course of the eruption, it follows that kinetic methods can be used to evaluate the constants of the rates of interchange of the various elements between the two types of melt. Results of the calculation (Table 4, Fig. 7) of diffusion for spherical volumes (Lykov, 1967) show that this process can be satisfactorily described in terms of a mechanism of diffusional mixing of components. The values of the constants of the rates of interchange for K, Na, Ca, Mg, Al, Ti, P, Rb, Li and Be coincide within limits of error and furnish a single constant value for D/r^2 of 5.4×10^{-8} s^{-1}.

The existence of a single constant of rate of interchange for different chemical elements seems at first sight unexpected, since it is well known from experiments that in silicate melts at temperatures of 1100–1250 °C there is a great difference in the diffusion coefficients of the elements involved. However, we also know that in crystalline silicate structures the

Table 4. *Values of constants for the rate of diffusion exchange between various components during mixing*

Element	D/r^2 (24 h^{-1})	(D/r^2) (mean)
K	0.0042 ± 0.0016 (24)	
Na	0.0055 ± 0.0029 (23)	
Ca	0.0065 ± 0.0032 (17)	
Mg	0.0046 ± 0.0030 (18)	
Al	0.0059 ± 0.0037 (16)	0.0047 ± 0.0028 (172), 24 h^{-1}
Ti	0.0032 ± 0.0016 (14)	$5.4 \times 10^{-8} \pm 3.2 \times 10^{-8}$ s^{-1}
P	0.0028 ± 0.0012 (17)	
Rb	0.0024 ± 0.0010 (16)	
Li	0.0065 ± 0.0054 (10)	
Be	0.0054 ± 0.0040 (17)	

In brackets, the number of analyses. D/r^2 is a constant consisting of diffusion coefficient D, and a characteristic dimension r (the radius of the sphere registering the diffusion flow).

displacement of the mobile components can be held back by the displacement of other components with higher activation energies for diffusion. Therefore the results of our calculation could indicate that at least one of the basalt melts had a quasi-crystalline structure at the time of mixing.

The change in the F and B contents, which do not fit into the general pattern of change in the petrogenic and minor element contents, is evidence that at the mixing stage the volatile components behaved independently of the assumed structure of the melt, i.e. it confirms the presence of fluid during the interaction of the two melts.

On the basis of the D/r^2 constant calculated, it is possible to estimate the values of r. Assuming that the diffusion coefficient of the element limiting the mobility of the other components in the coalescing melts could have values from $n \times 10^{-6}$ cm²/s to $n \times 10^{-12}$ cm²/s, then for r we obtain values in the range of 10 cm to 0.01 cm. This result is in itself interesting, since it implies considerable mechanical intermixing of the melts in the process of coalescence.

Altogether, a kinetical analysis of the mixing of the basalt melts produces the following characteristics for a geochemical model of the Great Tolbachik Fissure Eruption. Initial conditions consisted of the

Fig. 7. Alteration in the relative concentrations of K, Na, Ca, Mg, Al, Ti, P, Rb, Li and Be in basalts in the mixing stage. Continuous curve, theoretical dependence corresponding to the diffusion constant $D/r^2 = 5.4 \times 10^{-8}$ s⁻¹; Broken curve, measurement error $(\Delta = \pm 3.2 \times 10^{-8}$ s⁻¹); Circles, empirical data.

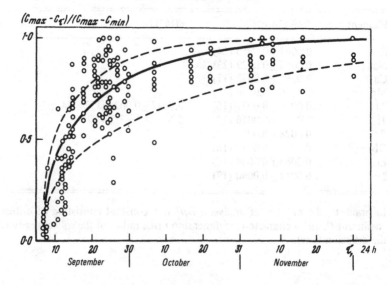

existence of two basalt melts of different composition (of which at least one had a quasi-crystalline structure), and of a fluid which, among other things, contained F and B. Conditions during the mixing were that there was some degree of contact between the mechanically intermixed basalt melts. The mixing mechanism was a diffusional exchange of components, limited by the rate of displacement of one of the chemical elements (with maximum diffusion activation in these conditions) in the quasi-crystalline structure of the melt.

In conclusion, it is important to emphasise that within the Holocene Tolbachik regional zone of cinder cones the association of moderately alkaline magnesian basalts and aluminous subalkaline basalts is consistently reproduced in the individual volcanic centres. The characteristics of the change in composition and facies of the rocks in the course of the activity of such centres are similarly reproduced: first the magnesian then the aluminous basalts come to the surface, in the former pyroclastics predominate over lavas, in the latter it is the other way round. All this indicates that eruptions like the Great Tolbachik Fissure Eruption of 1975–76 are not random processes, and consequently testifies in favour of the proposed model.

References

Balesta, S. T., Gorshkov, A. P., Zubin, M. I., Kargopol'tsev, A. A., Farberov, A. I. & Tarakanovskiy, A. A. 1976. Glubinnoye stroyeniye rayona Bol'shogo treshchinnogo Tolbachinskogo izverzheniya (po geofizicheskim dannym). (The deep structure of the region of the Great Tolbachik Fissure Eruption (from geophysical data).) In: *III Sovetsko–Yaponsk. simpozium po geodinamike i vulkanizmu zony perekhoda ot Aziatskogo kontinenta k Tikhomu okeanu (tezisy dokladov). (The 3rd Soviet–Japanese symposium on the geodynamics and volcanism of the transition zone from the Asian continent to the Pacific Ocean (theses of papers).)* Novoaleksandrovsk, p. 31.

Fedotov, S. A. 1976. O pod"yeme osnovnykh magm v zemnoy kore i mekhanizme treshchinnykh bazal'tovykh izverzheniy. (On the ascent of basic magmas in the crust and the mechanism of basalt fissure eruptions.) *Izv. AN SSSR*, Ser. geol., no. 10, 5–23.

Fedotov, S. A., Gorel'chik, V. I. & Stepanov, V. V. 1976a. Seysmicheskiye dannyye o magmaticheskikh ochagakh, mekhanizme i razvitii bazal'tovogo treshchinnogo Tolbachinskogo izverzheniya v 1975 g. na Kamchatke. (Seismic data on the magma chambers, mechanism and development of the Tolbachik basaltic fissure eruption of 1975 in Kamchatka.) *Dokl. AN SSSR*, **228**(6), 1407–10.

Fedotov, S. A., Khrenov, A. P. & Chirkov, A. M. 1976b. Bol'shoye treshchinnoye Tolbachinskoye izverzheniye 1975 g., Kamchatka. (The Great Tolbachik Fissure Eruption of 1975 in Kamchatka.) *Dokl. AN SSSR*, **228**(5), 1193–6.

Khedzh, K. Ye. & Gorshkov, G. S. 1977. Izotopnyy sostav strontsiya vulkanicheskikh porod Kamchatki. (The isotope composition of strontium in the volcanic rocks of Kamchatka.) *Dokl. AN SSSR*, **233**, 1200–3.

Kirsanov, I. T. & Ponomarev, G. P. 1974. Izverzheniya vulkana Ploskiy Tolbachik i nekotoryye osobennosti ikh produktov. (The eruptions of Ploskiy Tolbachik volcano and some features of their products.) *Byul. vulkanol. stantsiy*, No. 50, 53–63.

Korshak, V. V. & Vinogradova, S. V. 1968. *Ravnovesnaya polikondensatsiya. (Equilibrial polycondensation.)* M., Nauka, 444 pp.

Leonova, L. L. & Kirsanov, I. T. 1974. Geokhimiya bazal'tov Klyuchevskogo vulkana (Kamchatka). (The geochemistry of the basalts of Klyuchi volcano (Kamchatka).) *Geokhimiya*, No. 6, 875–83.

Leonova, L. L. & Ogorodov, N. V. 1975. Geokhimiya chetvertichnykh bazal'tov Sredinnogo khrebta Kamchatki. (The geochemistry of the Quaternary basalts of Sredinnyy Khrebet in Kamchatka.) *Geologiya i geofizika*, 102–8.

Leonova, L. L., Volynets, O. N., Yermakov, V. A., Kirsanov, I. T., Dubik, Yu. M. & Popolitov, E. I. 1974. Tipy chetvertichnykh bazal'tov Kamchatki v svyazi s problemoy pervichnykh magm. (Types of Quaternary basalts from Kamchatka and the problem of primary magma.) In: *Geodinamika vulkanizma i gidrotermal'nogo protsessa.* (*The geodynamics of volcanism and the hydrothermal process.*) Petropavlovsk-Kamchatskiy, 221 pp.

Lykov, A. V. 1967. *Teoriya teploprovodnosti.* (*The theory of thermal conductivity.*) M., Vysshaya shkola, 599 pp.

Menyaylov, I. A., Nikitina, L. P., Guseva, R. V. & Shapar', V. N. 1976. Rezul'taty otbora i analiza vulkanicheskikh gazov na Tolbachinskom treshchinnom izverzhenii v 1975 g. (Results of sampling and analysis of volcanic gases during the Tolbachik fissure eruption of 1975). *Dokl. AN SSSR*, **230**(2), 440–2.

Murata, K. J. & Richter, D. H. 1966. The settling of olivine in Kilauean magma as shown by lavas of the 1959 eruption. *Am. J. Sci.*, **264**(3), 194–203.

Popolitov, E. I., Leonova, L. L., Znamenskiy, Ye. B. & Tsykhanskiy, V. D. 1976a. Raspredeleniye Nb i Ta v vulkanicheskikh porodakh Kurilo–Kamchatskoy ostrovnoy dugi. (The distribution of Nb and Ta in volcanic rocks of the Kuril–Kamchatka island arc.) *Geokhimiya*, No. 1, 29–35.

Popolitov, E. I., Volynets, O. N. & Selivanova, G. I. 1967b. Geokhimicheskiye osobennosti chetvertichnykh bazal'tov Kamchatki. (Geochemical features of the Quaternary basalts of Kamchatka.) In: *Yezhegodnik 1974. Institut geokhimii SO AN SSSR.* (*Yearbook for 1974 of the Institute of Geochemistry of SO AN SSSR.*) Novosibirsk, pp. 92–7.

Shou, D. M. 1969. *Geokhimiya mikroelementov kristallicheskikh porod.* (*The geochemistry of micro-elements in crystalline rocks.*) L., Nauka, 207 pp.

Sirin, A. N. & Timerbayeva, K. M. 1971. O dvukh tipakh bazal'tov i sostave iskhodnoy magmy Klyuchevskoy gruppy na Kamchatke. (Two types of basalt and the composition of the primary magma of the Klyuchi group in Kamchatka.) In: *Vulkanizm i glubiny Zemli.* (*Volcanism and the Earth's depths.*) M., Nauka, pp. 147–50.

Turekian, K. K. & Wedepohl, K. N. 1961. Distribution of the elements in some major units of the earth's crust. *Bull. Geol. Soc. Am.*, **12**(2), 175–92.

Volynets, O. N., Yermakov, V. A., Kirsanov, I. T. & Dubik, Yu. M. 1976b. Petrokhimicheskiye tipy chetvertichnykh bazal'tov Kamchatki i ikh geologicheskoye polozheniye. (Petrochemical types of Quaternary basalts of Kamchatka and their geological position.) *Byul. vulkanol. stantsiy*, No. 52, 115–26.

Volynets, O. N., Flerov, G. B., Khrenov, A. P. & Yermakov, V. A. 1976a. Petrologiya vulkanicheskikh porod treshchinnogo Tolbachinskogo izverzheniya 1975 g. (The petrology of the volcanic rocks of the Tolbachik fissure eruption of 1975.) *Dokl. AN SSSR*, **228**(6), 1419–22.

Wright, T. L. & Fiske, R. S. 1971. Origin of the differentiated and hybrid lavas of Kilauea volcano, Hawaii. *J. Petrology*, No. 1, 1–65.

Yermakov, V. A. 1971. Megaplagiofirovyye lavy – veroyatnyy analog anortozitovykh porod. (Megaplagiophyre lavas are a probable analogue of anorthosite rocks.) *Izv. AN SSSR*, Ser. geol., No. 10, 56–72.

Yoder, G. S. & Tilli, K. E. 1965. *Proiskhozhdeniye bazal'tovykh magm.* (*The origin of basalt magmas.*) M., Mir, 248 pp.

First results of a study of ore minerals in eruption products of the Southern Breakthrough of the Great Tolbachik Fissure Eruption

V. M. Okrugin, V. K. Garanin, G. P. Kudryavtseva and V. N. Sokolova

This paper presents the first results of a preliminary study of ore minerals in the solid products (lavas, bombs and plagioclase lapilli) from the eruption of the Southern Breakthrough, samples of which were taken during monitoring observations from 8 to 20 October 1975.

During this time aluminous subalkaline basalts were pouring out at a great rate, accompanied by ejections of considerable quantities of pyroclastic material (Fedotov *et al.*, 1976; Volynets *et al.*, 1976). For comparison single samples of plagioclase lapilli from the lavas of Ploskiy Tolbachik (the eruption of the summit crater in 1970) and of the first bombs from the Northern Breakthrough (1975) were collected.

Together with the traditional classical procedures of petrography and mineragraphy in transmitted and reflected light, we used electron microprobe analysis, laser microanalysis, scanning electron microscopy and thermobarogeochemical methods. The shape, size and optical properties of the ore minerals were characterised using the most up-to-date optical microscopes (Polam P-312, NU-2E). An IXA-50A (Geol.) electron microprobe, equipped with a computer and a set of standard programs, was used for qualitative analysis of the varieties of ore minerals that had been separated. The most representative of these minerals were then subjected to quantitative chemical analysis, the results of which are given in Table 1. Very small phases ($\approx$ 2–3 µm), for example border phases, skeletal dendritic aggregates, and various sulphides and native metals, were subjected to semi-quantitative analysis.

A Carl Zeiss LMA-1 laser microanalyser was used in some cases to obtain a preliminary diagnosis of individual ore phases of high reflectivity. Although the size of the crater formed by detaching the specimen by laser far exceeded ($\approx$ 60 µm) the size of the individual ore grain itself ($n \times 1$ µm),

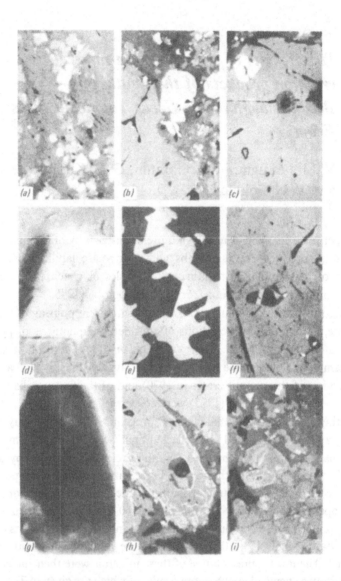

Fig. 1. Forms of spinellids: (*a*) hypidiomorphic granular texture; white, spinellids; grey, subphenocrysts of plagioclase, pyroxene; lava, sample 12; (*b*) relatively large isometric individuals of spinellids in glass of groundmass; lava, sample 12; (*c*) polygonal spinellid grain in the marginal zone of a plagioclase phenocryst, alongside a two-phased inclusion of solidified magmatic melt; lava, sample 9; (*d*) rhomboidal grain of spinellid in lapilli (plagioclase); sample 16, photo from SEM screen, ×25000; (*e*) sub-graphic texture formed by spinellid (white) and plagioclase (dark) in plagioclase lapilli, sample 12; (*f*) idiomorphic spinellid crystal (light grey) trapped in inclusions of

nevertheless by comparing the spectra of the silicate phases proper, without ore phases, with the spectra of sections containing an ore phase in the centre, it was possible to obtain quite definite information.

To examine the micromorphology of the ore minerals, and the structure of the inclusions of solidified magmatic melts, especially the ore minerals trapped in them, for the first time in Soviet practice a scanning electron microscope (SEM) of the CWIKSCAN-107 type was used. This instrument has a guaranteed resolution of 60 Å and offers a wide range of magnification: from optical (× 10) to electron-optical (× 200000). The use of an electronic gun with field emission in this particular model has a number of advantages over a scanning electron microscope with thermo-electronic emission: (1) the depth of focus is increased by more than 20 times, and at low voltage this is already measured in centimetres; (2) there are almost limitless possibilities in terms of clarity and definition of the image (Welter & McKee, 1972). The use of rapid television rasters considerably lowers the chance of charging the image and means that one can study uncoated and, in particular, thermally unstable products. It is by using these advantages that we have been able for the first time to obtain direct authentic photomicrographs of ore minerals that easily oxidise and disintegrate under an electron probe.

The sample of the eruption product to be studied was $1.0 \times 1.0 \times 0.5$ cm in size. Its surface was carefully cleaned of dust and the effects of mechanical disintegration by blowing with a stream of dry gaseous nitrogen. It was then attached to the microscope stage using electro-conductive adhesive. Because of the 'highly dissected relief' on the specimen's surface, the latter acquired a considerable charge. By careful application of the adhesive to the sides of the specimen it was possible to reduce this charge considerably. After this operation a thin film of gold

Caption to Fig. 1 (*cont.*)
solidified magmatic melt in a plagioclase phenocryst (grey); white – pyrite, black (isometric) – trace of gas bubble in amorphous glass base (dark grey) of the vacuole; lava, sample 14; (*g*) spinellids on the walls of vacuole, droplike inclusion of solidified magmatic melt in plagioclase lapilli, sample 14, photo from SEM screen × 20000; (*h*) texture of marginal rims caused by separation of spinellid (white) on the boundary between plagioclase subphenocryst (grey) and groundmass glass (dark grey); (*i*) distribution of spinellids (white) in pyroxene subphenocrysts (grey); lava, sample 16. (*a*), (*b*), (*c*), (*e*), (*g*), (*h*) and (*i*) are polished mounts, × 150. The number of the sample corresponds to the time the eruption product appeared at the surface (for example, sample 14 –14 October 1975).

(≈ 100 Å) was applied to the specimen by means of vacuum sputtering. This enabled the coefficient of the secondary emission to be increased. In order to obtain maximum resolution and to improve the quality of the photomicrographs, the SEM was used in a high voltage mode (accelerating the voltage 16 kW). In order to minimise distortions of the micro-images the specimen was placed almost normal to the electron beam.

The following ore minerals were determined: spinellids with a variable Cr_2O_3 content, sulphides (pyrrhotite, pyrite, pentlandite), native copper (?) and iron (?), and a number of intermetallic compounds containing copper, iron and nickel. Minerals with a spinel structure are the most common. They form: (1) relatively large (30–150 µm) crystals of isometric, occasionally idiomorphic shape occurring independently or forming intergrowths with subphenocrysts of plagioclase, pyroxene and olivine (Fig. 1(*a*), (*b*)); (2) a variety of poorly developed associations confined to the interstices of the rock-forming silicates (Fig. 1(*b*)–(*i*)); (3) idiomorphic, and less frequently, isometric to polygonal inclusions confined both to the central and peripheral zones of the phenocrysts of plagioclase and olivine (Fig. 1(*c*), (*d*)) and containing in their turn micro-inclusions of the latter (Fig. 1(*c*), (*d*)); (4) various rather sub-graphic formations in plagioclase lapilli, similar to decomposition structures (Figs. 1(*e*) and 2(*a*), (*b*)); (5) idiomorphic and occasionally xenomorphic crystals, which occur as minerals imprisoned in inclusions of solidified magmatic melts in the plagioclase phenocrysts (Fig. 1(*f*), (*g*)); (6) thin fringes (margin structures) along the periphery of subphenocrysts, and emulsion dissemination along the cleavage planes of the phenocrysts of the rock-forming minerals (Fig. 1(*h*)), or complex dissociation structures within them (Fig. 1(*i*)); (7) spherical to regular individuals restricted to the zoning planes of plagioclase plates in the eponymous lapilli (Fig. 3(*a*), (*b*)); (8) conglomerations of skeletal dendritic aggregates with amazingly complicated patterns (Fig. 4(*a*)–(*f*)). The overall quantity of spinellids is 2–3%, and their size varies from 1 µm to 200 and even, in exceptional cases, to as much as 1000 µm.

Four varieties of spinellids can be identified, according to reflectivity and colour: (1) a dark-grey variety with a faint dark-bluish tinge which has the lowest reflectivity; (2) a creamy-grey variety; (3) a light bluish-grey; and (4) a light-grey variety with a faint yellowish tinge. All these phases are isotropic but in phase 1 traces of anisotropy can sometimes be distinguished. Intermediate varieties can also be distinguished between phases 2 and 4. Phase 1 occupies a special position because of its optical properties. These varieties occur as single minerals (Fig. 1(*b*)–(*g*), Figs. 2 and 3) and complex two- and three-phase aggregates (Fig. 5). In the majority of cases phase 1 occurs as independent grains forming poikilitic idiomorphic and isometric intergrowths in the silicate phenocrysts. Phase 1 contains maximum quantities of Cr_2O_3 and can be ascribed to chrompicotite (see Table 1).

Fig. 2. Sub-graphic intergrowth of spinellid and plagioclase. (*a*), (*b*) polished mount, plagioclase lapillus; sample 19 (analysis No. 5), × 250; (*c*)–(*h*) photo from screen of X-ray microanalyser IXA-50A; (*c*), (*d*) back scatter electron (composition) (*c*) is × 500, (*d*) is × 1000; (*e*)–(*h*) in X-ray beam; (*e*) Fe, K; (*f*) Si; (*g*) Mg, K; (*h*) Al, K.

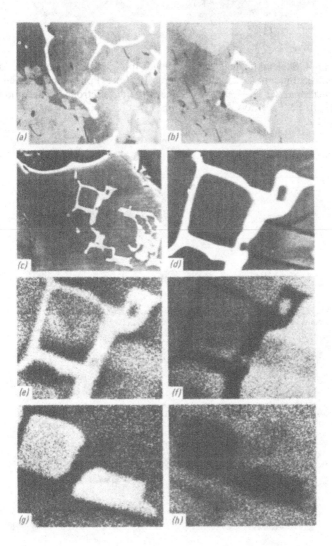

Fig. 3. Isometric grains of spinellids oriented along individual parallel growths of plagioclase plates. Photo from screen of microanalyser IXA-50A in back scatter electrons, sample 14; plagioclase lapilli; (a) × 300, (b) × 600.

Fig. 4. Skeletal aggregates of spinellids in a groundmass of glass: (a) polished mount, × 250; lava, sample 10; (b)–(f) photo from screen of microanalyser IXA-50A; (b)–(e) back scatter electrons (composition); (f) in Fe, K X-rays. Magnification: (b) × 300; (c)–(f) × 600.

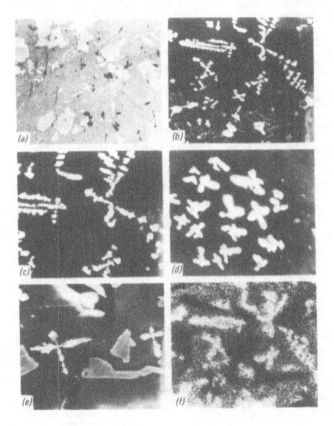

Fig. 5. Two-phase separations of spinellids: (*a*) polished mount, ×250; lava, sample 13; (*b*)–(*h*) photo from screen of X-ray microanalyser IXA-50A; (*b*)–(*c*) back scatter electrons, ×600; (*b*) relief; (*c*) composition; (*d*)–(*h*) in X-rays; (*d*) Cr, K; (*e*) Al, K; (*f*) Mg, K; (*g*) Fe, K; (*h*) Ti, K.

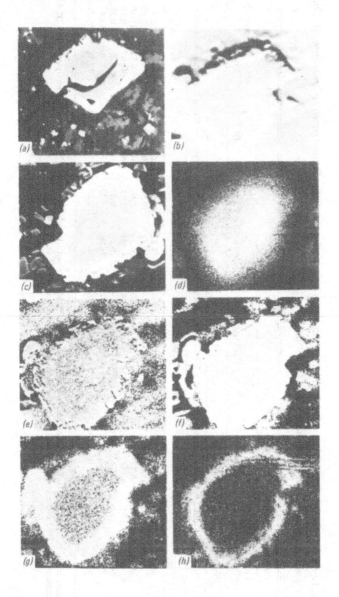

Table 1.—*Chemical composition of spinellids in the eruption products of the Southern Breakthrough of Tolbachik volcano (X-ray spectral analysis by microanalyser IXA–50A)*

Oxides	Plagioclase lapilli		Lava									Bomb		
	Ploskiy Tolbachik (1970)		9 Oct. 1975		19 Oct. 1975	13 Oct. 1975		16 Oct. 1975		9 July 1975		Northern Breakthrough, Cone 1		
	\[No. of grain; 1–4 number of analysis\]													
	Grain 1	Grain 2	Grain 3	Grain 4	Grain 1	Grain 1	Grain 2	Grain 1	Grain 2	Grain 1	Grain 2	Grain 3	Grain 4	Grain 5
	1	2	3	4	5	6	7	8	9	10	11	12	13	14
MgO	5.31	5.02	12.88	5.07	4.66	4.46	4.31	8.62	10.51	15.46	16.05	14.84	2.75	3.37
Al_2O_3	5.50	5.36	13.64	5.17	7.22	4.41	4.61	10.51	10.56	10.46	10.53	10.93	2.66	2.88
TiO_2	9.30	12.48	0.87	10.52	11.26	10.20	10.89	0.55	0.60	0.87	0.89	0.82	11.38	10.45
Cr_2O_3	3.53	2.82	48.44	5.04	2.62	6.97	6.21	48.44	48.11	46.92	47.61	44.97	0.12	0.10
MnO	0.34	0.43	0.16	0.33	0.29	—	—	0.38	0.41	0.13	0.14	0.21	0.38	0.37
Fe_2O_3	75.28	73.28	25.62	75.04	75.94	72.71	72.32	31.80	30.23	27.89	27.27	28.96	83.42	83.44
Total	99.26	99.96	101.61	101.17	102.00	98.75	98.34	100.30	100.23	101.73	102.45	99.58	100.71	100.61

Plagioclase lapilli collected by Yu. M. Stefanov near the summit crater of Ploskiy Tolbachik, belong to the 1970 eruption (see analyses 1 and 2); dates in the upper part of the table (9 Oct. 1975 and so on) give the time when the eruption products appeared on the surface.

Phases 2, 3 and 4 are found both together and separately. In the former case the structure of the intergrowths is as follows: the central part (core) consists of phase 2, the outer zone (or rim) of phase 3 (Figs. 5, 6 and 7). These outer fringes are obviously the least stable and it is to them that the lamellar and occasionally more complex forms of phase 4 belong. In several cases lamellae of phase 4 have been observed in the core itself (phase 2, Fig. 7). Phase 2 is interpreted as a chromium-bearing titanomagnetite, in which the Cr_2O_3 content varies between 2.62 and 6.9% (see Table 1, analyses 4–7). The width of the two- and three-phase rims that constitute

Fig. 6. Two-phase spinellids both phases of which are practically indistinguishable by reflectivity: (*a*) polished mount; lava, sample 13 (analysis 6, Table 1), × 350. (*b*)–(*e*) photo from screen of IXA-50A; (*b*) back scatter electrons (composition) on which are located lines of distribution for Chromium I and Titanium II; (*c*)–(*e*) in X-rays; (*c*) Cr, K; (*d*) Ti, K; (*e*) K; × 700.

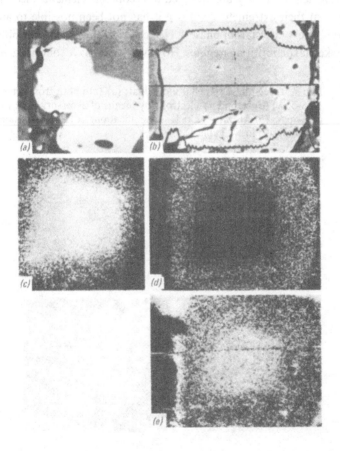

phase 3 is measured in several micrometres ($\approx$ 8 µm). The rims contain microscopic intergrowths of silicates that prevented us from obtaining a reliable analysis by electron microprobe. On the basis of a semi-quantitative microprobe analysis of these rims we can assume that the MgO and Al_2O_3 content does not exceed 2% and the Cr_2O_3 is less than 1%. The change in composition from the rim to the core occurs gradually, judging by the distribution curves of the chromium, titanium, aluminium and magnesium (Figs. 6 and 7). Optically, the extreme members of phases 2 and 3 are practically indistinguishable, nor can they be differentiated in the back scatter electron images, where the high intensity of the mineral is a function of its mean atomic number (the greater the atomic number of the mineral, the more the electrons are reflected and the brighter the mineral appears on the micro-analyser screen). In other words, these members have the same mean atomic number and the differences in their composition are only revealed when they are analysed into concrete elements (see Fig. 6). Because of their extremely small size it has not been possible to analyse the thin lamellar crystals of phase 4. Nor could probe analyses be obtained of the skeletal dendritic aggregates. However, we do have some idea of their

Fig. 7. Three-phase spinellid individual; (a)–(c) photo from screen of IXA-50A; (a) secondary electrons by means of accessory SEI; (b), (c) in X-rays; (b) Al, K; (c) Cr, K. Magnification: × 1500, lava, sample 13 (analysis No. 7, Table 1).

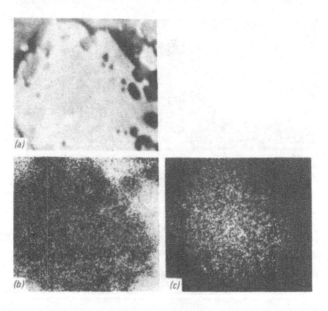

composition. Independent occurrences of phase 3, in the form of isometric and sometimes idiomorphic grains are common in the glass and the interstitial spaces. Chemically, they correspond to titanomagnetites, containing almost no Cr_2O_3 (see Table 1, analyses 13, 14). Phase 4 is almost indistinguishable from these formations in its optical properties. The rims of the subphenocrysts, the skeletal structures and the emulsion disseminations are optically similar to phases 3 and 4. The spinellid minerals imprisoned in inclusions of solidified magma melts consist in the majority of cases of phase 1 (Fig. 1(f)), as do the idiomorphic crystals in the phenocrysts (Fig. 1(c), (d)).

Sulphides, native and intermetallic compounds are extremely rare. Their ratio to spinellids is $\approx 1:1000$. Among the sulphides pyrite, pyrrhotine, chalcopyrite and pentlandite have been recorded. The pyrite and pyrrhotine form independent minerals as inclusions in the phenocrysts of plagioclase, olivine, pyroxene and large multi-phase aggregates of spinellids (Fig. 8(a)–(c)). They are no larger than a few micrometres, rarely a few tens of micrometres. Pyrite has been found in the form of imprisoned mineral in a multi-phase inclusion of solidified magma melt in plagioclase lapilli. Chalcopyrite, some pentlandite and pyrrhotine occur as spherical globule aggregates of up to 30–40 micrometres in size, orientated along the boundaries of individual plagioclase plates that form both phenocrysts and lapilli. The identification of these minerals is based on their optical properties and on the results of interpretation of spectra obtained using the Zeiss laser spectrum analyser. Intermetallic compounds of copper, iron and nickel, as well as native copper and iron have been found in isolated cases in the form of independent crystals in phenocrysts of olivine, pyroxene and occasionally plagioclase. They are of isometric form, have a high reflectivity and low relief, and are up to 10–15 μm in size. Their identification was also based on optical properties and a comparison with standards and the results of interpretation of spectra obtained on the LMA-1 laser microanalyser, and needs further refinement.

Because a number of ore minerals have been trapped in inclusions of melts in minerals, a general description is given of such inclusions in phenocrysts of plagioclase and in plagioclase lapilli.

The following types of inclusions of solidified magma melts have been identified: (1) single-phase: (a) vitreous inclusions, the amorphous vitreous base being either colourless or a pale-yellowish colour; (b) gaseous, occurring in vitreous cinder incrustations and tending to be concentrated in the outer zones of individual plagioclase plates in lapilli (Fig. 9(c));

Fig. 8. Iron sulphides: (a)–(c) photo from screen of SEM; (a) in plagioclase lapillus, × 25000; (b), (c) in rock-forming subphenocrysts; (b) general view, × 3000; (c) detail, × 25000.

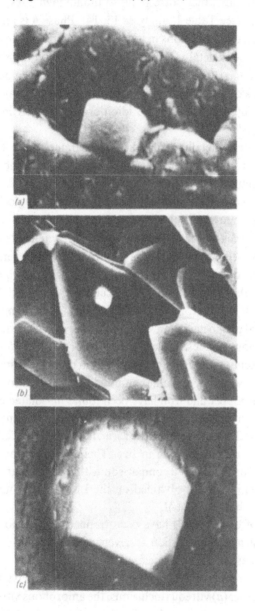

Fig. 9. Morphology of inclusions in plagioclase. Photograph from SEM screen: (a) oval vacuoles confined to the zone of parallel growth of two plagioclase plates (lapillus, general view); sample 16, × 100; (b) structural details of vacuoles in a plagioclase phenocryst; lava, sample 16, × 20000; (c) microrelief in the parallel growth zone of two plagioclase plates (outer part of the zone), caused by conjunction of vacuoles which have kept their gaseous inclusions; sample 16, × 500.

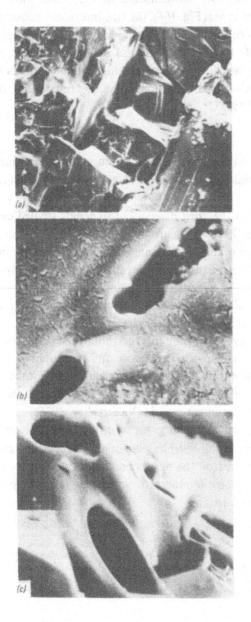

(2) two-phase inclusions, isometric to droplike in shape, the gas phase being separated into spherical bubbles occupying 5–60% of the volume, the size being 5–50 μm (Fig. 1(*c*) and Fig. 9(*a*), (*b*)); (3) multi-phase inclusions, consisting of gas, a vitreous base, and ore and non-ore enclosed minerals (the ore minerals are spinellids of phases 1 and 2, and pyrite; the non-ores have not been determined). They are isometric in shape and include droplike and more complex types (Fig. 1(*f*), (*h*), (*i*)). Individual measurements of the temperatures of homogenisation showed that the first signs of melting appear at 1010–1050 °C, and the gas bubble disappears at 1080–1100 °C, despite the fact that complete melting of the inclusions is observed at 1135–1140 °C. Some inclusions have a temperature of homogenisation of 1280–1350 °C.

Conclusions

1. Ore minerals have been identified in solid products of the eruption. The shapes, sizes and spatial position of the different varieties have been described in detail.

2. It has been shown that the great diversity of spinellids studied can be reduced to four phases. Phase 1 is chrompicotite; phases 2, 3 and 4 are titanomagnetites with a variable Cr_2O_3 content and relatively consistent TiO_2 content. The chrompicotites display some variations in Cr_2O_3, MgO and Al_2O_3 with a constant TiO_2 content and do not resemble anything described in the literature. Chrompicotites are present in the products of both the Northern and Southern Breakthroughs.

3. Among the sulphides pyrite, chalcopyrite, pyrrhotite and pentlandite have been identified. The chalcopyrite, pyrrhotite and pentlandite form characteristic sulphide globules similar to those from inclusions of ultrabasic rocks in explosion pipes (Vakhrushev & Prokoptsev, 1969; Vakhrushev, 1973) and in samples from the upper mantle (Vakhrushev, 1976).

4. A qualitative description has been given of inclusions of mineral-forming systems in minerals exemplified by plagioclase phenocrysts, and some data have been presented on the phases present and temperatures of homogenisation; these point to the minerals having undergone a complex thermal evolution of many stages.

References

Fedotov, S. A., Khrenov, A. P. & Chirkov, A. M. 1976. Bol'shoye treshchinnoye Tolbachinskoye izverzheniye 1975 g. na Kamchatke. (The Great Tolbachik Fissure Eruption of 1975 in Kamchatka.) *Dokl. AN SSSR*, **228**(5), 1193–6.
Vakhrushev, V. A. 1973. *Rudnyye mineraly izverzhennykh porod i ikh znacheniye pri*

petrologicheskikh issledovaniyakh. (*Ore minerals of extruded rocks and their significance for petrological studies.*) Novosibirsk, Nauka, pp. 5–27, 83–121.

Vakhrushev, V. A. & Kutolin, V. A. 1976. Sul'fidy v verkhney mantii. (Sulphides in the upper mantle.) *Dokl. AN SSSR*, **229**(2), 438–41.

Vakhrushev, V. A. & Prokoptsev, N. G. 1969. Pervichno-magmaticheskiye sul'fidnyye obrazovaniya v bazal'takh okeanicheskoy kory i vo vklyucheniyakh ul'traosnovnykh porod. (Primary magma sulphide formations in ocean crust basalts and in inclusions of ultrabasic rocks.) *Geologiya rudn. mestorozhd.*, No. 6, 14–26.

Volynets, O. N., Flerov, G. B., Khrenov, A. P. & Yermakov, V. A. 1976. Petrologiya vulkanicheskikh porod treshchinnogo Tolbachinskogo izverzheniya 1975 g. (The petrology of volcanic rocks of the Tolbachik fissure eruption of 1975.) *Dokl. AN SSSR*, **228**(6), 1419–22.

Welter, L. W. & McKee, A. N. 1972. Observations on uncoated, nonconducting or thermally sensitive specimens using a fast scanning field emission source SEM. *50th Ann. IITRI SEM Symposium*, pp. 161–8.

Gases from the basalt flows of the Tolbachik Fissure Eruption 1975–1976

Ye. K. Markhinin, V. A. Urakov, N. Ye. Podkletnov
and V. V. Ponomarev

Volcanism is the main process by which gas is released from the Earth's mantle. Volcanic eruptions at the Earth's surface of magma originating at depths of 150–200 km (Markhinin, 1967) result in the release of considerable volumes of magmatic gas. The accumulation of the components of this gas on the Earth's surface and their chemical evolution during the geological history of the Earth led ultimately to the formation of the Earth's air and water layers and the appearance of life. The formation of many useful ore and non-ore minerals is connected with the rise of magmatic gases to the surface and to the subsurface layers of the Earth's crust. Thus the importance of studying magmatic gases is obvious. However, collecting samples of actual magmatic gas, which is the first stage of the research, presents exceptional difficulties and so far only a few cases are known of such samples being taken.

In May 1912 A. Day and E. Shepherd, from the geographical laboratory of the Carnegie Institute, introduced an iron pipe into a lava bubble that had formed on the lava lake in the Halemaumau crater of Kilauea volcano on Hawaii (Day & Shepherd, 1913). They demonstrated that the magmatic gas contained up to 70% water, the presence of water being disputed at the time. But the results of the analysis of their sample were not entirely satisfactory. The iron pipe used to collect the sample distorted the true composition of the gas. The possibility of pollution by atmospheric air was not excluded. In 1917 Shepherd successfully collected several samples from the surface of Lake Halemaumau. More representative samples were collected in 1919, all from the same lake (Jaggar, 1940). In 1926 Shepherd again collected samples of gas from liquid lava, this time on Mauna-Loa (Hawaii) (Shepherd, 1938). In 1959 Chaigneau, Tazieff and Fabre extracted gas from the lava lake of the Nyiragongo volcano in Africa (Chaigneau *et al.*, 1960), plunging the collecting tube 15 cm into the red-hot lava. Gases were first collected from the sources of lava flows by Chaigneau in 1961 during the eruption of Etna. He took the samples by introducing a metal pipe into the molten lava (Chaigneau, 1962). In the period 1964–67, during the eruption of Surtsey in Iceland, G. Sigvaldason and G. Elisson collected more than ten samples from liquid lava flows (Sigvaldason & Elisson, 1968).

Examples of analyses of the most representative samples collected by foreign scientists are given in Table 1.

In the Soviet Union the first successful extraction of samples of magmatic gas directly from lava flows, their sources and middle course was achieved by Ye. K. Markhinin and V. A. Urakov in 1975–76 on the Southern Breakthrough of the Great Tolbachik Fissure Eruption.

Collecting tubes of ceramics, metal (stainless steel, titanium and vanadium), and later quartz glass (the most successful), varying in length from 1.5 to 4 m, were submerged in flowing liquid lava to a depth of 10–50 cm. In some cases gas samples were taken from bubbles of magmatic gas that had only just appeared on the surface of the lava flow. The lava

Table 1. *Composition of volcanic gases, vol. %*

Components	1	2	3	4	5
CO_2	21.4	46.2	40.9	4.6	18.2
CO	0.8	0.4	2.4	0.3	0.3
H_2	0.9	0.016	0.8	2.8	10.4
SO_2	11.5	7.6	4.4	4.1	21.7
S	0.7	0.0	—	—	—
SO_3	1.8	20.6	—	—	—
Cl_2	0.1	0.0	—	—	—
F_2	0.0	0.0	—	—	—
O_2	0.0	0.0	—	—	—
HCl	—	—	—	0.6	—
N_2 + rare gases	10.1	9.8	8.3	4.5	47.5
CH_4	—	—	—	—	0.7
H_2O	52.7	38.0	43.2	83.1	?

1, Kilauea, Hawaiian Islands; average value of 10 gas samples from the lava lake of Halemaumau taken by Jaggar, 1917–18; 2, Mauna Loa, Hawaiian Islands; average of 2 samples from molten lava, taken by Shepherd in 1926; 3, Nyiragongo, Africa; sample from lava lake taken by Chaigneau and Tazieff in 1959; 4, Surtsey, Iceland; average value of 11 samples from liquid lava taken by Sigvaldason in 1967; 5, Etna, Sicily; average value of 2 samples from liquid lava, taken by Chaigneau in 1961.

had a vesicular structure and the magmatic gas was concentrated in its interstitial bubbles. When the mouth of the collecting tube entered the gas pore it did not become clogged up with lava. This was evidently because where the tube and the lava came into contact a fall in pressure occurred which increased the flow of gas bubbles towards the tube and into the collecting apparatus. The temperature of the lava was constantly being measured by a platinum/platinum–rhodium thermocouple type PP-1 and was found to be 1000 ± 65 °C.

The apparatus for collecting high-temperature gases consisted of a specially constructed channel for the gas (Fig. 1) made of quartz glass, a cooling chamber with many outlets to condense the vapours of the magmatic gas, a thermometer to measure the temperature of the gas entering the apparatus, burettes with absorbents for the various components, a tap-pipette or vessels for collecting the gases, and an aspirator

Fig. 1. Gas conduit for extracting high-temperature magmatic gas. 1, quartz tube; 2, condensate collector; 3, condensate; 4, capillary condensate collector.

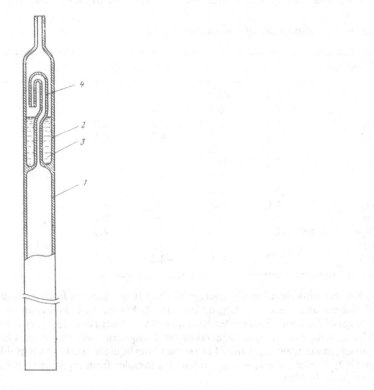

tap-pipette or a pump depending on the purpose for which the gas was being collected (Fig. 2). Before the gases were collected the apparatus was washed through several times by the magmatic gas; moreover, as a rule the gas entered the apparatus under pressure greater than atmospheric, which precluded air being sucked in from the atmosphere. After the condensation of vapour contained in the gas, the gas itself was collected into special containers and analysed in the laboratory. In storing the gas every possible precaution was taken to prevent it being contaminated by atmospheric air.

The gas was analysed on an LKhM-8MD chromatograph (model V and 'Luch'). For the LKhM-8MD chromatograph the carrier gas was helium, the immobile phase 'poropak', the temperature of the columns 60 °C, the length 3 m and the detector a catarometer. On the 'Luch' chromatograph the carrier gas was argon, the immobile phase molecular sieves 5A, the temperature of the columns 80 °C, the length 3 m, and the detector a catarometer.

Analysis of 36 gas samples gave the following results (vol. %):

Fig. 2. Diagram of extractor of gaseous samples of volcanic gas. 1, gas conduit; 2, vessel with cooling liquid; 3, vessel for outflow of cooling liquid; 4, multiple cooling jacket; 5, vessel for condensate collection; 6, thermometer; 7, pipette stopcock; 8, aspirator; 9, absorber.

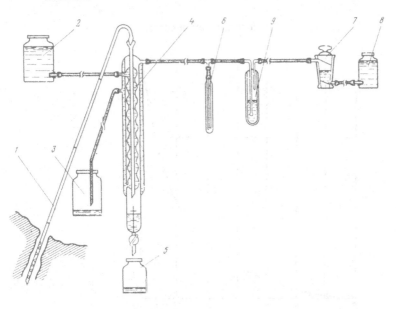

H_2 – 0.002–25.5; O_2 – 0.6–19.7; N_2 – 38.6–83.4; He + Ne – 0.02–0.17; CO_2 – 0.2–41.0; CO – 0.06; CH_4 and other hydrocarbons – 10^5–0.56; HF – 0.005–0.06; HCl – 0.05–0.7; SO_2 – 0.002–0.006; H_2S – 0.003–0.008 (analyses carried out by S. P. Levshunova and R. V. Guseva).

Fig. 3. Dependence of viscosity of lavas on their water content.

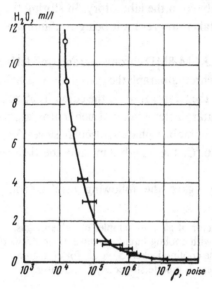

Fig. 4. Hydrogen–water correlation curve.

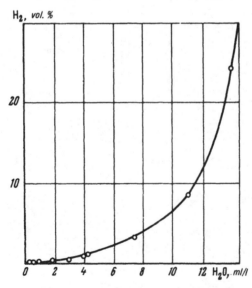

Compared with samples collected by foreign scientists, the gas from the liquid lava of the Tolbachik eruptions has a noticeably low carbon monoxide content. The amount of carbonic acid gas can be anything up to 41% vol. Small quantities of helium are present in all samples. The Moore–Lepape index $(100 \times$ rare gases$)/N_2$ is 0.547 for the average sample, whereas for atmospheric air it is 1.177. Consequently, one can assume that the nitrogen and rare gases in the samples are juvenile. The relation N_2/O_2 is much higher than in air. Depending on the degree of degasification of the lava the composition of the magmatic gas varied greatly.

Besides the overall composition of the gas samples, the composition of the condensate obtained by cooling the magmatic gas was also studied. It comprised on average up to 11.2 ml for every litre of gas obtained, which corresponds to about 90% by volume. It has been established that the viscosity of the lava is dependent on the amount of water dissolved in it (Fig. 3). This is in good agreement with laboratory experiments (Kadik *et al.*, 1971).

The correlation between the quantity of condensed water vapour and the hydrogen content of the magmatic gas is shown in Fig. 4. In the samples of condensate from the magmatic gas, fluoride, chloride, sulphate and metallic ions were discovered. When fluorine was determined by direct potentiometry using a fluorine-selective electrode of lanthanum fluoride after alkalisation only 660 mg/l were found. When complexing agents were added to the solution, a further 199 mg/l were found. This means that approximately 60% of the fluorine in the condensate was bound up in a metal complex.

Two methods were used to identify the chlorine: direct potentiometry with a silver chloride electrode, and argentometric titration. Both methods showed a very high chloride content – up to 23 800 mg/l of condensate.

The metals in the condensate were identified by atomic absorption on an S-302 spectrophotometer in an air–acetylene flame.

The following are the mean results from an analysis of the condensate at pH 0.66–0.85 in mg/l gas: H_2O – 11 200; Cl – 23 800; F – 1900; SO_4^{2-} – 220.4; H_2SO_4 – 1327.3; NH_4^+ – 6.2; Na^+ – 770.0; Al^{3+} – 250.6; Zn^{2+} – 190.3; Cu^{2+} – 93.5; Ca^{2+} – 68.0; Fe^{2+} – 37.5; Mg^{2+} – 18.3; Cd^{2+} – 24.79; Li^+ – 1.5; Pb^{2+} – 1.41; Mn^{2+} – 1.23; Sn^{2+} – 0.05; As^{2+} – 0.05; Ag^{2+} – 0.001; Co^{2+} – traces.

Thus the analysis showed that the main ingredient of volcanic emanations is water. Almost all the samples contained high concentrations of zinc, copper and cadmium, and in some cases these elements even predominated over the rock-forming ones. It is also extremely interesting that relatively little potassium was found in the samples – approximately the same amount as lithium.

The condensate has a comparatively low magnesium and calcium content. In general one can say that the magmatic gases carry relatively small quantities of petrogenic elements (with the exception of sodium), although they are rich in ore elements.

As well as general analysis, a more detailed analysis was also carried out on the hydrocarbon part of the magmatic gas, using gas chromatography.

Table 2. *Hydrocarbon content of magmatic gas*

Sample	Hydrocarbon group composition,						
	CH_4	C_2H_6	C_2H_4	C_3H_8	C_3H_6	iC_4H_{10}	nC_4H_{10}
Magmatic gas from liquid lavas	66.17	2.56	4.06	2.10	6.39	1.28	3.01
Magmatic gas from liquid lavas	41.33	8.98	21.56	4.49	10.78	0.52	2.25
Magmatic gas from liquid lavas	33.45	4.49	15.30	28.85	4.77	0.86	2.06
Chilled lava from liquid flow	71.93	12.42	9.34	2.74	2.30	0.04	0.23
Chilled lava from liquid flow	61.29	12.88	15.07	3.27	2.62	0.20	1.28
Chilled lava from liquid flow	69.40	6.94	14.90	2.25	2.04	0.12	0.66
Cinder bomb	55.37	5.70	25.00	3.50	3.80	0.34	1.90

Sample	Hydrocarbon group composition, (vol. %)						
	C_4H_8	iC_5H_{12}	nC_6H_{12}	iC_6H_{14}	nC_6H_{14}	nC_6H_{16}	Σ
Magmatic gas from liquid lavas	3.16	Trace	Trace	Σ 11.28		—	100.00
Magmatic gas from liquid lavas	1.17	5.75	1.08	Σ 2.08		—	99.98
Magmatic gas from liquid lavas	6.68	0.81	3.81	Σ 2.86		—	101.01
Chilled lava from liquid flow	0.25	0.09	0.09	0.15	0.34	0.08	100.00
Chilled lava from liquid flow	2.18	0.15	0.22	0.50	0.28	0.06	100.00
Chilled lava from liquid flow	1.40	0.16	0.20	1.26	0.36	0.37	100.06
Cinder bomb	3.30	0.44	0.53	Σ 0.12		—	100.00

Analyst S. P. Levshunova.

The results of this, showing the contents of hydrocarbon groups in the magmatic gas, are set out in Table 2. The hydrocarbon part of the gas is dominated by methane and its homologues up to hexane and higher. Unsaturated hydrocarbons are also present (ethylene, propylene, butylene, etc.) This last circumstance means that we can assume the hydrocarbons had been subject to thermal cracking. A direct relationship was discovered between the hydrogen content and the range of hydrocarbons in the magmatic gas. This tends to suggest that the magmatic hydrocarbons were of abiogenic origin.

As the gas and condensate were being collected from lava flows, samples of lavas, volcanic bombs and ash were also collected, and the hydrocarbon content determined of the gases adsorbed by the volcanic products.

Analyses of these gases are also given in Table 2. Methane is the predominant hydrocarbon constituent of the gases from these solid products (of the order of 70%). As well as methane, a series of complex hydrocarbons were discovered, up to C_6H_{14} and heavier. The overall quantity of adsorbed hydrocarbons is about 0.08–0.3 cm³/kg of lava. It is interesting to note that the composition of the gases collected from the lava is identical to that of the adsorbed gases of the solid products.

References

Chaigneau, M. 1962. Sur les gaz volcaniques de l'Etna (Sicile). *Compt. rend. hebdomadaires des séances de l'Acad. des Sci.*, No. 23, 254.

Chaigneau, M., Tazieff, H. & Fabre, R. 1960. Composition des gaz volcaniques du lac de lave permanent du Nyiragongo (Congo belge). *Compt. Rend. Acad. Sci.* Paris, **250**, 2482–5.

Day, A. L. & Shepherd, E. S. 1913. Water and volcanic activity. *Bull. Geol. Soc. Am.*, **24**, 573 pp.

Elskens, I., Tazieff, H. & Tonani, F. 1964. A new method for volcanic gas analysis in the field. *Bull. Volcanol.*, Ser. 2, **27**, 347–50.

Jaggar, R. A. 1940. Magmatic gases. *Am. J. Sci.*, **238**, 313–53.

Kadik, A. A., Lebedev, Ye. B. & Khitarov, N. I. 1971. *Voda v magmaticheskikh rasplavakh.* (*Water in magmatic melts.*) M., Nauka, 268 pp.

Markhinin, Ye. K. 1969. *Rol' vulkanizma v formirovanii zemnoy kory.* (*The role of volcanism in the formation of the Earth's crust.*) M., Nauka, 256 pp.

Shepherd, E. S. 1938. The gases in rock and some related problems. *Am. J. Sci.*, Ser. 5, **35-A**, 311–51.

Sigvaldason, G. E. & Elisson, G. 1968. Collection and analysis of volcanic gases at Surtsey, Iceland. *Geochim. Cosmochim. Acta*, **32**, 797–805.

Chemistry, metal content and reaction products of gases from the New Tolbachik Volcanoes of 1975

I. A. Menyaylov, L. P. Nikitina, L. P. Vergasova,
R. V. Guseva, V. N. Shapar' and A. M. Rozhkov

The Tolbachik fissure eruption that began on 6 July 1975 was a very suitable subject for the study of volcanic gases. During the eruption cinder cones were formed and liquid basalt lavas poured out at the Northern and Southern Breakthroughs. In 1975 more than 30 gas vents were sampled on the New Tolbachik Volcanoes, as a result of which about 100 gas samples, 30 condensate samples and 200 samples of sublimates were collected. In the present article only a portion of the analyses and condensates are described, although the graphs have been constructed from all the data available at the present time.

The technique for collecting the gases and condensates (Fig. 1) corresponded to modern usage (Sigvaldason & Elisson, 1968; Mizutani & Matsuo, 1959; Le Gern, 1972). The method used to collect sublimates was one of areal sampling and separation of the minerals in the field.

Samples were analysed by various methods. Gases were identified using chromatography (R. V. Guseva, A. M. Rozhkov) and mass spectrometry, and aqueous solutions by 'wet chemistry' methods (N. A. Peretolchina, S. N. Litasova), atomic absorption (N. V. Reznikov, L. P. Nikitina, G. N. Anoshin), polarography (A. N. Nevzorov, L. P. Nikitina), flame photometry (Yu. D. Kuz'min, S. N. Litasova) and adsorption chromatography with complexing agents (L. P. Nikitina). This not only meant that the composition of the exhalations could be identified with the greatest accuracy and a high degree of sensitivity, but also enabled us to compare different methods on the same very specific and hard to identify samples of these ultra-acidic complex solutions of volcanic origin.

Volcanic exhalations were studied at the explosive–effusive stage of the eruption at the Northern Breakthrough. Gas vents on the lava flows of Cone I were sampled at the consolidation stage and those on the lava flow of Cone II while mobile. At the Southern Breakthrough gases were studied during the effusive stage of the eruption. Gases and condensates were collected from the lava stream at the outpouring stage. The gases of the Southern Breakthrough are magmatic discharges.

Magmatic gases. Samples of magmatic gases were collected from fissures in a thin crust (Table 1, Nos. 8, 9) or from beneath the rim (Table 1, Nos. 10, 11) of congealed lava above the sources of the lava rivers. As sample No. 11 was being collected (see Table 1) the source of the lava river moved 5–6 m down the relief and the gas-collecting tube was caught up by the lava crust in such a way that its lower end was left in the lava river considerably higher than the source.

It was in some of the samples of magmatic gases released from liquid lava at varying distances from the South Cone that the highest HF, HCl, SO_2, H_2S and H_2O contents of all the samples were found. Some samples (see Table 1, No. 9) corresponded to Ellis' (1957) theoretically derived composition at a similar temperature and a pressure of 1 atm., i.e. where the molecular ratios for the system $H_2O–CO_2–H_2–S_2$ were $100:10:2:1$. Like Ellis, Krauskopf (1957), Matsuo (1960), and Volkov & Ruzaykin (1974) took as their basis for equilibrium calculations the samples collected by Shepherd and Jaggar from the lava lake of Kilauea. Gases from the Southern Breakthrough and Surtsey in Iceland differed from the Kilauea gases in having more hydrogen. Obviously gases released from liquid lavas some distance from the conduit of a volcano, as with Surtsey and the Southern Breakthrough, should differ from those released above the conduit (the Kilauea, Nyiragongo and Erta-Ale lava lakes).

Fig. 1. Diagram of gas collection system.
1, ceramic tube; 2, 'fluorplastic' tubing; 3, wash bottle with metal-free water; 4, wash bottle with cadmium acetate solution; 5, three-way stopcock; 6, syringe; 7, gas receiver with saturated solution of NaCl; 8, rubber tubing.

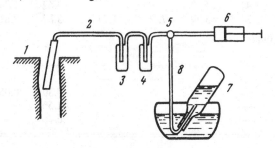

Systematic sampling of gases from flowing lava as the lava vents migrated further and further from the main conduit (Fig. 2) showed that their average composition was changing: the further from the main conduit, the less CO_2 the gases contained and the more H_2O and especially H_2, whilst the HF, HCl and SO_2 contents remained approximately the same

Table 1. *Composition of the volcanic gases of the Tolbachik Fissure Eruption (vol. %) in 1975*

No.	$t(°C)$	Date of sampling	Volcanic gases apart from H_2O, O_2, N_2				
			HF	HCl	SO_2	H_2S	CO_2
Northern Breakthrough							
1	500	16 Aug.	0.94	2.55	0.03	0.29	89.19
2	800	17 Aug.	2.37	2.46	0.18	—	73.55
3	485	26 Aug.	1.71	4.42	0.01	—	87.10
4	400	14 Aug.	1.61	6.92	0.07	—	68.32
5	900	22 Aug.	2.61	4.84	0.02	—	36.12
Southern Breakthrough							
6	750	18 Oct.	0.69	3.17	0.38	—	90.57
7	800	13 Oct.	1.32	3.65	0.59	—	92.82
8	990	15 Dec.	8.56	43.18	5.54	—	28.18
9	980	16 Dec.	7.50	27.30	28.00	0.18	36.65
10	1020	21 Dec.	0.04	21.00	5.32	0.54	30.51
11	1020	22 Dec.	5.54	30.06	5.34	0.36	12.54

No.	$t(°C)$	Volcanic gases apart from H_2O, O_2, N_2					
		CO	CH_4	C_2H_6	C_3H_8	H_2	NH_3
Northern Breakthrough							
1	500		1.65	—	—	4.18	1.18
2	800	2.10	12.08	0.48	—	1.75	5.08
3	485	1.13	4.78	—	—	0.42	0.43
4	400	3.42	14.10	—	—	3.62	1.94
5	900	25.84	15.29	7.16	1.74	3.87	2.61
Southern Breakthrough							
6	750	—	—	—	—	4.31	0.82
7	800	—	—	—	—	—	0.29
8	900	—	—	—	—	14.53	0.01
9	980	—	—	—	—	0.30	0.05
10	1020	—	—	—	—	42.56	—
11	1020	—	—	—	—	46.16	—

Table 1. (*cont.*)

No.	$t(°C)$	Total composition of gases				N_2/O_2	H_2O/H_2 $\times 10^{-4}$
		H_2O	Σ_B^*	O_2	N_2		
Northern Breakthrough							
1	500	81.83	4.30	0.04	13.83	345.5	0.04
2	800	83.10	1.95	2.10	12.85	6.1	0.3
3	485	87.63	2.62	0.48	9.27	19.1	0.9
4	400	87.59	3.83	0.89	7.68	8.6	0.06
5	900	75.39	2.54	1.40	20.67	14.1	0.07
Southern Breakthrough							
6	750	66.03	0.42	4.03	29.52	7.3	0.3
7	800	66.40	0.50	5.81	26.79	4.6	—
8	990	82.06	1.65	2.57	12.82	4.6	0.03
9	980	88.39	0.40	2.24	8.97	4.0	—
10	1020	93.73	0.93	0.58	4.76	8.2	0.03
11	1020	94.02	1.18	0.89	3.91	4.4	0.01

Sample Nos. 1, 2, gases from fumaroles on surface of lava flow at the southern base of Cone I; 3, 4, gases from fumaroles on surface of lava flow at the northern base of Cone I; 5, gases from a cleft between lava blocks on the front of lava flowing from Cone II; 6, 7, gases from fissures in the lava covering the source of a lava river (200 m from cone); 8, 9, gases from source of lava river 400 m from cone; 10, 11, gases from source of lava river 1000 m from cone. Temperatures of 1, 3, 4 were measured by thermometer; of 8, 9, 10, 11 by thermocouple; of 2, 5, 6, 7 from the colour of incandescent lava material. The dash means that the component was determined, but not found. Solutions from the absorption vessels were analysed by N. A. Peretolchina. V. A. Urakov took part in the gas sampling. Σ_B^* = total volcanic gases.

(see Table 1, Nos. 6–11). This change in the composition of the gases agrees with the pattern of degasification in the course of eruption of basalt lavas (Sigvaldason, 1974). In collecting samples of gases from one and the same source of a lava river it was noticed that several hours before the source was exhausted the H_2O content of the gases increased and H_2S appeared in considerable amounts (see Table 1, Nos. 8, 9), whilst colloidal sulphur was formed in the condensates. Sample No. 8 (see Table 1) was taken forty-eight hours before the exhaustion of a lava river, and No. 9 several hours before. It is possible that the change in the composition of the gases was connected with a reduction or increase in the intensity of the gas release from the magma in the conduit, since the exhaustion of some lava rivers and the appearance of others depended to a certain degree on the intensity of the eruption.

Fumarole gases on both the Southern and Northern Breakthroughs differed from the magmatic gases in containing less HF, HCl, SO_2, H_2S and H_2, but more H_2O, CO_2, CO and CH_4 (+ other hydrocarbons; see Table 1). The composition of the gases of the New Tolbachik Volcanoes is typical of the basalt volcanoes of Kamchatka and the Kuril Islands, where the source of the magma is known to lie at some depth (Menyaylov, 1976). The main factor determining the composition of the fumarole gases was temperature. Fluctuations in the composition of the fumarole gases depending on temperature (Fig. 3) corresponded to known patterns (Basharina, 1963).

However, the composition of the gases also depended on the location

Fig. 2. Diagrammatic section of the Southern Cone, showing gas sampling locations. Vertical scale not accurate.

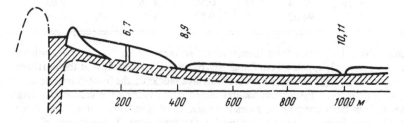

Fig. 3. Dependence of gas composition on temperature. The curves are for maximum concentrations.

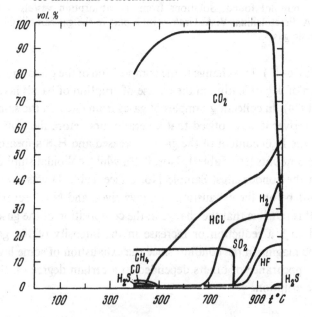

of the fumaroles. There are quite conspicuous differences between the gases of the Northern and Southern Breakthroughs (Fig. 4) at similar temperatures. The N_2/O_2 ratio in the volcanic gases can be considered an indicator of the physical and chemical conditions in the magma melt and of its depth. The N_2/O_2 ratio was high in the gases of the Northern Breakthrough and low in those of the Southern (Fig. 5), i.e. in the magma melt of the Northern Breakthrough the conditions were more reducing, indicating greater depths, whereas in that of the Southern Breakthrough, especially in the initial stages of the eruption, they were more oxidising, indicating lesser depths. In gases of the Northern Breakthrough CO, CH_4 (+ other hydrocarbons) and H_2 were recorded, and high H_2O/H_2 ratios (see Fig. 5). It can be assumed that the magma of the Northern Breakthrough came to the surface from a greater depth and with greater speed than that of the Southern (Fedotov *et al.*, 1976).

The He content in the gases of the Northern Breakthrough was greater

Fig. 4. Average composition of the gases from the Northern (I) and Southern (II) Breakthroughs from analyses of 35 samples.

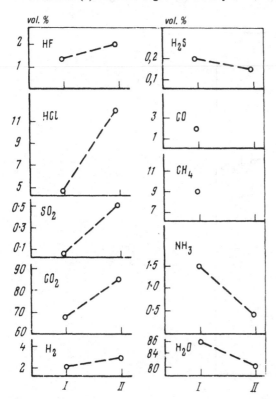

than in those of the Southern (Table 2), which is further evidence that the source of the magma of the Northern Breakthrough was deeper.

Condensates of volcanic gases. Condensates were collected at the same time as gas samples. The following were discovered in the magmatic gas condensates: chlorine (a few tens of grams per litre), fluorine and sulphate (several grams per litre), aluminium, sodium and potassium (in hundreds of milligrams), and calcium, magnesium and zinc (tens of milligrams) (Table 3). The molar ratio of water:gas:metals in the gas samples was 1000:63:0.1. Magmatic gases of the Southern Breakthrough contained much greater quantities of ore elements than the fumarole gases of the Northern Breakthrough (Fig. 6).

Fumarole minerals. The composition of sublimates from the lava flows of the Northern and Southern Breakthroughs, their abundance, and their

Fig. 5. Average values of the ratios of N_2/O_2 and H_2O/H_2 in gases of the Northern (I) and Southern (II) Breakthroughs from analyses of 35 samples.

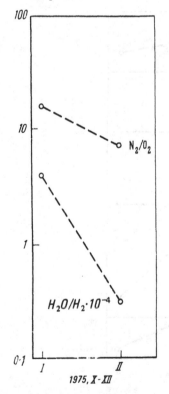

trace element composition are all closely related to the composition of the gases and condensates. When the fumarole gases from lava flows of the Northern Breakthrough contained relatively large amounts of HCl and NH_3, sublimates of ammonium chloride and Na, K and Fe chlorides formed around them. The chief fumarole mineral of the sublimates around the fumaroles of the Northern Breakthrough was ammonium chloride. The

Table 2. *He in gases of the Tolbachik Fissure Eruption 1975–beginning of 1976*

Sampling site		He (vol. %)	Samples analysed for He	Number of samples in which He was found	% occurrence of He in gases
Northern Breakthrough	Lava flow from Southern Bocca, Cone I	1×10^{-3}	13	3	23.1
	Cone II	6.43×10^{-3}	—	—	—
	Cone II	5.45×10^{-3}	—	—	—
Southern Breakthrough	Lava river	1.8×10^{-3}	26	1	3.8
—	Content in the atmosphere	5.4×10^{-4}	—	—	—

I. A. Markov, V. V. Ponomarev and C. A. Urakov also took part in the sampling.

Table 3. *Composition of condensates of gas from the New Tolbachik Volcanoes (p.p.m.)*

Components	1	2	Components	1	2
pH	0.43	1.55	Fe	9.0	7.0
Cl	21347	2615	Al	202.5	24.3
F	2700	540	Zn	9.250	0.384
S	564	5.8	Pb	0.300	0.040
Na	700	29.2	Cu	14.000	3.500
K	240.5	10.2	Sn	0.030	0.013
Ca	3.4	7.8	Ag	0.002	0.006
Mg	9.3	7.5	As	0.020	0.010

1 corresponds to No. 11, 2 to No. 3 in Table 1.

trace element composition of the sublimates showed that they were enriched with copper, zinc, lead and silver (Fig. 7). The ammonium chlorides and Cu, Na, K and Fe chlorides of the Southern Breakthrough were richer in trace elements than those of the Northern (Fig. 8). When H_2S was present in the gases (see Table 1, No. 1), native sulphur formed around the fumaroles. The gases of the Southern Breakthrough were on the whole significantly richer in SO_2 (see Table 1), and the main types of sublimate at the Southern Breakthrough were high temperature sulphates, also enriched with trace elements (see Fig. 8). Among the sublimates collected from the lava flows of the Northern and Southern Breakthroughs the following were discovered: native sulphur (S); halogen compounds – ralstonite $(Na_x(Mg_xAl_{2x}) \cdot (F, OH)_6 \cdot y(H_2O)$, sylvite (KCl), halite (NaCl), sal ammoniac (NH_4Cl), kremersite $(NH_4KFeCl_5 \cdot H_2O)$, atacamite $(Cu_2Cl(OH)_3)$, paratacamite $(CuCl_2 \cdot 2H_2O)$, melanothallite $(CuCl_2)$, molysite $(FeCl_2)$; oxides – tenorite (CuO), hematite (Fe_2O_3); oxy-salts – aphthitalite $(K, Na)_3Na(SO_4)_2$, euchlorine $(KNa_2) \cdot SO_42CuSO_4CuO)$, chalcocyanite $(CuSO_4)$, chalcanthite $(CuSO_4 \cdot 5H_2O)$; sulphides – realgar (AsS), orpiment (As_2S_3). The trace element content of the sublimates was much higher than that of the condensates and aqueous extracts from the ashes (Fig. 9).

Aqueous extracts from the ashes. The study of aqueous extracts from the ashes adds to our understanding of the exhalations of volcanoes, especially

Fig. 6. Average content of microelements in condensates of gases from the Northern (i) and Southern (ii) Breakthroughs from analyses of 9 samples. Temperature of gases 485 and 980–1020 °C respectively.

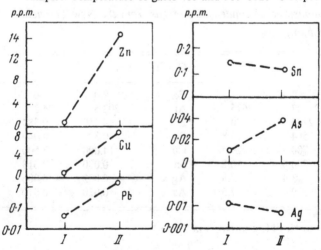

Fig. 7. Maximum contents of microelements in sublimates of the Northern (I) and Southern (II) Breakthroughs relative to their Clarkes in basalts (from 50 samples). The abscissae indicate the Clarke of the element according to Vinogradov (1962).

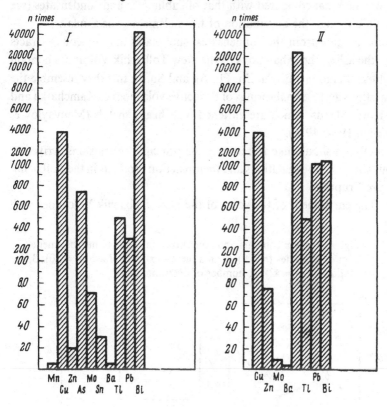

Fig. 8. Average contents of microelements in sublimates of the Northern (I) and Southern (II) Breakthroughs (from 50 samples). 1, in ammonium chloride, temperature (t) at sampling site = 150–350 °C; 2, chlorides of Na, K, Cu, Al, Fe, t = 100–< 500 °C; 3, sulphates of Na, K, Cu, t = 400–500 °C.

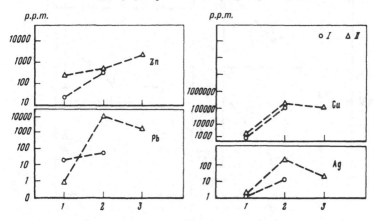

at the explosive stage. The trace element content in aqueous extracts from ash was minimal compared with that of sublimates and condensates (see Fig. 9), although the distribution of trace elements was similar.

Thus judging from the condensates, sublimates and aqueous extracts from the ashes, the exhalations of the New Tolbachik Volcanoes contain heightened amounts of Cu, Zn, Pb, Ag and Sn. In this they resemble the exhalations and thermal springs of the active volcanoes of Kamchatka and the Kuril Islands, which are enriched with heavy metals (Menyaylov & Nikitina, 1974; 1976).

Finally we should like to reiterate the principal conclusions arrived at in our study of the exhalations and their reaction products in the Tolbachik Fissure Eruption.

1. The enrichment of the gases of the New Tolbachik Volcanoes with

Fig. 9. Average contents and concentration limits of microelements. 1, in condensates ($n = 9$); 2, in aqueous extracts of ash ($n = 20$); 3, in sublimates ($n = 42$); n, number of determinations.

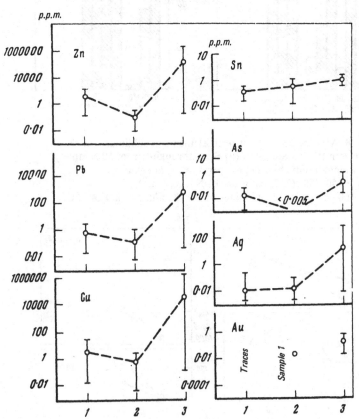

HF, HCl, H_2 and CH_4, and the high N_2/O_2 and H_2O/H_2 ratios, suggest that the magma originated from a deep source.

2. The nature of the N_2/O_2 and H_2O/H_2 ratios and of the H_2, CH_4 (+other hydrocarbons), CO and He content in gases of the Northern and Southern Breakthroughs suggests a deeper source for the magma of the Northern Breakthrough than for that of the Southern.

3. The metallogenic specialisation of the exhalations of the New Tolbachik Volcanoes is typical of exhalations and thermal springs of active volcanoes.

References

Basharina, L. A. 1963. Eksgalyatsii pobochnykh kraterov Klyuchevskogo vulkana na razlichnykh stadiyakh ostyvaniya lavy. (Exhalations of the subordinate vents of the Klyuchi volcano at different stages in the cooling of the lava.) In: *Vulkanizm Kamchatki i nekotorykh drugikh rayonov SSSR.* (*The volcanism of Kamchatka and some other regions of the USSR.*) M., Izd-vo AN SSSR, pp. 169–227.

Ellis, A. Y. 1957. Chemical equilibrium in magmatic gases. *Am. J. Sci.*, **255**(6), 416–30.

Fedotov, S. A., Gorel'chik, V. I. & Stepanov, V. V. 1976. Seysmicheskiye dannyye o magmaticheskikh ochagakh, mekhanizme i razvitii bazal'tovogo treshchinnogo Tolbachinskogo izverzheniya v 1975 g. na Kamchatke. (Seismic data on the magma chambers, mechanism and development of the Tolbachik basalt fissure eruption of 1975 in Kamchatka.) *Dokl. AN SSSR*, **228**(6), 1407–10.

Krauskopf, K. B. 1957. The heavy metals content of magmatic vapour at 600 °C. *Econ. Geology*, **52**, 786–807.

Le Gern, F. 1972. *Etudes dynamiques sur la phase gazeuse éruptive.* Rapport CEA-R4383. Paris, 169 pp.

Matsuo, S. 1960. On the origin of volcanic gases. *J. Earth Sci.*, Nagoya Univ., **8**(2), 222–45.

Menyaylov, I. A. 1976. Sostav gazov fumarol vulkanov Klyuchevskoy gruppy v zavisimosti ot sostoyaniya ikh aktivnosti. (The composition of the fumarole gases of the volcanoes of the Klyuchi group in relation to the latter's activity.) In: *Glubinnoye stroyeniye, seysmichnost' i osobennosti sovremennoy deyatel'nosti Klyuchevskoy gruppy vulkanov.* (*The deep structure, seismicity and features of current activity of the Klyuchi group of volcanoes.*) Vladivostok, Dal'nevostochn. knizhnoye izd-vo. 15 pp.

Menyaylov, I. A. & Nikitina, L. P. 1974. Tsink i svinets v gazakh i vodakh vulkana Ebeko i Pauzhetskogo mestorozhdeniya. (Zinc and lead in the gases and waters of Ebeko volcano and the Pauzhets deposit.) In: *Gidrotermal'nyye mineraloobrazuyushchiye rastvory oblastey aktivnogo vulkanizma.* (*Hydrothermal mineral-forming solutions in regions of active volcanism.*) Novosibirsk, Nauka, pp. 103–10.

Menyaylov, I. A. & Nikitina, L. P. 1976. Zn, Cu, Pb, Cd v fumarol'nom i gidrotermal'nom pare. (Zn, Cu, Pb and Cd in fumarole and hydrothermal vapour.) *Byul. vulkanol. stantsiy*, No. 52.

Mizutani, Y. & Matsuo, S. 1959. Successive observation of chemical components in the condensed water from a fumarole of volcano Showa-Shinzan. *Bull. Volcanol. Soc. Japan*, **3**(2), 119–27.

Sigvaldason, E. 1974. Chemical composition of volcanic gases. In: *Phys. Volcanol.* Amsterdam e.a., pp. 215–40.

Sigvaldason, E. & Elisson, G. 1968. Collection and analysis of volcanic gases at Surtsey, Iceland. *Geochim. Cosmochim. Acta*, **32**(8), 797–805.

Volkov, V. P. & Ruzaykin, G. I. 1974. *Matematicheskoye modelirovaniye bazovykh ravnovesiy v vulkanicheskom protsesse.* (*Mathematical modelling of basic equilibria in the volcanic process.*) M., Nauka, 150 pp.

The influence of volcanic eruption on the chemical composition of surface and underground waters. (*As in the eruption area of the Tolbachik volcanoes, Kamchatka*)

T. P. Kirsanova, G. F. Pilipenko and L. M. Yurova

The Great Fissure Eruption of the Tolbachik basalt volcanoes lasted from 6 July 1975 to 10 December 1976. The first-formed group of three new cinder cones at the Northern Breakthrough was active until 16 September 1975. The eruption was basically an explosive one. On 17 September 1975 the second – Southern – Breakthrough was formed, and was active until 10 December 1976. In this case effusive activity predominated. The total quantity of juvenile material extruded during the eruption is estimated at 2.78×10^9 t, including 1.44×10^9 t (0.72 km^3) of lava, 1.28×10^9 t (1.085 km^3) of pyroclastic material and 0.06×10^9 t (72.28 km^3) of volatile material (see Markhinin *et al.*, this volume).

A large proportion of the ash and gaseous products was scattered over great distances, and its effect on the hydrosphere is difficult to quantify. The present paper discusses the influence of the abyssal material on the chemical composition of surface and underground waters in the immediate vicinity of the eruption.

The eruption area – the southern foot of Tolbachik volcano – is a mantle of lavas formed over the last 2000–3000 years by a series of fissure eruptions similar to the most recent one (see Braytseva *et al.*, this volume). This mantle is 800 km^2 in area and up to 100 m thick. It is made up of flows of sheet and blocky basalt lava, highly fissured, with numerous cavities and caverns hundreds of metres in extent. These lavas are extremely permeable. The actual rates of movement of underground waters in complexes of volcanic rocks like these may be as much as 800 m per day, according to Manukhin (1972). The Holocene lava mantle lies above the local base drainage level and is a source and transit area for a formidable flow of groundwater that discharges along the front of the mantle to form the rivers Levyy Tolbachik and Ozernaya. The outlets of the springs are linear and concentrated and the discharge is as much as 100 l/s. One may assume that deeper aquifers also discharge their waters here.

Within the lava mantle itself there are no constant water courses, but seams of thin redeposited pyroclastics form localised water resistant layers. At such points springs appear with discharges of up to 2 l/s, streams 10 to 20 m long, and small lakes (see Fig. 1). As a rule they are fed by snow patches, which persist until the end of summer.

In 1976 we tested all manifestations of water within the Tolbachik lava shield: springs, rivers, lakes, snow patches and atmospheric precipitation (rain). The observations were made over a period when the effusive activity of the volcano was already greater than its explosive activity, explosions in the crater were infrequent and relatively low-power, and the quantity of pyroclastics and gases was declining sharply. The results of the most typical analyses (major components) are summarised in Tables 1 and 2. Table 2 shows that the influence of the eruption primarily took the form of increased mineralisation of atmospheric precipitation, as well as a change in the proportions of cations and anions. Precipitation in this region usually has a mineralisation of between 20 and 60 mg/l, with a sodium–calcium hydrocarbonate composition. However, during the eruption mineralisation of the precipitation reaches 300 mg/l, the pH drops, the relative and absolute quantities of calcium and magnesium increase and chlorine appears among the anions. High proportions – unusually high for Kamchatka – of fluorine are recorded, with as much as 20 mg/l in the snow near the South Cone. The distribution of F^- is shown schematically in Fig. 1.

Mineralisation of precipitation and snow patch water decreases with distance from the centre of the eruption (Table 2). The similarity in the chemical composition of the waters and atmospheric precipitation within the lava mantle (Table 3) indicates that the surface and groundwater is fed by atmospheric precipitation. This is borne out by the ratio rF^-/rCa. As Table 4 shows, the most typical rF^-/rCa ratio for atmospheric precipitation applies to all forms of water on the lava mantle, in contrast to the waters of high-discharge sources on its periphery, in particular those of Ozernaya and Amrok (Table 1, No. 4 and Fig. 1). In terms of composition these waters, which are fed from the lava mantle region, are typical hydro-carbonate waters produced by the leaching of young basalt lavas, but the mineralisation is slightly higher: up to 500 mg/l. Their low F^- content can probably be explained by formation of complex salts with increased pH

Fig. 1. Hydrochemical sketch map of the Southern sector at the base of the Tolbachik volcano. Compiled by T. P. Kirsanova, L. M. Yurova, G. F. Pilipenko, July and August 1976.

1, cones of the Northern and Southern Breakthroughs; 2, cones of breakthroughs of the Great Tolbachik Fissure Eruption and Ploskiy Tolbachik; 3, cones of previous eruptions; 4, springs: a, sampled, b, not sampled; 5, sites of extraction of water samples from rivers; 6, sampled snowfields; 7, lakes; 8, caves in lava flows; 9, lava flows from the Northern and Southern Breakthroughs; 10, boundaries of *kekur*'s (conical monoliths); 11, number of the analysis in Table 1.

Content of F in water bodies and sediments (mg/l): 12, 1.5; 13, from 1.5 to 5; 14, from 5 to 10; 15, from 10 to 20; 16, maximum content of F (mg/l) in individual water bodies.

Areal distribution of hydrochemical types of water: 17, Cl–Ca waters; 18, HCO$_3$–Mg–Ca waters.

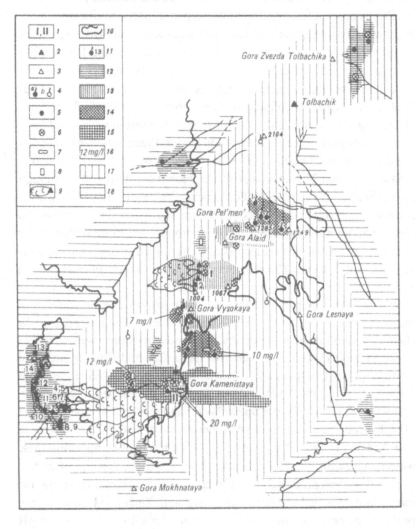

Table 1. *Chemical composition of waters in the Tolbachik eruption area*

Components	1 mg/l	1 mg equ. (%)	2 mg/l	2 mg equ. (%)	3 mg/l	3 mg equ. (%)	4 mg/l	4 mg equ. (%)	5 mg/l	5 mg equ. (%)	6 mg/l	6 mg equ. (%)	7 mg/l	7 mg equ. (%)
Cations														
NH_4^+	0.4	0.41	None	—	0.7	0.66	—	—	0.2	0.14	—	—	0.1	0.16
Na^+	30.0	26.42	44.0	23.01	42.0	30.24	40.0	23.75	42.5	25.41	42.0	29.4	42.0	28.71
K^+	4.6	2.44	3.4	1.08	6.1	2.66	—	—	4.6	1.65	—	—	4.1	1.74
Ca^{2+}	44.0	44.72	97.0	58.43	40.0	33.22	22.0	16.8	30.0	20.60	23.0	18.2	25.0	19.72
Mg^{2+}	15.6	26.01	17.7	17.48	24.4	33.22	47.4	59.45	46.4	52.20	40.4	52.5	38.4	49.68
Total	94.6	100.00	162.1	100.00	113.2	100.00	109.4	100.00	123.7	100.00	105.4	100.00	109.6	100.00
Anions														
F^-	10.0	8.05	2.7	1.40	5.0	3.86	0.6	0.18	1.0	0.70	0.5	0.32	1.0	0.78
Cl^-	142.0	60.79	248.5	69.93	156.4	65.28	24.4	9.28	34.1	13.43	27.1	11.6	34.1	15.08
SO_4^{2-}	80.0	25.38	85.0	17.68	80.0	24.78	17.3	4.84	2.0	0.56	19.2	6.08	3.0	0.94
HCO_3^-	18.3	4.56	67.1	10.99	19.5	4.75	387.9	85.7	327.1	85.31	329.4	82.0	323.3	83.20
S total	1.3	1.22	—		1.5	1.33	—		—		—		—	
Total	251.7	100.00	403.3	100.00	262.4	100.00	430.3	100.00	409.2	100.00	376.3	100.00	361.4	100.00
Total mineralization	346.3		565.4		375.6		555.4		532.9		498.2		471.0	
pH	6.0		7.0		6.3		8.05		7.3		7.5		7.1	

Table 1 (cont.)

Components	8 mg/l	8 mg equ. (%)	9 mg/l	9 mg equ. (%)	10 mg/l	10 mg equ. (%)	11 mg/l	11 mg equ. (%)	12 mg/l	12 mg equ. (%)	13 mg/l	13 mg equ. (%)	14 mg/l	14 mg equ. (%)
Cations														
NH_4^+	—	—	None	—	None	—	None	—	None	—	None	—	None	—
Na^+	38.2	27.10	45.0	26.45	16.6	25.36	41.2	29.51	41.2	30.37	44.2	35.56	30.0	27.31
K^+	—	—	5.8	2.02	—	—	3.9	1.64	3.5	1.56	3.1	1.48	2.4	1.26
Ca^{2+}	22.0	17.90	30.0	20.24	20.0	36.23	20.0	16.39	22.0	19.03	22.0	20.37	24.0	26.21
Mg^{2+}	41.3	55.00	46.4	51.29	12.9	38.41	39.0	52.36	34.2	48.44	28.1	42.59	26.9	46.22
Total	101.5	100.00	127.2	100.00	49.5	100.00	104.1	100.00	100.9	100.00	97.5	100.00	83.3	100.00
Anions														
F^-	0.5	0.25	1.0	0.72	1.1	2.13	1.1	0.95	1.0	0.88	—	—	—	—
Cl^-	20.3	0.63	39.8	16.12	11.4	11.35	35.5	15.82	34.1	16.87	34.1	15.73	25.6	14.88
SO_4^{2-}	17.3	5.42	4.0	1.15	3.0	1.42	3.0	0.95	4.0	1.41	2.0	0.66	4.0	1.65
HCO_3^-	346.5	85.70	347.7	82.01	136.4	85.10	317.2	82.28	280.6	80.84	311.1	83.61	246.5	83.47
S Total	—	—	—	—	—	—	—	—	—	—	—	—	—	—
Total	384.5	100.00	392.5	100.00	150.9	100.00	356.8	100.00	319.7	100.00	347.2	100.00	276.1	100.00
mineralization	486.0		519.7		200.4		460.9		420.6		444.7		359.4	
pH	7.5		7.3		7.1		7.1		7.3		7.4		7.4	

Table 1. (cont.)

Components	15 mg/l	15 mg equ. (%)	16 mg/l	16 mg equ. (%)	17 mg/l	17 mg equ. (%)
Cations						
NH_4^+	0.2	0.11	None	—	None	—
Na^+	44.2	21.45	18.0	12.87	35.5	30.6
K^+	8.5	2.46	7.0	2.97	4.0	1.38
Ca^{2+}	90.0	50.48	60.0	49.50	40.0	39.7
Mg^{2+}	28.0	25.70	26.6	34.66	17.1	27.72
Total	171.0	100.00	110.6	100.00	95.6	100.00
Anions						
F^-	19.0	11.01	15.0	10.72	6.0	5.72
Cl^-	227.2	70.43	191.7	73.27	10.65	54.4
SO_4^{2-}	52.0	11.89	44.0	12.48	96.0	36.2
HCO_3^-	36.6	6.61	15.8	3.53	12.2	3.62
S total	—	100.00	—	100.00	—	100.00
Total	334.8		266.6		220.7	
Total mineralisation	505.8		337.2		316.3	
pH	6.0		6.0		—	

1, spring in a crater on the Kleshnya lava flow (August); 2, spring (ground seepage); Gora 1004 (July); 3, Magus'kin's spring (August); 4, 5, Amrok springs, south bank (4, in April, 5, in August); 6, 7, Amrok springs, north bank (6, in April, 7, in August); 8–13, Ozernaya springs (8, in April, 9–13, in August); 14, R. Ozernaya, 12 km from its source (August); 15, lake and cave at the Southern Breakthrough (August); 16, 17, lake at Gora Kamenistaya (16, in July, 17, in September) (the F^- content was determined by an ion selective electrode method with LaF_3 suggested by V. P. Roze and A. V. Gordiyevskiy).

Table 2. *Composition of atmospheric precipitates within the confines of the lava shield of the southern foot of Tolbachik volcano*

Components	1 mg/l	1 mg equ. (%)	2 mg/l	2 mg equ. (%)	3 mg/l	3 mg equ. (%)	4 mg/l	4 mg equ. (%)
Cations								
NH_4^-	1.30	24.14	0.5	13.64	1.2	2.27	2.0	3.46
Na^+	Traces	—	0.3	4.55	7.0	9.74	15.4	21.07
K^+	0.6	6.90	Traces	—	4.1	3.57	—	—
Ca^{2+}	3.0	51.72	3.0	68.18	20.0	32.47	30.0	47.17
Mg^{2+}	0.6	17.24	0.4	13.63	19.5	51.95	10.9	28.30
Total	5.50	100.00	4.2	100.00	51.8	100.00	58.4	100.00
Anions								
F^-	0.6	6.94	0.6	4.76	5.0	9.08	8.3	15.60
Cl^-	11.4	44.44	7.1	31.75	78.1	76.92	56.8	56.74
SO_4^{2-}	None	—	4.0	12.70			20.0	14.89
HCO_3^-	15.9	36.11	14.6	38.10	19.5	11.19	29.0	12.77
S total	1.4	12.51	1.3	12.69	1.4	2.80		
Total	29.4	100.00	27.7	100.00	104.0	100.00	107.1	100.00
Total	34.9		31.9		155.8		165.5	
mineralisation								
pH	4.7		6.5		5.3		6.0	

1, the NE foot of Ploskiy Tolbachik, height 1700 m, 9 Aug. 1976; 2 the Peski level, southern base, height 1400 m, 19 Aug. 1976; 3, Gora Vysokaya, 25 July 1976; 4, Southern Breakthrough, Volcanologists' Camp, 21 July 1976.

and Mg and Ca contents as the water moves from the supply regions towards the points of discharge (Kraynov & Petrova, 1976). None of the characteristic signs of addition of thermal water (such as increased temperatures and noticeably higher concentrations of Cl, Na, K) were discovered in any of the sources on the periphery of the lava mantle. However, in this specific region (fissure type basalt volcanism) perhaps we should take as indicators of thermal waters such parameters as the HCO_3^- ion and H_4SiO_4 in water and H_2 in gas, which are characteristic of the thermal waters of Iceland (also with high energy parameters).

It is interesting to compare the chemical composition of water from one of the sources feeding Lake Amrok – a typical source for the Ozernaya and Amrok groups (see Table 1) – with that of water from the front of the lava flow of the Southern Breakthrough, which came right up to the lake in December 1975 (Table 5 and Fig. 1). As a result of leaching from fresh

Table 3. *Types of chemical composition of waters in the Tolbachik area (from 104 analyses)*

Site of water sampling	Type of water body	Total mineralisation (g/l)	Formula of the chemical composition of the waters	pH
Southern Breakthrough	Atmospheric precipitates (rain)	0.17–0.30	$Cl_{57-65}SO_{4, 15-20}F_{9-16}$ <hr> $Ca_{44-47}Mg_{28-30}Na_{21-25}$	5.4–6.0
Northern Breakthrough (Gora Vysokaya)	Rain	0.03–0.16	$Cl_{23-28}SO_{4, 0-61}HCO_{3, 0-26}F_{2-14}$ <hr> $Ca_{32-77}Mg_{6-63}NH_{4, 2-22}Na_{2-10}$	3.8–5.7
Lava shield	Snowfields	0.05–0.13	$Cl_{39-56}HCO_{3, 10-45}F_{1-32}SO_{4, 2-21}$ <hr> $Ca_{31-99}Mg_{0-51}Na_{1-23}NH_{4, 0-5}$	4.4–6.2
Lava shield	Lakes	0.19–0.38	$Cl_{66-77}HCO_{3, 7-20}SO_{4, 2-20}F_{3-11}$ <hr> $Ca_{33-74}Mg_{5-35}Na_{13-26}$	6.0–6.8
Lava shield	Caves	0.36–0.50	$Cl_{61-71}SO_{4, 12-25}F_{10-11}$ <hr> $Ca_{46-63}Mg_{12-25}Na_{21-33}$	6.0–6.4
Lava shield	Springs	0.13–0.6	$Cl_{54-90}SO_{4, 17-34}HCO_{3, 5-13}F_{1-13}$ <hr> $Ca_{33-73}Mg_{17-48}Na_{3-30}$	5.7–6.7
Lake Amrok	Springs	0.46–0.89	$HCO_{3, 82-91}Cl_{8-15}$ <hr> $Mg_{49-53}Na_{21-29}Ca_{20-21}$	7.2–7.3
River Ozernaya	Springs	0.31–0.52	$HCO_{3, 61-68}Cl_{11-31}SO_{4, 1-18}$ <hr> $Mg_{33-51}Na_{25-36}Ca_{16-36}$	7.2–7.4

lava, the amount of mineralisation was doubled, as was the content of Mg^{2+}, Ca^{2+}, HCO_3^- and K, whereas the chlorine content was unchanged.

Thus the chemical composition of the waters of the lava mantle is a direct indication of both the atmospheric origin of the waters themselves and the volcanic source of their mineral component. Soluble substances might have been captured by the waters from the gas cloud of the volcano or from pyroclastic material (both in the air and on the ground).

Water extracted from ash and cinder shows that every kilogram of pyroclastic material contains from 300 to 3400 mg of highly soluble substances (see Table 5 and Menyaylov *et al.*, this volume).

We can thus give a first approximation of the quantity of soluble substances entering the underground waters with the atmospheric precipitation in the immediate vicinity of the eruption. For this purpose we shall take the area of 200 km^2 in which observations were carried out. The annual precipitation in this region is 900 mm and the mineralisation of the precipitation (rain, snow patches, see Tables 2 and 3) is on average close to 100 mg/l. The total quantity of the dissolved material in the atmospheric precipitation during the eruption (19 months) will therefore be 2.2×10^4 t, including $Cl - 1 \times 10^4$, $F - 0.2 \times 10^3$, $S - 0.4 \times 10^3$ t.

Only an approximate comparison of these figures with the overall amount of water-soluble substances ejected during the eruption can be made, since the composition of the eruptive gases remains unknown. The available analyses of gases taken from lava flows remote from eruptive centres can give only a qualitative idea of the composition of the volcanic gases. However, estimates of the most general kind can be made. If we take

Table 4. *rF/rCa ratios in waters of the area of the Tolbachik volcanoes*

Type of water	Place of occurrence	rF/rCa	Most characteristic rF/rCa ratios
Atmospheric precipitation (rain)	Northern Breakthrough, Gora Vysokaya	0.08–0.35	0.20–0.22
Snowfields	Northern Breakthrough	0.80–0.90	—
Snowfields	Southern Breakthrough	0.90	—
Snowfields	Lava shield	0.17–0.62	—
Lakes	Lava shield	0.14–0.26	0.18–0.25
Waters of lava caves	Lava shield	0.22–0.25	0.22–0.25
Springs	Amrok waters	0.03–0.04	0.03–0.04
	Ozernaya waters	0.04–0.06	0.04–0.06

Table 5. *Composition of waters extracted from eruption products of the Tolbachik volcanoes*

	1		2		3	
Components	mg/l	mg equ. (%)	mg/l	mg equ. (%)	mg/l	mg equ. (%)
Cations						
NH_4^+	0.2	0.56	0.01	0.01	2.6	1.22
Na^+	31.5	79.2	46.9	45.0	55.6	21.25
K^+	9.0	13.3	13.7	7.7	13.0	2.90
Ca^{2+}	2.4	6.94	37.1	40.7	50.0	21.95
Mg^{2+}	None	—	3.7	6.6	73.2	52.68
Total	43.1	100.00	101.4	100.00	194.4	100.00
Anions						
F^-	1.5	4.55	15.2	14.4	1.2	0.51
Cl^-	10.6	17.05	56.7	28.90	34.1	8.12
SO_4^{2-}	28.8	34.10	134.5	50.4	2.0	0.34
HCO_3^-	47.6	44.3	21.3	6.3	656.4	91.03
Total	88.5	100.00	212.5	100.00	694.9	100.00
Total mineralisation	131.6	—	313.9	—	889.3	—
pH	7.84	—	6.4	—	7.3	—

1, cinder at the Southern Breakthrough cone, 24 Sept. 1976; 2, ash from tent in the Volcanologists' Camp, Southern Breakthrough, 6 Sept. 1976; 3, water at the front of the lava flow that reached Lake Amrok, 15 Aug. 1976.

Table 6. *Amount of easily soluble material entering the hydrosphere, T*

Mechanism by which components entered the hydrosphere	Components, T		
	Cl	F	S
Came to surface with gaseous products	2×10^5	7×10^4	4×10^4
Fell out with precipitates	1×10^4	2×10^3	$4 \times 10^*$
Fell out with precipitates and was absorbed from pyroclastics	1.1×10^4	2.1×10^3	4.2×10^3
Discharged by the Amrok and Ozernaya springs	9×10^3	3×10^2	1.2×10^3

The discharge of water in the River Ozernaya (August 1976) was ≈ 5 m³/s.

the quantity of gaseous products 6.0×10^7 t (see Markhinin *et al.*, this volume), and ascribe to them the average gas content (% volume) of the lava flows from the Tolbachik volcanoes (Menyaylov *et al.*, 1976): $H_2O - 95$; $HF - 0.04$; $HCl - 0.2$; $SO_2 - 0.05$; $H_2S - 0.005$; $CO_2 - 0.5$; $N_2 - 0.4$; $H_2 - 0.4$; we obtain the relationships given in Table 6.

Similar estimates may be made for other elements. Although the quantitative estimates are approximate, the following conclusions can be made.

1. The eruption has a major influence on the composition of surface and underground waters only over a small area.

2. Most gaseous products are dispersed far from the site of the eruption and cause no sharp deviations in the chemistry of the hydrosphere.

3. The region at the southern foot of the Tolbachik volcano can be considered potentially rich in thermal waters, which should be sought by geophysical methods. Geochemical criteria appropriate to the geothermal regions of Iceland are difficult to apply here because of intensive and extensive dilution by groundwater (the modulus for surface supply is 25.8 $l/s \times km^2$ and for underground run-off 50 $l/s \times km^2$).

References

Gidrogeologiya SSSR. Kamchatka. Kuril'skiye i Komandorskiye ostrova, t. XXIX. (*The hydrogeology of the USSR. Kamchatka. The Kuril and Komandor Islands, Vol. XXIX.*) 1972. M., pp. 120–1.

Kraynov, S. R. & Petrova, N. T. 1976. Ftoronosnyye podzemnyye vody, ikh geokhimicheskiye osobennosti i vliyaniye na biogeokhimicheskiye protsessy. (Fluoric underground waters, their geochemical features and influence on biogeochemical processes.) *Geokhimiya*, No. 10, 1533–41.

Manukhin, Yu. F. 1975. *Osobennosti vodoobmena i gidrodinamicheskaya zonal'nost' geotermal'nykh rayonov oblasti sovremennogo vulkanizma (na primere Kamchatki).* (*Features of water circulation and the hydrodynamic zoning of geothermal areas in a region of present-day volcanism (Kamchatka).*) Avtoref. kand. dis. L., pp. 1–29.

Menyaylov, I. A., Nikitina, L. P., Guseva, R. V. & Shapar', V. N. 1976. Rezul'taty otbora i analiza vulkanicheskikh gazov na Tolbachinskom treshchinnom izverzhenii v 1975 godu. (Results of sampling and analysis of volcanic gases during the Tolbachik fissure eruption of 1975.) *Dokl. AN SSSR*, **230**(2), 440–2.

Part II

Geophysical studies

The development of the Great Tolbachik Fissure Eruption in 1975 from seismological data

S. A. Fedotov, V. I. Gorel'chik, V. V. Stepanov and V. T. Garbuzova

The Great Tolbachik Fissure Eruption (GTFE) which began on 6 July 1975 (Fedotov *et al.*, 1976b) was preceded and accompanied by a large number of volcanic earthquakes and intensive volcanic tremor. Incomplete seismological data on the eruption have already been published in condensed form (Fedotov *et al.*, 1976a). The present paper sets out and discusses the results of detailed seismological studies of the mechanism and development of the eruption.

The strongest earthquakes before and during the eruption were registered by all fifteen regional seismic stations of the Institute of Volcanology of the DVNTs AN SSSR. However, the bulk of the earthquakes were recorded by the six stations closest to the site of the eruption: Klyuchi, Kozyrevsk, Apakhonchich, Esso, Kronoki and Krutoberegovo. Further observations were made by four temporary stations: Tolbachik, Plotina, Levyy Tolbachik and Proryv (Fig. 1). Amplitude–frequency characteristics of the seismic channels were tabular within periods of 0.2–0.9 s and amplification from 10000 to 100. The amplification was reduced owing to the severe volcanic tremor, which at the stations nearest to the new volcanoes exceeded 10 μm.

In the analyses of the earthquakes pallets of isochrones were used that were constructed using the time field calculated by V. I. Gorel'chik (Gorel'chik & Stepanov, 1976) for a velocity section in the vicinity of the Klyuchi volcanic group with geoseismic sounding (GSZ) data (Anosov *et al.*, 1974). For verification, the site of the severest earthquakes was identified separately by near and distant stations, by combinations of three stations and graphically by the Vadati method. The depth of the focus was calculated from theoretical travel–time curves, the true angles of emergence and converted waves. Errors in determining coordinates were mainly of

190   *Fedotov, Gorel'chik* et al.

the order $\pm 5 - \pm 10$ km for the epicentre and $\pm 5 - \pm 15$ km for depth. The actual accuracy of determination of their coordinates was higher for earthquakes with distinct arrivals of body waves and small $S - P$ distances at the nearest field station. Thus at Tolbachik, Proryv and Levyy Tolbachik stations the time difference for the transmission of transverse and longitudinal waves generally varied from 1 to 4 s (sometimes up to 5 or 6 s), which was in itself sufficiently clear evidence of the position of the earthquake foci.

From 1973 to 1974 a certain increase in seismic activity in the crust and upper mantle was noted in the area of the future eruption, compared with the years 1964–71. To the south-west of Ploskiy Tolbachik, in the zone of areal volcanism, earthquakes were recorded of energy classes $K_{S1.2}^{\phi 68} = 10$–12 at depths of 0–25 km and more than 150 km. $K_{S1.2}^{\phi 68}$ is the energy class of the earthquakes as defined by S waves according to

Fig. 1. Sketch map of seismic stations in the eruption area.
1–2, seismic stations: 1, permanent, 2, temporary; 3, active volcanoes: I, Klyuchi, II, Bezymyannyy, III, Ploskiy Tolbachik, IV, Kizimen, V, Komarova, VI, Gamchen, VII, Kronotskiy; 4, the 1975 eruption centres: N, Northern Breakthrough, S, Southern Breakthrough.

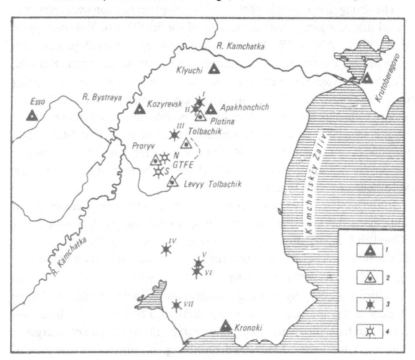

Fedotov's nomogram (Fedotov, 1972). $K_{S1.2}^{\phi 68} = 4.6 + 1.5M$, where M is the magnitude determined from surface waves.

Data from the Tolbachik field station, which operated periodically at the south-eastern foot of Ploskiy Tolbachik, show that in this region volcanic earthquakes of Type 1 occurred, with a depth of focus of probably not more than 10 km.* Thus, in 1971, over 60 days of operation, the station recorded six earthquakes with $K_S \leqslant 6.5$ and $S - P \leqslant 4.0$ s, in 1972 twelve such earthquakes over 70 days, in 1973 ten over 60 days and in 1974 seventeen over 40 days. On 15 August 1974, 10 km south-west of Ploskiy Tolbachik crater an earthquake of $K_S = 7$ ($M = 1.5$) was recorded at a depth of 5–10 km. On 18 May 1975, 7 km south-east of the crater another occurred, with $K_S = 7$ and a depth of 5–10 km.

From June 1975 onwards, during the preparatory stages of the GTFE, seismic activity around Ploskiy Tolbachik, to its south-west, south and east increased by several orders compared with the preceding years.

From June to September 1975 three large volcanic earthquake swarms and severe volcanic tremor were observed in the area of the eruption. In the first swarm (27 June–5 July) there were 300 earthquakes of energy classes $K_S \geqslant 7$ ($K_S^{\max} = 11.5$, $M = 5$), 2–17 August nearly 1400 ($K_S^{\max} = 10$) and 1–17 September 280 such earthquakes ($K_S^{\max} = 10$). Fig. 2 shows the nature of the change in time of seismic activity preceding (period A) and accompanying (period C) the eruption. The graphs showing the distribution of the number, N, and maximum energy class $K^{\max}$ of volcanic earthquakes were constructed at 12 hour intervals using data from the Klyuchi station (distance = 75–80 km), and that showing the change in average amplitude of the volcanic tremor $\bar{A}$ in micrometres is based on data from the Tolbachik station (distance = 15 km). As Fig. 2 shows, the formation of the fissures on which the large cinder cones I, II, III and VIII (with relative heights of 330, 290, 140 and 110 m)† grew during the eruption, was preceded by earthquake swarms which began 5–10 days before magma and gases broke through onto the surface. The earthquake swarms appeared against a background of reductions, or complete absence

---

* At the moment most scientists term earthquakes volcanic when they are spatially and genetically connected with volcanic activity. Volcanic earthquakes of Type 1 are connected with processes in the magma chamber and the volcanic conduit. They correspond to earthquakes of Type A in T. Minakami's classification (Minakami, 1960) and resemble in their recording pattern ordinary tectonic earthquakes with distinct arrivals of P and S body waves, and may occur not only just before, but also between, eruptions (Tokarev, 1966; Gorel'chik, 1973).

† The height of Cone VIII had increased to 160 m by the end of 1976.

of volcanic tremor energy, whilst the breakthroughs themselves originated when the number (N) and energy (K^{max}) of the volcanic earthquakes had dropped and volcanic tremor was again on the increase. The formation of Cone I (commencement of eruption – period A) was preceded by the longest and seismically most intensive preparation. From this swarm accurate predictions were made of the time and place of the beginning of the eruptions. Subsequently, the characteristic drop in the number of earthquakes at the end of the swarms enabled us to predict the formation of Cones II, III and VIII.

The dynamics of the eruption correlated well with the graph for the variation in average amplitude of volcanic tremor (Fig. 2).* Period '*a*' is the intensification of the eruption at its inception, accompanied by an intensification of volcanic tremor. From 9 to 23 August a mighty, continuous blast of gas from Cone I was observed; it ejected incandescent pyroclastics up to 2.5 km into the air. Tremor at this time was moderate, up to 1.5 μm. The 26 and 27 July (period '*b*') were the turning point in the eruption: the first pauses occurred, alternating with heavy explosions, and Cone I and its adjacent extinct cone to the south were torn apart by fissures. On 29 July lava began to flow from these (the southern bocca, in Fig. 2), and the eruption became an explosive–effusive one. This stage is marked by sharp fluctuations in the level of volcanic tremor from 0.2 to 5–6 μm. Period '*c*' was marked by an intensification of volcanic tremor which reflects the brief explosive activity in the pass between Cone I and its extinct neighbour. Periods '*d*' and '*e*' were the result of strong explosive activity in Cones II and III, occurring after their formation. Surges of tremor in the period 'f'–'g' accompanied the final stage of eruption of Cone II, when powerful explosions became intermittent. The period '*h*' is one of intensified tremor which accompanied the formation of Cone VIII.

After lava had broken through, tremor diminished. This was observed during the breakthrough of the southern and northern boccas of Cone I and the main flow of Cone II (Fig. 2).

The activity of the Southern Breakthrough (small uniform explosions and continuous effusion of fluid basalt lavas) was accompanied by comparatively weak volcanic tremor up to 1.0–1.2 μm.

Fig. 2 distinguishes two periods when volcanic tremor was practically absent for two to three days. At this time none was recorded even at the Proryv seismic station 4 km from Cone I. Both pauses in volcanic tremor

---

* For details of volcanic tremor see Gorel'chik *et al.*, this volume.

preceded events which radically altered the course of the eruption. From 6 to 9 August the activity of Cone I gradually died down and the seismic preparation (earthquake swarm) for the formation of Cone II proceeded. The second pause in tremor (from 15 to 17 September) coincided with the end of the eruption at the Northern Breakthrough and with intensive seismic activity (an earthquake swarm) to the south of the Northern Breakthrough, at the point where Cone VIII was later to form. Evidently magma no longer flowed along existing channels during these periods (the emission of lava from the southern and northern boccas of Cone I at this time was most probably from magma cauldrons near the surface), and new routes to the surface were being formed (earthquake swarms), initiating new eruption centres.

The map in Fig. 3, which shows the seismic situation immediately before the eruption, depicts all the earthquakes of swarm I, from energy class 8 upwards, as well as isolated weaker earthquakes. Altogether 217 earthquakes were analysed in this swarm. As Fig. 3 shows, the epicentres

Fig. 2. Variation with time of the amplitude of volcanic tremor (data from Tolbachik seismic station), of the number N and of the maximal energy class of earthquakes K^{max} (data from Klyuchi seismic station) during the Tolbachik eruption. Explanation of conventional signs in text. I–VIII are cones.

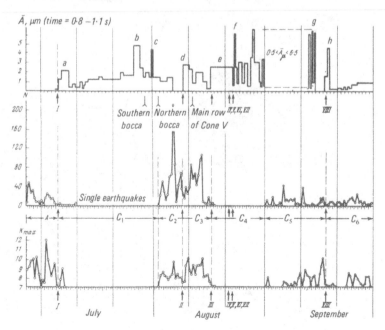

194   *Fedotov, Gorel'chik* et al.

of earthquakes with $K_S \geqslant 8$ form an area $\approx 20 \times 10$ km with its long axis running north–south within which the original three cones of the eruption are situated. The depth of earthquake foci varied from 15 to 20 km (possibly 30 km) to sea level and above. This evidently indicates a general tendency for the foci to become shallower with time.

Fig. 3. Map of epicentres of earthquakes preceding the eruption (swarm I).
1, the volcanic masses of Ploskiy and Ostryy Tolbachik; 2, crater collapse of Ploskiy Tolbachik; 3, zone of areal volcanism; 4, 1975 eruption centres; 5, seismic stations.

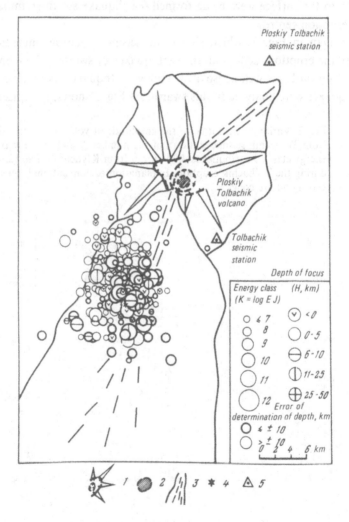

The distribution of the earthquakes of swarm I with $K_S \geqslant 8$ (7.5) in terms of energy and focus depth is given in Table 1.

The most powerful earthquakes, with $K_S = 11.5$ and 11.1 ($M \approx 5$) and depth of focus 10–20 km, were recorded on 2 August. The epicentre of the strongest one occurred on the axis of the zone of cinder cones extending south-west of the summit caldera of Ploskiy Tolbachik, and practically coincided with Cone I, i.e. the point where the eruption began. Together with the earthquakes with $K_S = 12.5$ which preceded the eruption of Shiveluch in 1964, these are the most powerful events recorded in volcanic earthquake swarms since seismological records of the Kamchatka volcanoes began in 1949. The extent of the fractures at the foci of such earthquakes is as much as 5–8 km, according to V. M. Zobin (this volume). The feeder dyke may have had approximately the same horizontal length on 2 July, though its size could have increased in the days following.

Thus, according to seismological data the period of preparation of the New Tolbachik volcanoes was one of intensive tectonic movements and fissuring, which may have involved the lower horizons of the crust. If we assume that the level from which magma began to move up to the surface corresponds approximately to the lower estimates of earthquake focus depths, and since we know the duration of swarm I, we can attempt a first approximation of the rate of rise of the magma. At a maximum depth of earthquake focus of 10–20 or 15–30 km the values for this rate are in the range of 50–100 or 100–150 m/h respectively.

After earthquake swarm I there was a lull, during which isolated

Table 1. *The distribution of earthquakes of swarm I with $K_S \geqslant 8$ (7.5), in terms of energy, depth and time*

Period of observation	Energy class $K = \log E$ (J)	Depth of focus, H (km)			
		$H \leqslant 0$	$0 < H \leqslant 5$	$5 < H \leqslant 10$	$10 < H \leqslant 20$ (30?)
27 June–	8	18	38	2	2
2 July (07.43)	9	5	13	2	1
	10	3	4	0	2
	11	0	2	2	1
	12	0	0	0	1
2 July (07.52)	8	16	9	2	0
–5 July	9	13	5	1	0
	10	1	1	0	0
	11	0	1	0	0
	12	0	0	0	0

earthquakes were recorded (period *C*, see Fig. 2). The severest earthquakes during this period, with $K_S \geqslant 9$, were located in the areas of Cone I at a depth of not more than 5 km (Fig. 4). The end of this phase coincided with the breakthrough of the southern bocca of Cone I on 28 July, and of the northern bocca of Cone I on 2 August. At the same time (28 July onwards) the first earthquakes with $K_S \geqslant 8$ and focus depth 0–5 km were recorded not directly from the vicinity of the eruption but to the east and south-east of the Ploskiy Tolbachik crater. It was also on 28 July that the bottom of Ploskiy Tolbachik crater began to shatter (see Farberov, this volume), heralding the collapses that ensued from the end of August and the formation of a new summit caldera.

Fig. 4 plots the epicentres of earthquake swarm II, which lasted from 2 to 17 August. All earthquakes with $K_S \geqslant 9$ and some of those in energy classes 8 and 7 have been fully analysed for this swarm.

The great majority had a focus depth of 0–5 km and only a few at the beginning had 5–10 km. The swarm may be regarded as a juxtaposition of two swarms, those of 2–9 August (C_2) and 10–17 August (C_3), at the end of which Cones II and III formed on 9 August and 17 August respectively, 750 m apart. In all probability, the feeder channels of Cone II began to form at a depth of not more than 10 km, and those of Cone III even closer to the surface.

The intensity of the seismic preparation of Cones II and III was approximately the same: in the swarm of 2–9 August two earthquakes were recorded with $K_S = 10$ and thirteen with $K_S = 9$, while in the swarm of 10–17 August there was one earthquake with $K_S = 10$ and twenty-two with $K_S = 9$. But in both cases the seismic energy released was more than one order less than that of swarm I. As in swarm I, the most intense earthquakes (all those in energy class 10 and the majority in class 9) tended, in August, towards the region of the Northern Breakthrough. The foci of weaker earthquakes with $K_S \leqslant 8$ and sometimes $K_S = 9$ was dispersed over the considerable area of ≈ 400 km² and occurred between the eruption centres and the summit of Ploskiy Tolbachik or to its south and south-east. The number of earthquakes from this region (known as the Tolud epicentral zone) was approximately 5–10 times greater than the number of those immediately preceding the formation of Cones II and III. Thus the swarm of 2–17 August was complex, consisting both of earthquakes connected with the formation of the feeder channels of Cones II and III at depths of 0–10 and 0–5 km, and of earthquakes caused by numerous movements around the summit of Ploskiy Tolbachik and south of it, at depths of 0–5 km and above sea level. Moreover, the degree of seismicity in the Tolud epicentral zone varied with the progress of the eruption: just before Cones II and III were formed seismic activity was intensified, to diminish again after their appearance.

During the second half of August the summit crater of Ploskiy Tolbachik was collapsing on a grand scale, resulting in a pit which by

Fig. 4. Map of epicentres of earthquakes accompanying the eruption of 6 July to 31 August 1975. Conventional signs 1–5, energy class and depth of earthquake focus signs as on Fig. 3. I, II, III, the 1975 eruption centres.

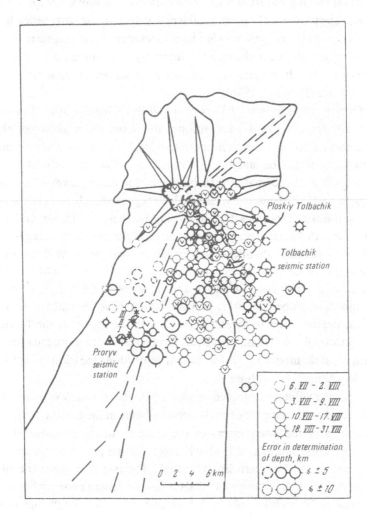

September had a diameter of 1600 m, a depth of about 5000 m, and a volume of 0.3 km³. The volume of ash, cinder and lava emitted by Cones II and III exceeded 0.6 km³ (Fedotov *et al.*, 1976b). Judging from seismological data, the earthquake swarm of 2–17 August, with its tectonic movements south of Ploskiy Tolbachik, the formation of the pit, and the change in the course of the eruption (cessation of Cone I's activity on 9 August, and the appearance of Cones II and III alongside on 9 and 17 August) were interconnected events which may be explained as follows.

Ploskiy Tolbachik has features characteristic of the Hawaiian type of basalt volcano: a collapsed summit caldera with a lava lake, and 'rift' zones spreading out from it in which numerous subsidiary eruptions take place. The presence of a summit caldera usually indicates that beneath such volcanoes there are shallow (a few kilometres) peripheral magma reservoirs, connected with larger chambers in lower layers of the crust, and in these reservoirs basalt magma accumulates as it advances from its zone of generation (Fedotov, 1976).

During the eruptions of Hawaiian volcanoes basalts often flow along the rifts from the lava lakes and possibly from the shallow peripheral chambers into the eruption centres, which are up to several tens of kilometres from the summit of the volcano. The structure of Ploskiy Tolbachik and the spatial distribution of the earthquakes of swarm II suggest that the seismic activisation which began in late July – early August to the south and south-east of the Ploskiy Tolbachik crater was connected with the withdrawal of magma from a shallow peripheral chamber beneath Ploskiy Tolbachik. Parallel to and evidently connected with the change in the nature of Cone I, the termination of its activity and the formation of the new feeder dykes of Cones II and III, some of the magma from this peripheral chamber in Ploskiy Tolbachik could have been moved along the rift at depths of up to several kilometres to the region of the Northern Breakthrough. It remains unclear, however, whether magma from a peripheral chamber could in fact flow at depth along the Ploskiy Tolbachik feeder channel at this time.

Since the volume of erupted products of Cones II and III was more than twice that produced by the simultaneous collapse of the caldera, the magma chambers in the lower layers of the crust or in the transitional layer between crust and mantle probably remained the principal source of the magmas of the Northern Breakthrough. The injection along the rift of characteristic high-aluminium subalkaline lavas from a peripheral chamber of Ploskiy Tolbachik could have been one reason for the change in lava composition which occurred during the final stage of activity of the Northern Breakthrough.

On 22–23 August (during the quiet C_4 period), 700–150 m west of the craters of Cones II and III, the small cones IV, V, VI and VII appeared. The last three were lava cauldrons situated on the gaping tension fissures that had appeared a day earlier. The absence of earthquakes indicates that the phenomenon originated locally and close to the surface. Possibly it was caused by the injection of a shallow sill which had branched out from the feeder channel of Cone II.

On 1 September 1975 a new earthquake swarm commenced, which culminated in the formation of the Southern Breakthrough. All the earthquakes of this swarm from $K_S \geqslant 8$ onwards, as well as some of the lesser events, were analysed (Fig. 5). As we can see, the epicentral zone of earthquake swarm III (like that of swarms I and II) covers a wide area, from Ploskiy Tolbachik to the eruption zone, and the first half of the swarm coincides in time with the earthquakes in the Tolud epicentral zone. The earthquakes are shallow as before, 0–5 km, and only isolated ones to the east of the Ploskiy Tolbachik crater had a focus depth of 5–10 km, $K_S^{max} = 9$. The first earthquakes in the eruption area, to be more precise in the area of the Northern Breakthrough, occurred on 9–12 September, i.e. when lavas of mixed composition had begun to flow from Cone II. It is instructive that the activity of earthquake swarm III reached its peak on 16–17 September, immediately after Cone II had ceased activity on 15 September (see Fig. 5). The foci of the shallow earthquakes ($K_S^{max} = 10$) spread south-west of the Northern Breakthrough on these two days, and in the far end of the new epicentral zone on 17 August at 20.00 hours GMT Cone VIII of the Southern Breakthrough formed. This location of the foci of preliminary earthquakes and the site of the breakthrough differs from that of the two preceding swarms, when the sources of the strongest events were grouped around Cones I, II and III. This leads us to suppose that in the case of the Southern Breakthrough the injection of magma occurred not upwards from below, as in the Northern Breakthrough, but through an array of shallow faults and fissures in the domed part of the zone of cinder cones to the south of Cones I, II and III, at depths of 0–5 km.

After the formation of Cone VIII, seismic activity shifted (October–December 1975) from the eruption area to that of the central crater of Ploskiy Tolbachik, as well as to the east and south-east of it, and also extended northwards and north-westwards of the crater, a manifestation that had not been observed previously (Fig. 6). Evidently there was a revival of deep magmatic activity and faulting throughout the region of areal volcanism, as Piyp (1956) calls the region of localised zones of cinder cones around Ploskiy Tolbachik.

On the basis of the seismological data, the probable mechanism of the GTFE may be described as follows. Between 27 June and 6 July 1975, 18 km south-west of the summit of Ploskiy Tolbachik, high-magnesium basalts rose from the lower layers of the crust and possibly from the transitional layer between crust and mantle, at an average speed of 100–150 m/h, and this led to the formation of the new Cone ɪ of 1975.

Fig. 5. Map of epicentres of earthquakes in September 1975, that preceded the formation of the Southern Breakthrough (1–17 Sept. 1975). Conventional signs 1–5, energy class and earthquake focus signs as on Fig. 3. ɪ, ɪɪ, ɪɪɪ, vɪɪɪ, the 1975 eruption centres.

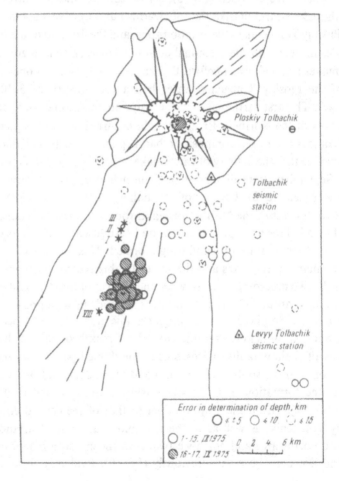

In August the feeder channels of Cone I became extinct and feeder channels formed for Cones II and III of the Northern Breakthrough. This was accompanied by the collapse of the summit caldera and the revival of seismic activity in the region of the Ploskiy Tolbachik crater, to its east and south-east. The obvious interdependence of these events would suggest a connection between the magma source of the eruption and Ploskiy Tolbachik, and that in August 1975 the basalts were transported from the shallow chamber at depths of 0–5 km beneath Ploskiy Tolbachik to the area of the Northern Breakthrough. Between 1 and 17 September

Fig. 6. Map of epicentres of earthquakes during the period 17 Sept. to 31 Dec. 1975. Conventional signs 1–5, energy class and earthquake focus signs as on Fig. 3. I, II, III, VIII, the 1975 eruption centres.

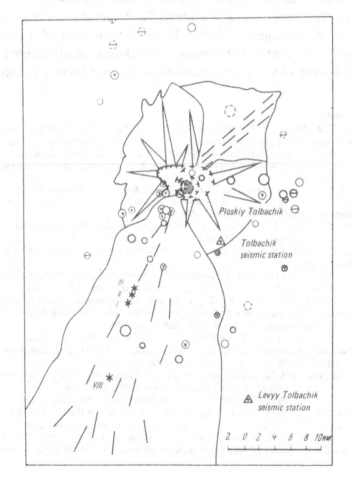

202   *Fedotov, Gorel'chik* et al.

movements and fissures occurred at a depth of 0–5 km south of Cones I, II and III and the eruption was concentrated in the Southern Breakthrough. Before December 1976 its cone, VIII, was pouring out the fluid high-alumina basalt lava typical of the summit eruptions of Ploskiy Tolbachik.

In spite of the great diversity of seismological data, a number of important aspects of the eruption mechanism remain obscure. For example, it is not known what direction the main basalt withdrawal of 0.3 km³ took from the new summit caldera of Ploskiy Tolbachik – whether it occurred at shallow depth along a rift into the eruption area, into the thicknesses of the volcanic structure to the south-east and east or downwards into an intermediate reservoir, the existence of which may be expected in the lower layers of the crust or the transitional layer (at a depth of 20–30 or 20–40 km). Possibly the basalts spread out in all these directions.

The GTFE ceased on 10 December 1976. A careful study of complete data on the earthquakes of 1975–76, and a correlation of them with changes in the course of the eruption, would undoubtedly improve our understanding of the magma chambers of this great fissure eruption.

References

Anosov, G. A., Balesta, S. T., Ivanov, B. V. & Utnasin, V. K. 1974. Osnovnyye cherty tektonicheskogo stroyeniya Klyuchevskoy gruppy vulkanov (Kamchatka) v svyazi s yeye glubinnoy strukturoy. (Basic features of the tectonic structure of the Klyuchi group of volcanoes (Kamchatka) in respect of its deep structure.) *Dokl. AN SSSR.* **219**(5) 1192–5.

Fedotov, S. A. 1972. *Energeticheskaya klassifikatsiya Kurilo–Kamchatskikh zemletryaseniy i problema magnitud. (Energy classification of the Kuril–Kamchatka earthquakes and the problem of magnitudes.)* M., Nauka, 116 pp.

Fedotov, S. A. 1976. O mekhanizme glubinnoy magmaticheskoy deyatel'nosti pod vulkanami ostrovnykh dug i skhodnykh s nimi struktur. (The mechanism of deep magmatic activity beneath the volcanoes of island arcs and similar structures.) *Izv. AN SSSR*, Ser. geol., No. 5, 25–37.

Fedotov, S. A., Gorel'chik, V. I. & Stepanov, V. V. 1976a. Seysmologicheskiye dannyye o magmaticheskikh ochagakh, mekhanizme i razvitii bazal'tovogo treshchinnogo Tolbachinskogo izverzheniya v 1975 g. na Kamchatke. (Seismological data on the magma chambers, mechanism and development of the Tolbachik basalt fissure eruption of 1975 in Kamchatka.) *Dokl. AN SSSR*, **228**(6), 1407–10.

Fedotov, S. A., Khrenov, A. P. & Chirkov, A. M. 1976b. Bol'shoye treshchinnoye Tolbachinskoye izverzheniye 1975 goda (Kamchatka). (The Great Tolbachik Fissure Eruption of 1975 (Kamchatka).) *Dokl. AN SSSR*, **228**(5), 1193–6.

Gorel'chik, V. I. 1973. *Seysmicheskiye proyavleniya vulkanicheskoy deyatel'nosti. (Seismic manifestations of volcanic activity.)* Avtoreferat kand. dis., M., 23 pp.

Gorel'chik, V. I. & Stepanov, V. V. 1976. Seysmichnost' rayona Severnoy gruppy vulkanov Kamchatki v 1971–1972 gg. (Seismicity of the area of the Northern Group of volcanoes in Kamchatka, 1971–1972.) In: *Glubinnoye stroyeniye, seysmichnost' i sovremennaya deyatel'nost' Klyuchevskoy gruppy vulkanov. (Deep structure, seismicity and present-day activity of the Klyuchi group of volcanoes.)* Vladivostok, pp. 108–18.

Minakami, T. 1960. Fundamental research for predicting volcanic eruptions (p. 1). *Bull. Earthqu. Res. Tokyo Univ.*, **38**, part 4, 497–544.

Piyp, B. I. 1956. *Klyuchevskaya sopka i yeye izverzheniya v 1944–1945 gg. i v proshlom.* (*Klyuchevskaya Sopka and its eruptions of 1944–45 and in the past.*) Trudy Labor. vulkanol., **11**, 308 pp.

Tokarev, P. I. 1966. *Izverzheniya i seysmicheskiy rezhim vulkanov Klyuchevskoy gruppy.* (*Eruptions and seismic regime of the volcanoes of the Klyuchi group.*) M., Nauka, 118 pp.

Volcanic tremor during the Great Tolbachik Fissure Eruption of 1975

V. I. Gorel'chik, V. V. Stepanov and V. P. Khanzutin

One of the many interesting seismic phenomena observed during the Tolbachik fissure eruption was volcanic tremor. At certain periods it reached such intensity that at Proryv, Tolbachik and Levyy Tolbachik, the seismic stations closest to the eruption, amplification had to be brought down to 100.* The amplitude of the tremor at Proryv station (distance = 4 km) reached 30–50 μm.

This paper gives a detailed analysis of the energy change in volcanic tremor at Tolbachik and Levyy Tolbachik stations compared with visual observation of the progress of the eruption. Pending analysis of all the field data on the tremor, our conclusions here are of a preliminary nature.

Seismic stations and instrumentation. The Tolbachik station began registering approximately 24 hours before the eruption and was situated 15 km north-east of the point where Cone I of the Northern Breakthrough formed and 22 km from the Southern Breakthrough. It was equipped with standard seismological instruments (VEGIK – GB-IV) with tabular frequency characteristic at intervals of 0.2–0.9 s. The range of variation in amplification was 500–5000.

Levyy Tolbachik station, 20 km south-east of the Northern Breakthrough and 19.5 km east of the Southern Breakthrough, began registering on 2 September 1975. Its instruments and frequency characteristic were the same as at Tolbachik, with a range of variation in amplification of 100–5000.

Description of the phenomenon. The graph of the change in average amplitude of volcanic tremor for July–September 1975 at Tolbachik and

---

* The location of these stations is shown on Fig. 1 of Fedotov *et al.* (The development of the Great Tolbachik Fissure Eruption...), this volume.

for September 1975 at Levyy Tolbachik is shown in Fig. 1. The standard deviation in determining the amplitude and period of volcanic tremor at maximum amplification was $\pm 0.15\,\mu m$ and $\pm 0.15\,s$ respectively. On lowering the amplification, the error in amplitude determination increased to $\pm 0.5\,\mu m$ and more ($\approx$ up to $\pm 2\,\mu m$).

As Fig. 1 shows, the nature of the change in intensity of the volcanic tremor correlates well with the principal stages of the eruption. A slight volcanic tremor (0.25–$0.3\,\mu m$) was recorded by the Tolbachik station on the morning of 5 July (GMT), i.e. just before the eruption, which began on 5 July at 21.45. In the first hours the new eruptive centre consisted of four explosive vents which were establishing themselves on a newly-formed fissure of north-west strike (Fedotov *et al.*, 1976b). Twenty-four hours later one crater was active. In this period, termed the 'initial regime', the amplitude of the volcanic tremor gradually increased, first to $1.5\,\mu m$ then to 3.5–$4.5\,\mu m$.

Up to 9 July there was a build-up in eruptive force and Cone I of the Northern Breakthrough began to form. The tremor at this time (the 'transitional regime') was characterised by sharp fluctuations in amplitude – from 1–$4.5\,\mu m$, with periods of increase and decrease lasting on average 2–3 hours.

After 9 July three principal stages can be identified in the eruption: a purely explosive one (9–26 July), which was connected with a powerful gas blast and ejections of pyroclastic material from Cone I of the Northern Breakthrough; an explosive–effusive phase (26 July–15 September), in which lava poured out and new eruption centres were formed on the Northern Breakthrough; and a predominantly effusive stage (after 17 September), connected with the formation and activity of the Southern Cone.

The emission of the gas stream from Cone I (the explosive stage), and the steady, low explosive activity and continuous effusion of liquid basalt lavas at the Southern Breakthrough (the predominantly effusive stage) were characterised by moderate volcanic tremor. In the first instance the average amplitude of the tremor at Tolbachik station did not exceed $1.5\,\mu m$, and in the second its maximum was $1.2\,\mu m$ at Tolbachik station and 2–$3\,\mu m$ at Levyy Tolbachik.

Between 23 and 26 July the first pauses in the eruption occurred, alternating with powerful explosions. On 28 July lavas began to pour from the southern bocca of Cone I, and the eruption took on an explosive–effusive nature. The dynamics of the eruption in this period (from 26 July to

15 September) varied dramatically: Cone I was torn apart, lava boccas formed and lava poured out, Cones II and III appeared, they were intensely active explosively, eruption centres IV, V, VI, and VII were initiated, and powerful explosions occurred in the final phase of eruption of Cone II. These events were reflected in considerable drops in the intensity of volcanic tremor – from 6–8 to 0.5–0.8 μm amd right down to complete cessation. The connection between volcanic tremor and the dynamics of the eruption was clearest in the periods preceding the formation of new eruption centres and during the effusion of lava. As already noted (Fedotov *et al.*, 1976a), the formation of the major eruption Cones I, II, III and VIII was preceded by earthquake swarms. These occurred when volcanic tremor was decreasing or completely absent. At the Tolbachik station volcanic tremor ceased for the three days preceding the appearance of Cone II of the Northern Breakthrough (from 6 August at 10.45 to 9 August at 07.00), and six hours before Cone III broke through.

It is noteworthy that the duration of the earthquake swarms did not entirely coincide with that of the pauses in the volcanic tremor. The earthquake swarm preceding the formation of Cone II began on 2 August. Up to 12.00 on 6 August earthquakes occurred with $K_{S1.2}^{\phi\,68} \leqslant 9$, the foci of which were mainly to the east of Cone I, in the region of the Ploskiy Tolbachik crater and to its south. In the breakthrough area during this time (3 August) only six earthquakes of energy classes 8 and 9 were recorded. The average amplitude of the volcanic tremor at this time was 2.8 μm. At Cone I, bomb showers, blow-outs and lava-fountains from the boccas, etc., continued. A sharp decrease in volcanic tremor followed at 10.45 on 6 August, the beginning of which coincided with a 9-hour pause in the eruption. Thenceforth, although Cone I resumed activity, no further volcanic tremor was recorded at the Tolbachik station. Its absence corresponds to the time of the second half of the swarm, when (from 14.00, 6 August onwards) some of the epicentres with $K_S \leqslant 9$ shifted to the vicinity of the breakthrough. The formation of Cone II during the

Fig. 1. Variation of mean amplitude (*A*) of volcanic tremor, July–September 1975, for Tolbachik seismic station (*a*), (*b*), (*c*) and September 1975 for Levyy Tolbachik seismic station (*d*). Volcanic tremor during: 1, formation of eruption centres; 2, formation of lava boccas: southern bocca and northern bocca of Cone I, and of the main flow of Cone II; 3, initiation of explosive activity; 4, abatement of eruptive activity; 5, intense bomb falls; 6, pauses in eruption.

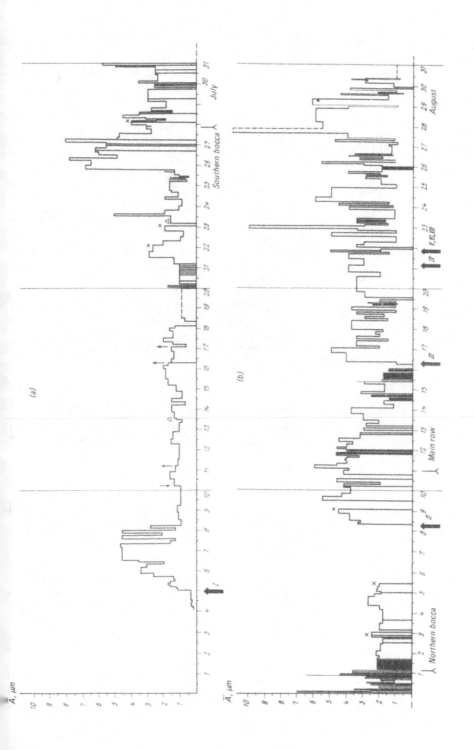

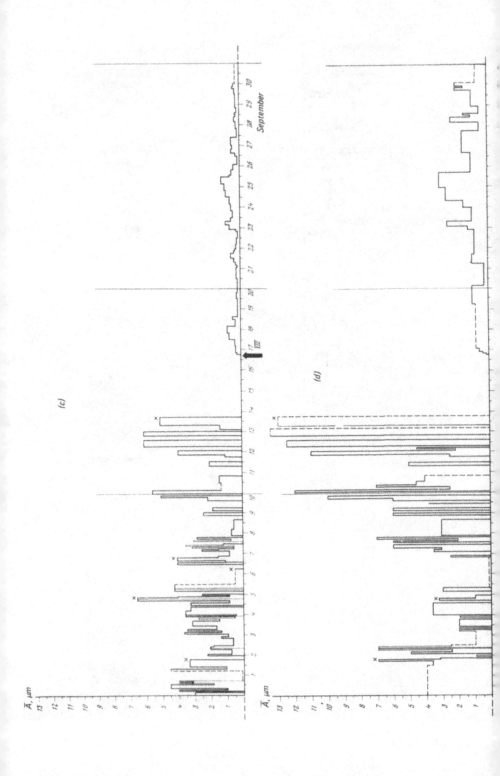

(c)

(d)

$\bar{A}$, μm

September

eruption of 9 August was accompanied by a sharp rise in volcanic tremor. Less than 24 hours later tremor ceased for 10 hours. The pause followed an earthquake of $K_S = 10$ in the eruption area. It is quite likely that this and subsequent pauses in tremor during the continuing earthquake swarm were related to subsequent events such as the formation of Cones III and IV and the opening up of a system of fissures (V, VI, VII). The nature of the eruptions in periods when there was no volcanic tremor varied. On 6, 7 and 8 August, when no tremor was recorded at the Tolbachik station, the effusive–explosive activity of Cone I continued and although noticeably less intense it did not cease until several hours before Cone II erupted. Just before Cone III was formed, during a 6-hour absence of volcanic tremor on 17 August, Cone II did not cease activity once. The eruption of Cone II ended on 14 September 1975 at 16.30 GMT. This was the end of the activity of the Northern Breakthrough and coincided with a sharp fall in the amplitude of volcanic tremor at Tolbachik station from 5 μm, and at Levyy Tolbachik station 13 μm, to one commensurate with the background. Subsequently volcanic tremor reappeared only on 17 September, when the eruption resumed at the Southern Breakthrough.

As a rule, when a lava flow began to form, accompanied by deformations of the cone structures, the volcanic tremor intensified, and after lava had broken through and movement ceased, it would subside. This was the case when the southern and northern boccas of Cone I opened up, and lava began to flow from the central vent of Cone II (main flow on Fig. 1) and from Cone VIII.

Thus, all the principal turning points in the eruption were marked by corresponding changes in the intensity of volcanic tremor. The connection between tremor and lesser events is less clear and straightforward. In a number of cases increases in explosive activity (blow-outs) are seen to coincide with a decrease in the energy of volcanic tremor (usually by not more than 0.5 μm at the Tolbachik station) and vice versa. Intensive bomb showers are sometimes accompanied by an increase in volcanic tremor. The sharpest decline, or indeed absence, of tremor was observed when explosive activity in the craters momentarily ceased (pauses in the eruption, see Fig. 1).

It is interesting to compare the results of recording volcanic tremor simultaneously at the Tolbachik and Levyy Tolbachik stations in September 1975 (see Fig. 1). Ground conditions were not tested at either station, but a comparison of earthquake recordings suggests that the amplitudes of ground displacement at Levyy Tolbachik were overestimated two or three times compared with those of the Tolbachik station. With relationships between source and receiver constant, the relation between amplitudes of volcanic tremor at the two stations should remain the same. This was the case throughout the active period of the Southern Breakthrough, from 17 September onwards, and in the culminating stage of activity of Cone II at the Northern Breakthrough from 8 to 14 September. In these cases the ratio of mean amplitudes of volcanic tremor at Levyy Tolbachik and Tolbachik stations was 2–2.5. Between 2 and 7 September this ratio changed from 2.2 (3 September) to 0.6–1.0 (3–7 September), i.e. from 3 to 7 September volcanic tremor at the Levyy Tolbachik station was weaker than or equal in intensity to that recorded at the Tolbachik station.

The predominant identifiable periods of volcanic tremor at these two stations were between 0.8 and 1.2 s and did not change substantially in the course of the eruption.

Discussion of results. Volcanic tremor is a phenomenon intimately connected with the dynamics of the volcanic process but comparatively little studied. It is difficult to study, partly because observations have to be carried out in the immediate vicinity of the source of the tremor (often in complicated conditions), and partly because it may have many causes. A review of a number of studies of the nature of volcanic tremor has been made by Farberov & Balesta (1966) and Gorel'chik (1976). We shall confine ourselves to a few principal aspects of the subject as it stands today. It is now known that both body and surface waves may be generated during volcanic tremor. The existence of several sources of tremor on one and the same volcano is considered practically proven. The depth of the source may vary from hundreds of metres to several tens of kilometres. As a rule, periods of volcanic tremor coincide in every case with magmatic activity either on the surface or at depth. For example, during the eruption of the Hawaiian volcano Kilauea in 1963 (Shimozuru *et al.*, 1966) volcanic tremor was generated not only during fountaining and during the formation of a lava lake, but also during a time when surficial activity was not observed – while the magma was advancing from a deeper reservoir towards a near-surface chamber and migrating along fairly extended channels from the crater to the eastern rift zone. At various stages in the eruption, the activity of some sources of volcanic tremor may abate, while that of others may intensify, and this gives rise to various types of volcanic tremor, differing in intensity, duration and wave type. The interference picture obtained when several sources are active at once, makes it difficult to pinpoint the sources and may obscure the connection between the tremor and the observable features of the eruption.

From July to September 1975 during the GTFE we were dealing in all probability with not one but several fairly extensive sources of tremor which were functioning either alternately or simultaneously. Apart from volcanic tremor with predominant periods of 0.8–1.1 s, from the very first days of the eruption the Tayga and Poisk seismographs registered from a distance of over 10 km volcanic tremor in the range of 0.1–0.15 s (see Balesta *et al.*, this volume), which consisted, in the authors' opinion, of body waves generated presumably at a depth of $\approx$ 5 km from the zone of both the Northern and Southern Breakthroughs. This type of tremor was also observed during pauses in the eruption and when there were no prolonged periods of tremor.

The volcanic tremor registered by standard seismological instruments, which is the subject of this article, consists of surface waves, judging from the way their amplitude attenuates with distance, and their predominant periods of oscillation. The rise in the energy of the tremor the closer it is to the eruption centres and the high correlation between the change in tremor intensity and the stages of the eruption suggest that the source of the oscillations is connected with the Northern and Southern Break-throughs. Possibly volcanic tremor with periods of 0.8–1.1 s is the result of magma moving towards the surface along channels that formed during eruption. The advent of fresh amounts of magma causes more intense volcanic tremor, with ensuing effusion of lava (see Fig. 1). The formation of the feeder channels of new eruption centres is marked by several series of earthquakes, the foci of which are situated in immediate proximity to the location of future breakthroughs. Long-period volcanic tremor is always negligible at this time or absent altogether until a new eruptive centre becomes active, i.e. when a channel has formed along which magma can travel. Such seismic preparation is typical of lateral craters breaking through basalt volcanoes (Tokarev, 1966; Gorel'chik & Zobin, 1971) and evidently it basically reflects the mechanism by which magma is injected and conveyed to the surface.

The study of volcanic tremor connected with the GTFE continues. The problems of locating the sources of tremor and explaining the mechanism by which oscillations are generated still await solution. The preliminary results of research may briefly be formulated thus.

1. During the GTFE, intensive volcanic tremor was recorded with predominant periods of oscillation of 0.8–1.1 s. This type of tremor

212   *Gorel'chik* et al.

consists of surface waves and displays a close correlation with the principal stages of the eruption.

2. In the purely explosive stage (activity at Cone I of the Northern Breakthrough from 6 to 26 July 1975) and the chiefly effusive stage (activity at the Southern Breakthrough after 17 September 1975), volcanic tremor was moderate – up to 1.5 μm on average at the Tolbachik station. During the explosive–effusive stage of eruption (when lava was pouring from Cone I and the remaining centres of eruption at the Northern Breakthrough were active), an abruptly changing pattern of volcanic tremor is characteristic, with 'pauses and spurts' of up to 10–15 μm (on average 7–8 μm) according to data from the Tolbachik station.

3. The formation of new eruption centres was usually preceded by a decrease in the energy of volcanic tremor, and always accompanied by a sharp rise in the latter. The longest pauses in volcanic tremor were observed when Cone I's activity was dying down, just before Cone II was formed, and after eruption ceased at the Northern Breakthrough prior to the formation of the Southern Breakthrough.

4. The discharge of lava and the deformations of the cone structure were accompanied by a rise in volcanic tremor. Once lava had broken through and movements ceased, tremor abated.

References

Farberov, A. I. & Balesta, S. T. 1966. Ob issledovanii vulkanicheskogo drozhaniya. (The study of volcanic tremor.) *Byul. vulkanol. stantsiy*, No. 40, 45–60.
Fedotov, S. A., Gorel'chik, V. I. & Stepanov, V. V. 1976a. Seysmicheskiye dannyye o magmaticheskikh ochagakh, mekhanizme i razvitii bazal'tovogo treshchinnogo Tolbachinskogo izverzheniya v 1975 godu na Kamchatke. (Seismic data on the magma chambers, mechanism and development of the Tolbachik basalt fissure eruption of 1975 in Kamchatka.) *Dokl. AN SSSR*, **228**(6), 1407–10.
Fedotov, S. A., Khrenov, A. P. & Chirkov, A. M. 1976b. Bol'shoye treshchinnoye Tolbachinskoye izverzheniye 1975 goda (Kamchatka). (The Great Tolbachik Fissure Eruption of 1975 (Kamchatka).)
Gorel'chik, V. I. 1976. Seysmicheskiye proyavleniya vulkanicheskoy deyatel'nosti. (Seismic manifestations of volcanic activity.) In: *Glubinnoye stroyeniye, seysmichnost' i osobennosti sovremennoy deyatel'nosti Klyuchevskoy gruppy vulkanov. (Deep structure, seismicity and features of the present-day activity of the Klyuchi group of volcanoes.)* Vladivostok, pp. 25–35.
Gorel'chik, V. I. & Zobin, V. M. 1971. O kharaktere razvitiya royev zemletryaseniy v oblastyakh aktivnogo vulkanizma na Kamchatke. (On the nature of the development of earthquake swarms in areas of active volcanism in Kamchatka.) In: *Vulkanizm i glubiny Zemli. (Volcanism and the Earth's depths.)* M., Nauka, pp. 118–20.
Shimozuru, D., Kamo, K. & Kinoshita, W. T. 1966. Volcanic tremor of Kilauea volcano, Hawaii, during July–December, 1963. *Bull. Earthqu. Res. Inst. of Tokyo*, **44**, part 3, 1093–1148.
Tokarev, P. I. 1966. *Izverzheniya i seysmicheskiy rezhim vulkanov Klyuchevskoy gruppy. (The eruptions and seismic regime of the volcanoes of the Klyuchi group.)* M., Nauka, 118 pp.

A study of volcanic tremor during the Tolbachik eruption

E. I. Gordeyev, V. D. Feofilaktov and
V. N. Chebrov

Introduction

Of the many phenomena that accompany volcanic eruptions, one of the most interesting is volcanic tremor, the continuous seismic signal recorded during almost any eruption and sometimes even when there are no external signs of volcanic activity. The first mention of volcanic tremor relates to the eruption of the volcano Usu in Japan (Omori, 1911). Interest in tremor has increased as the seismological study of active volcanoes has developed, and there is now a wide range of works devoted to the problem (Sassa, 1935; Finch, 1949; Tokarev, 1963; Farberov & Balesta, 1966; Tanaka, 1968, 1969; Kubotera, 1974). However, there is still no single, comprehensive interpretation of the causes of tremor, or satisfactory description of its sources; at least ten different hypotheses exist. Some authors identify the source as auto-oscillatory processes in a magma chamber (Sassa, 1935; Kubotera, 1974); others consider its main causes to be either gas emission or disturbances created by magma as it moves along its complex conduits (Shimozuru et al., 1966; Guerra et al., 1976). It is difficult to answer these questions at present because of the wide variety of forms of volcanic tremor during an eruption. Even in one volcano several types with different frequency ranges and different wave composition can be distinguished (Shimozuru et al., 1966; Tanaka, 1968, 1969; Kubotera, 1974). However, there is definitely a type of volcanic tremor observable during the eruption of basaltic volcanoes. Its records resemble quasi-sinusoidal oscillations complicated by a process similar to pulsation. This is the type of tremor examined here.

Field experiments were conducted in the region of the Tolbachik eruption in 1975 and 1976. The Great Tolbachik Fissure Eruption (GTFE) commenced at the beginning of July 1975, 18 km south-south-west of the

213

214   *Gordeyev* et al.

central crater of Ploskiy Tolbachik volcano, which is situated in the southern part of the Klyuchi group. On 17 September, three days after the newly formed Northern Cones ceased activity, the eruption shifted 9 km south, where the Southern Breakthrough was initiated. During 1975 and 1976 a moderate eruption continued at the Southern Breakthrough. Using their observations of tremor at many points on the Southern Breakthrough, the authors have attempted to measure the dynamic characteristics of the wave fields of volcanic tremor; the direction of the source of tremor is determined by the method described below.

Experimental methods and the processing of results

In September and November 1975 observations of volcanic tremor were conducted using three-component seismic field stations at the points marked on Figs. 1 and 2 by triangles. The maximum distance from the site of the eruption was 23.5 km, the minimum 0.8 km. The application

Fig. 1. Diagrammatic map of the eruption area.
1, site of seismic field station; 2, azimuths to source; 3, angular error; 4, area bounded by azimuths, its centre denoted by a double circle.

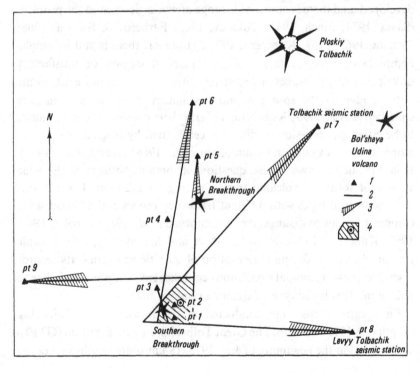

of PSR field seismic recorders (Dergachev *et al.*, 1974) recording on magnetic tape meant that we could reposition the stations and commence observations in other places at short notice. Each of the three seismic stations consisted of PSR recorders, SM-3 seismometers, an exact time signal receiver, an MKh-6 marine chronometer and an MGPA calibration generator (*Apparatura i metodika...(Instrumentation and methods...)*, 1974). The SM-3 seismic detectors were tuned to the period of their own oscillation, which was 2.0 s, and had complete cut-off above the critical period with constant D-2.

To get better conformity with the amplifiers, pendulums were used with rewound working coils ($R_{sg} = 1500$ cm). The transmission band of the seismic through-channel together with the arrangement of the transcription relative to the rate of ground displacement covers a frequency range from 0.3 to 3 Hz (Fig. 3). Maximum magnification could be varied from 1 to

Fig. 2. Diagrammatic map of the situation of observation points and of the statistical centre of volcanic tremor, from determinations on 12 September 1975.

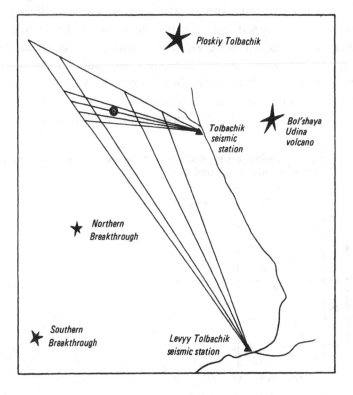

500 thousand, and in each specific case the best amplification for the level of signal was selected.

The recording of volcanic tremor was carried out in simultaneous sessions at two or three points, each session usually lasting several hours. After its reproduction from magnetic tape the recording of the tremor was presented in visual form on photographic paper. The fact that there were precise time readings meant that simultaneous recordings could be selected which were then encoded and selectively analysed using a Nairi-K computer. Errors in encoding were caused by the thickness of the recorded line and never exceeded 2–3%. The error in definition of amplitude–frequency characteristics of the apparatus using an MGPA calibration generator was 10%.

Normality and stationariness of volcanic tremor

Volcanic tremor is a continuous seismic signal which may be considered a random process. One is normally dealing with separate events that are selective functions of a random process. In order to analyse the final events, some of the terms of this random process must first be worked out, particularly its stationariness, ergodicity and normality. If the process has a normal density of distribution, then the proof of its low stationariness automatically entails stationariness in the strict sense. For ergodicity it is enough for the random process to be stationary, but if in addition it has gaussian or normal distribution, then this process is strictly ergodic. On this basis it is first necessary to check whether the random process in question obeys the law of normal distribution. Verification is carried out

Fig. 3. Amplitude–frequency characteristic of a continuous seismic channel relative to ground shift.

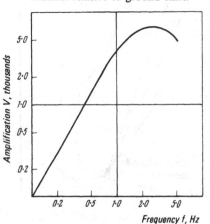

using the conformity criterion χ^2. The principle of this method is that to estimate the divergence between theoretical and experimental densities of distribution a statistic is used which is approximately described by the distribution χ^2. The extent of divergence is defined as follows:

$$X^2 = \sum_{i=1}^{K} \frac{(f_i - F_i)^2}{F_i}$$ (Bendat & Pirsol, 1974).

Here K is the number of categories over which the observation results are spread; f_i is the observed frequency, the number of values that fall into category i; F_i is the expected frequency, the number of values that could fall into category i if the true density of distribution coincided with that expected. X^2 has the same sort of distribution as χ_n^2 with n degrees of freedom. The number of degrees of freedom is determined by the number of categories $n = K - 3$.

To accept the hypothesis that there is agreement between the experimental and theoretical densities of distribution, it will be enough to satisfy the inequality $X^2 \leqslant \chi_{n;\alpha}^2$, where α is the level of significance.

To test the hypothesis concerning the normality of the process of volcanic tremor three selective events were studied. The average length of each sample was 50 s, where the total number of values was $N = 500$. The magnitudes of X^2 were 15.5, 29.8, 20.7, which satisfies the inequality $X^2 \leqslant \chi_{15;0.05}^2 = 32.8$ where the level of significance was 0.05. Thus the hypothesis of normal distribution of the random process of volcanic tremor is acceptable with 95% probability.

To test stationariness, the criterion of series was used, as described in detail by Dzh. Bendat & A. Pirsol (1974). In the majority of physical processes stationariness can be adequately tested by checking the average values of the square of individual intervals, for the absence of a trend. Two selected events of volcanic tremor were tested, each of 200 s duration; the duration of a single interval was 10 s. The criterion of series at a significance level $\alpha = 0.05$, was not infringed once. Consequently, for intervals of 200 s the process of volcanic tremor may be considered stationary in the loose sense, which, if distribution is normal, automatically means strict stationariness. Ergodicity is also automatically guaranteed.

The polarisation method of determining the direction of a quasi-stationary source of seismic signals

The polarisation of seismic waves, together with kinematic features (rate of distribution), is one of the most important characteristics of a wave form. If we know these characteristics we can define the wave type, but normally only kinematic features are used to identify seismic waves, since the determination of polarisation has its own difficulties and does not always achieve the desired accuracy. Nevertheless, the polarisation properties of seismic waves have been used by various authors studying both brief signals (earthquakes and explosions) (Kedrov & Bashilov, 1975), and continuous signals (microseismic storms) (Rykunov, 1967). The study of continuous seismic signals (microseismic storms, volcanic tremor, various seismic noises) is possible using statistical methods of processing the data, and these considerably enhance the accuracy of the final results. In the present work the authors have attempted to use the polarisation properties of seismic waves to develop a method for determining the direction of the quasi-stationary source. The quasi-stationariness of the source and hence of the wave field mean that statistical methods can be applied to selected events in this particular random process.

For an homogeneous and elastic environment we can consider two types of wave emanating from a primary source, longitudinal (P) and transverse (S). The diversity of wave forms engendered in the inhomogeneities of a real environment complicates the overall picture and means that more sophisticated methods are needed for detailed analysis than those applicable to an homogeneous environment. One can assume, however, that in a horizontally-stratified medium the polarisation vectors of primary longitudinal waves and of secondary waves that originated at inhomogeneities remain on a single vertical plane, and that this plane corresponds to the direction of the source. On the other hand, the polarisation vectors of transverse waves (SH) lie in a plane normal to the first. Projections of the vectors of the first and second planes onto a horizontal plane form two mutually perpendicular directions. An exactly similar picture is obtained if we look at the Rayleigh surface waves and the Love waves. When we are dealing with statistically independent waves of various types with polarisation in normal planes, we can find a system of coordinates in which events of two mutually perpendicular components have a maximum coefficient of correlation. The statistical independence of various types of waves is reached at some distance from the source, as is necessary if these waves, having different rates of propagation, are to be subject to different time delays. Let us attempt to calculate this distance.

On propagation from the primary source, the difference Δt in the time of arrival of the various types of wave must be sufficient for the decrease of the auto-covariational function, so that it should not introduce significant distortions into the mutual covariation function. For greater detail on the effect of auto-covariation on the function of mutual covariation, see Dzhenkins & Batts (1972). Let us assign a tenfold decrease to the auto-covariation, and let this decrease correspond to a delay of Δt s. With the ratio of the velocities of various wave types approximately $\sqrt{3}$ the minimum distance S to the source for this delay is calculated as: $S = (a \times \Delta t \sqrt{3})/0.7$, where Δt is the delay in seconds, and a is the velocity of propagation of the slower waves in km/s. A typical auto-correlational function for the volcanic tremor of the Tolbachik eruption is given in Fig. 4. A tenfold diminution in the auto-correlational function gives a delay of 2 s. Thus when velocity $a = 1.2$ km/s, S is approximately 6 km.

It should be noted that the value of S obtained in this way is known to be excessive, since the degree of decrease of the auto-covariational function is ten times greater than the level at which the error in evaluation of mutual covariation (due to the effect of auto-covariation) is comparable with the dispersion of a random process. Consequently it may be said with confidence that at distances from the source greater than S different types of waves are statistically independent.

Let us consider an ideal case where the source may be regarded as a single point within the observation system and waves polarised in normal planes are statistically independent. When registering oscillations on a three-component seismograph with one horizontal component set along the direction from the source and the other perpendicular to it, we obtain three independent recordings of the random function, describing the ground oscillations: $\{X_{\parallel}\}$, $\{X_{\perp}\}$ and $\{Z\}$.

Fig. 4. Typical auto-correlation function of volcanic tremor.

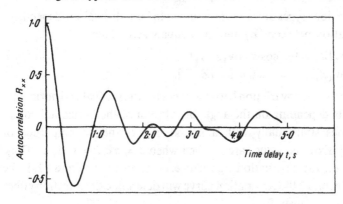

If the horizontal components are rotated through an angle α in the horizontal plane, the events $\{X_\parallel\}$ and $\{X_\perp\}$ are transformed as follows:

$$\{X'_\parallel\} = \{X_\parallel\}\cos\alpha + \{X_\perp\}\sin\alpha,$$
$$\{X'_\perp\} = \{X_\parallel\}\sin\alpha + \{X_\perp\}\cos\alpha.$$

The covariation between $\{X'_\parallel\}$ and $\{X'_\perp\}$ with zero time lapse is defined as:

$$\mathrm{cov}\,[X'_\parallel, X'_\perp] = -\sin\alpha\cos\alpha\,\mathrm{cov}\,[X_\parallel, X_\parallel] + \sin\alpha\cos\alpha,$$
$$\mathrm{cov}\,[X_\perp, X_\perp] + \cos^2\alpha\,\mathrm{cov}\,[X_\parallel, X_\perp] - \sin^2\alpha\,\mathrm{cov}\,[X_\parallel, X_\perp].$$

Since $\{X_\parallel\}$ and $\{X_\perp\}$ are independent, then

$$\mathrm{cov}\,[X_\parallel, X_\perp] = 0,$$
$$\mathrm{cov}\,[X'_\parallel, X'_\perp] = 1/2\,\sin 2\alpha(\mathrm{var}\,[X_\perp] - \mathrm{var}\,[X_\parallel]),$$

where $\mathrm{var}\,[X_\parallel]$ and $\mathrm{var}\,[X_\perp]$ are dispersions of the random processes $\{X_\parallel\}$ and $\{X_\perp\}$ respectively. Consequently, the covariation between the events $\{X'_\parallel\}$ and $\{X'_\perp\}$ varies sinusoidally with period π and amplitude $1/2\,(\mathrm{var}\,[X_\perp] - \mathrm{var}\,[X_\parallel])$. The maximum and minimum values are reached when:

$$\alpha = \pi/4(2k+1), \quad k = 0, 1, \ldots$$

When the horizontal components are rotated through an angle equal to $\pi/4$ from the direction of the source, either a maximum is reached, if $\mathrm{var}\,[X_\perp] > \mathrm{var}\,[X_\parallel]$, or a minimum if $\mathrm{var}\,[X_\perp] < \mathrm{var}\,[X_\parallel]$.

Thus an element of uncertainty arises in establishing the direction of the source. To overcome this we shall consider the covariation between events on the horizontal and vertical components.

The covariation between events $\{X'_\parallel\}$, $\{X'_\perp\}$ and $\{Z\}$ is defined as follows:

$$\mathrm{cov}\,[Z, X'_\parallel] = \cos\alpha\,\mathrm{cov}\,[Z, X_\parallel] + \sin\alpha\,\mathrm{cov}\,[Z, X_\perp],$$
$$\mathrm{cov}\,[Z, X'_\perp] = -\sin\alpha\,\mathrm{cov}\,[Z, X_\parallel] + \cos\alpha\,\mathrm{cov}\,[Z, X_\perp].$$

Since the various types of wave are assumed to be statistically independent, the covariation between $\{Z\}$ and $\{X_\perp\}$ equals zero. Then:

$$\mathrm{cov}\,[Z, X'_\parallel] = \cos\alpha\,\mathrm{cov}\,[Z, X_\parallel],$$
$$\mathrm{cov}\,[Z, X'_\perp] = -\sin\alpha\,\mathrm{cov}\,[Z, X_\parallel].$$

Consequently, the covariation between events of the vertical and horizontal components depending on the angle of rotation of the horizontal component, varies sinusoidally with a period of 2π, and an amplitude of $\mathrm{cov}\,[Z, X_\parallel]$. Extreme values are reached when $\alpha = \pi k, k = 0, 1, \ldots$ Thus, depending on the covariation sign between the events $\{Z\}$ and $\{X_\parallel\}$, the maximum value of the covariation curve will determine the direction either to the source or away from it.

An analysis of the covariation curves $\text{cov}[X'_\|, X'_\perp]$ and $\text{cov}[Z, X'_\|]$ obtained using an observation system consisting of several three-component seismographs, may provide a unique solution to the problem of locating the source.

In practice it is more convenient to use coefficients of correlation which are defined as normalised.

$$\rho_{X'_\| X'_\perp} = \frac{\text{cov}[X'_\|, X'_\perp]}{\sqrt{(\text{var}[X'_\|]\,\text{var}[X'_\perp])}}$$

where $\text{var}[X'_\|]$ is the dispersion of the random process $\{X'_\|\}$. It is not difficult to show that the coefficients of correlation vary according to the angle of rotation α in the same way as the covariations, i.e. extreme and zero values are reached when there are identical values of α.

Indeed, for the horizontal plane:

$$\text{var}[X'_\|] = \text{var}[X_\|]\cos^2\alpha + \text{var}[X_\perp]\sin^2\alpha,$$

$$\text{var}[X'_\perp] = \text{var}[X_\|]\sin^2\alpha + \text{var}[X_\perp]\cos^2\alpha,$$

$$\text{cov}[X'_\|, X'_\perp] = 1/2\sin 2\alpha(\text{var}[X_\perp] - \text{var}[X_\|]),$$

and accordingly:

$$\rho_{X'_\| X'_\perp} = \frac{1/2\sin 2\alpha(\text{var}[X_\perp] - \text{var}[X_\|])}{\sqrt{(\text{var}[X_\|]\,\text{var}[X_\perp] + 1/4\sin^2 2\alpha(\text{var}[X_\perp] - \text{var}[X_\|])^2)}}.$$

The expression obtained for the coefficient of correlation acquires zero values when $\alpha = \pi/2k, k = 0, 1, \ldots,$ and extreme values when $\alpha = \pi/4(2k+1), k = 0, 1 \ldots$. The maximum value:

$$\rho_{max} = \frac{\text{var}[X_\perp] - \text{var}[X_\|]}{\text{var}[X_\perp] + \text{var}[X_\|]},$$

is reached when k is even and the minimum $-\rho_{min} = -\rho_{max}$ when k is odd.

All the above observations relate to a point source; where there are two or more sources they remain valid but the extreme values of the covariation curves define an azimuth towards the centre of gravity of the sources according to the characteristics $(\text{var}[X_\perp] - \text{var}[X_\|])$ for the horizontal plane and $\text{cov}[Z, X_\|]$ for vertical planes. These statements may also be extended to a three-dimensional source if this is thought of as a totality of independent elementary sources.

In conclusion, it should be remembered that we have considered an ideal case, whereas real media and actual sources may present complications owing to the presence of focusing inhomogeneities such as lenses, vertical bedding in the form of wedges, or a plane-parallel vertical system of layers with different characteristics. All these may distort the picture and give a false direction for the source.

Locating the sources of volcanic tremor in 1975 and 1976

Using the method described above, selected events of volcanic tremor as registered by SM3-PSR three-component seismographs in September 1975 and November 1976 during the Tolbachik eruption were analysed.

The encoded events of horizontal components were transformed analytically to a new system of coordinates obtained by rotating the axes of coordinates in the horizontal plane through an angle α in fixed steps of $\Delta\alpha = 18°$.

For each successive pair of transformed events, values of the coefficients of correlation $r_{x'_\parallel x'_\perp}$, $r_{zx'_\parallel}$ and $r_{zx'_\perp}$ were calculated. Then curves were constructed showing the dependence of the coefficients of correlation on the angle of rotation α, and the direction of the source was determined from the extreme values of these curves. An example of this dependence is shown for horizontal and vertical planes in Fig. 5. This presents variation curves for coefficients of correlation, constructed in polar coordinates for selected events of volcanic tremor at the Tolbachik station on 12 September 1975.

Fig. 5. An example of the azimuthal correlation curves for 12 September at Tolbachik seismic station.
1, in the horizontal plane; 2, in vertical planes; 3, direction to Northern Breakthrough.

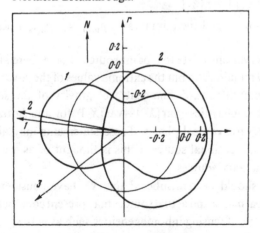

A unique determination of source location requires the use of at least two three-component seismographs. For the purpose of analysis, simultaneous recordings made on 12 September 1975 at Tolbachik and Levyy Tolbachik stations were chosen. Fig. 2 shows the siting of the stations, and also the azimuth directions of the source as determined by not less than three selected events of volcanic tremor for each station. All the events selected occurred within half an hour. A high correspondence of azimuths was observed for the various events for each station. The statistical centre of the source is located at the point denoted by a circle on Fig. 2.

The fact that one can distinguish vertical planes corresponding to extreme values of the coefficients of correlation suggests that in each of these planes there exists a certain predominant wave type. If we determine the functions of mutual correlation between the values of the vertical and horizontal constituents in the plane of maximum level of the correlation coefficient, we can discover the chief characteristics of these events. Fig. 6 shows the mutual correlation functions calculated for events of volcanic tremor on 12 September at the Levyy Tolbachik station.

In the main, functions of the correlation between events of the vertical component and the horizontal one directed towards the source display a symmetry of relatively zero displacement. Thus, between the vertical and horizontal constituents of this plane the phase displacement is most probably zero. The correlations between events of the vertical component and the horizontal component directed perpendicularly to the source lie within 95% confidence, and consequently these events have not been correlated.

The results obtained from observations at Levyy Tolbachik on other days are given in Table 1. This presents the results of determinations of

Fig. 6. An example of the mutual correlation function between events of the vertical and horizontal components. $\alpha = 95\%$ confidence range.

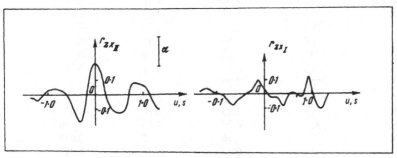

azimuths towards the source and of dispersions according to three components for all events of volcanic tremor from 12 to 21 September inclusive.

This period is interesting because it includes eruptions of both the Northern and Southern Breakthroughs. On 15 September the Northern Breakthrough ceased activity and on 18 September the activity of the Southern Breakthrough began. No noticeable changes were observed in the azimuths to the source, whereas the average amplitude of volcanic tremor from 12 to 21 September inclusive was ten times less.

A more accurate determination of the position of sources of volcanic tremor may be obtained using a system of observations at several points distributed over the area surrounding the eruption zone. Such observations were made in November 1976.

The simultaneous registration of volcanic tremor at the Southern Breakthrough was carried out at three points, using data from two groups of three locations for 12 and 18 November. At each location not less than five areas with tremor duration of about 20 s were determined. Then the average values of the azimuth towards the source and errors in determining direction were found. Fig. 1 is a diagram of the eruption zone marking observation points and average azimuths towards the source of each point. The location of the source according to the convergence of independent

Table 1. *Results of determining azimuths and dispersions for selected events of volcanic tremor at Levyy Tolbachik seismic station in September 1975*

Date	Azimuth(°)	Dispersion, $\sigma_{x\parallel}^2$ (mkm²)	Dispersion, $\sigma_{x\perp}^2$ (mkm²)	Dispersion, σ_z^2 (mkm²)
12	326	2.1	1.16	0.3
12	323	1.05	0.71	0.16
12	351	1.41	0.69	0.15
14	298	1.49	0.33	0.12
14	305	1.54	0.36	0.14
14	302	1.85	0.54	0.18
15	310	6.83	4.71	3.31
15	324	7.00	3.96	3.11
15	339	7.36	4.97	3.40
20	339	0.15	0.11	0.031
20	313	0.22	0.16	0.046
20	350	0.16	0.12	0.039
21	332	0.13	0.075	0.022

results obtained at six different points was fairly successful. The area bounded by the azimuths is 3 km across and is most likely determined not by the true position of the source of volcanic tremor, but by the area within which the formation of surface waves takes place. Proof that in the remote zone (beyond 10 km) the volcanic tremor of the Southern Breakthrough consists mainly of surface waves, is given below. The error in the determination of azimuths at all observation points is approximately 6°, and the eruption site given by each azimuth falls within the region of error.

Spectra of volcanic tremor

To assess the dynamic characteristics of the wave fields of volcanic tremor from the observations made in 1976, spectrum densities of ground displacement were used. Theoretically, spectrum density is a Fourier transformation from the real time function within infinite limits. In practice, when dealing with selected events of finite length, the following definition is usually adequate:

$$S(f) = \Delta t \sum_{k}^{N} F(k\Delta t) e^{-ik\Delta t 2\pi f},$$

$$F(t) = \int_{-\infty}^{\infty} S(f) e^{i2\pi ft} df,$$

where $F(t)$ is the real time function; $S(f)$ is the spectrum density of this function; Δt is a step in discretisation; and $N\Delta t = T$ is the length of the selected event.

The length of the event T determines the separation of peaks with frequency $\Delta f = 1/T$ in a Fourier transformation and the discretisation step determines the maximum of the different frequencies $f = 1/2\Delta t$, which is called the Nyquist frequency.

In our case the length of selected events was approximately 50 s, and the Nyquist frequency $f = 7$ Hz at a discretisation step of $\Delta t = 0.07$ s. The final spectrum densities were smoothed out using a triangular window. Altogether the spectrum densities were calculated for the vertical and horizontal components of ground displacement at seven different observation points. The minimum distance from the eruption point was 0.8 km, the maximum 23.5 km.

The relative variation of spectra with distance is shown in Fig. 7, which gives the spectrum densities of vertical displacement components for five different distances. The respective observation points are marked in Fig. 1. Spectrum density decreases with increasing distance from the eruption point. The points closest to the source in terms of spectrum density are points 1 and 3 in Fig. 1. This accords with the results obtained earlier and confirms that the source of volcanic tremor is directly connected with the site of the eruption. With increasing distance from the source, the greater is the decrease in spectra that accompanies increase in frequency; this is caused by the different absorptions of the high and low frequency components of the tremor.

Fig. 7. Spectral density of ground shift for various distances from the eruption site. Points: 1, 0.8 km; 3, 2.5 km; 4, 8.0 km; 6, 19 km; 7, 23 km.

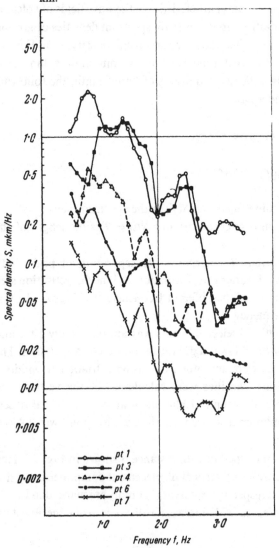

The ratio of spectrum densities seems somewhat unusual for the distances 0.8 and 2.5 km. In the frequency range of 1 to 2.5 Hz at a distance of 2.5 km the values of spectrum density are comparable with those for a distance of 0.8 km. In the zone close to the source we are most probably dealing with the process of formation of surface waves. Moreover, the difference in spectrum densities at these points may be due to different radiation from the source in various directions. In assessing the absorption of volcanic tremor, therefore, results of observations in the zone adjoining the source (less than 2 km) were not used.

The relative change in spectrum densities with increasing distance enables us to obtain absorption as function of frequency. If we assume that in the propagation of a harmonic wave in an absorbing homogeneous medium the amplitudes at recording points 1 and 2 are equal to A_1 and A_2, then the coefficient of absorption is defined as:

$$\alpha = \frac{\ln(A_1/A_2) - m \ln(r_2/r_1)}{r_2 - r_1},$$

where r_1 and r_2 are the distances from the recording points to the source; m is the degree of divergence ($m = 1$ for a spherical wave, $m = \frac{1}{2}$ for a cylindrical wave and $m = 0$ for a plane wave).

Where spectra are used, the amplitudes A_1 and A_2 are replaced by values of spectrum densities at a selected frequency f. The calculation of absorption coefficients over the entire frequency range of spectrum determination gives the dependence of absorption on frequency.

Fig. 8. Dependence of the absorption coefficient of volcanic tremor on frequency for degree of divergence $m = \frac{1}{2}$.
1, 2, curves for a source coinciding with the position of an active cone; 1, curve for horizontal components of shift; 2, curve for vertical components of shift; 3, curve for horizontal components when the source has shifted into the centre of the area bounded by the azimuths (see Fig. 1).

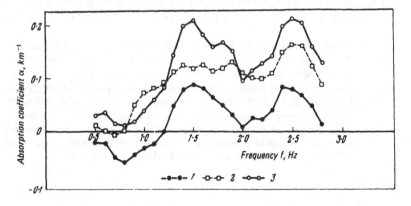

In determining the absorption of volcanic tremor, coefficients were calculated in the frequency range 0.5–2.8 Hz for seven different distances from the source and for two values of degree of divergence $m = 1$ and $m = \frac{1}{2}$. The steps in frequency were of 0.1 Hz. For the propagation of a spherical wave ($m = 1$) as it is postulated here, the absorption coefficients obtained do not have a physical value, since they assume negative values over the whole frequency range. It may thus be considered that volcanic tremor at the Southern Breakthrough in the more distant zone (beyond 5 km) consists mainly of surface waves.

Fig. 8 shows the dependence of absorption on frequency for surface waves ($m = \frac{1}{2}$), giving averaged values of the coefficients for all pairs of observation points. The analysis of averaged values is justified, because we do not know the ground conditions at each point where seismometers were installed and cannot make any corrections. Curve (2) is constructed for vertical channels, and curve (1) for horizontal. The dependence of absorption on frequency may be said to be the same for vertical and horizontal components of ground displacement. On both curves, where an increase in absorption coefficient values tends to occur as frequency rises, there are clearly marked minima at the frequencies 0.7–0.9 Hz and 2.0–2.2 Hz. These minima are most likely caused by the layered structure of the medium in which the surface waves are propagated. An attempt can be made to assess the effective thickness of a layer by the transmission path of the waves if one takes a particular wave type and particular rates of propagation. Let us suppose that the horizontal component consists mainly of surface Love waves. Then the conditions for the diffusion of a harmonic wave of frequency f in a layer of thickness H will be expressed as follows:

$$f = (2n+1)\,b/(4H \sin e),$$

$a = \arccos b/c$ (Savarenskiy, 1972), where n is the order of overtone; b is the velocity of propagation of the wave in the layer; and c is the velocity of propagation of the Love wave along the surface of the layer. The values of velocities b and c are 0.9–1.0 km/s and 1.2–1.3 km/s respectively.*

---

* The values of the velocities b and c are taken from: Balesta, S. T., Kargopol'tsev, A. A. & Grigoryan, G. B. *Seysmicheskiye dannyye o stroyenii rayona Bol'shogo treshchinnogo Tolbachinskogo izverzheniya.* (*Seismic data on the structure of the region of the Great Tolbachik Fissure Eruption.*) (In press) and Tokarev, P. I. & Lemzikov, V. K. *Predvaritel'nyye rezul'taty seysmologicheskikh nablyudeniy vulkanicheskogo drozhaniya Yuzhnogo proryva Bol'shogo treshchinnogo Tolbachinskogo izverzheniya v avguste 1976 goda.* (*Preliminary results of seismological observations of volcanic tremor at the Southern Breakthrough of the Great Tolbachik Fissure Eruption in August 1976.*) (In press.).

Assuming that the first minimum on the absorption curves corresponds to the main tone ($n = 0$) and the second to the first overtone ($n = 1$), we obtain the following effective values of the thickness of the layer: $H_0 = 430$ m; $H_1 = 480$ m. Doubtless these values are not highly accurate and for more authentic results one would need to have reliable values for the velocities of propagation and be certain of the wave type. It should be noted, however, that the study of wave fields of volcanic tremor for the purposes of studying the structure of the eruption medium, could be useful and have great potential; its main advantage over other methods is the continuity and quasi-stationariness of the process of volcanic tremor.

Fig. 8 shows that at frequencies below 1.2 Hz the values of absorption coefficients for the horizontal component of ground displacement are negative. The explanation of this may be that the true location of the source of surface waves does not correspond to a position exactly at the eruption centre. Fig. 8 presents an absorption curve calculated for a source that is situated at the statistical centre of an area bounded by the azimuths obtained in the preceding section of this paper (see Fig. 1). The curve is constructed for horizontal components, and even such an insignificant shift in the location of the source takes this curve completely into a positive area of values of absorption coefficients, which is more realistic. The absolute values of absorption coefficients agree quite reasonably with the data provided by Guerra *et al.* (1976) and Tanaka (1968). Tanaka gives coefficients of absorption of volcanic tremor for the volcano Makhara-yama as 0.09 to 0.3 km^{-1}, Guerra has $\alpha = 0.4$ km^{-1} at frequency 2 Hz for Etna, and in our case $\alpha = 0.1$–0.2 km^{-1}.

In conclusion, the reduction of spectra to a single level for different periods of observation was carried out using data from the Levyy Tolbachik station, where volcanic tremor was recorded continuously by the regional seismograph VEGIK-GB VI (*Apparatura i metodika...* (*Instrumentation and methods...*), 1974).

Conclusions

1. Volcanic tremor at the Southern Breakthrough of the Tolbachik eruption at intervals of approximately 200 s may be considered a stationary random process.

2. The method of separating polarisation planes of seismic waves gives results with good internal consistency.

3. The determination of the source of volcanic tremor in November 1976 suggests that tremor generation is directly linked to the site of the

eruption. This conclusion is supported by an analysis of variation of spectrum density with distance from the eruption site. The 1975 results suggest a source to one side of the active cone, which may be due either to the different nature of volcanic tremor or to limitations in the observation system.

4. The dependence of absorption coefficients of volcanic tremor on frequency, with a general tendency to increase as the latter increases, produces clearly marked minima which may be due to the diffusion of surface waves in a layered medium.

5. Apart from the obvious necessity of determining the characteristics of the source of volcanic tremor, an extensive study of tremor wave fields could be useful for studying the structure of the environment inside the eruption zone.

References

Apparatura i metodika seysmometricheskikh nablyudeniy v SSSR. (*Instrumentation and methods of seismometric observations in the USSR.*) 1974. M., Nauka, 244 pp.

Bendat, Dzh. & Pirsol, A. 1974. *Izmereniya i analiz sluchaynykh protsessov.* (*The measurement and analysis of random processes.*) M., Mir, 464 pp.

Dergachev, A. A., Dzhanov, S. M., Zhadin, V. V. & Savinov, V. A. 1974. Apparatura dlya registratsii seysmologicheskoy informatsii. (Apparatus for recording seismological information.) In: *Geofizicheskaya apparatura.* (*Geophysical instrumentation.*) L., Nedra, Issue 54, pp. 129–32.

Dzhenkins, G. & Batts, D. 1971. *Spektral'nyy analiz i yego prilozheniya.* (*Spectrum analysis and its applications.*) Issue 1. M., Mir, 316 pp.

Dzhenkins, G. & Batts, D. 1972. *Spektral'nyy analiz i yego prilozheniya.* (*Spectrum analysis and its applications.*) Issue 2. M., Mir, 288 pp.

Farberov, A. I. & Balesta, S. T. 1966. Ob issledovanii vulkanicheskogo drozhaniya. (Studying volcanic tremor.) *Byul. vulkanol. stantsiy*, No. 40, 45–60.

Finch, R. H. 1949. Volcanic tremor (pt. 1). *Bull. Seism. Soc. Am.* **39**(2), 356–72.

Guerra, I., Lo Bascio, A., Luongo, G. & Scarpa, R. 1976. Seismic activity accompanying the 1974 eruption of Mt Etna. *J. Volcan. Geotherm. Res.*, **1**(4), 347–62.

Kedrov, O. K. & Bashilov, I. P. 1975. O polyarizatsionnom sposobe vydeleniya ob"yemnykh voln ot udalennogo seysmicheskogo istochnika. (The polarisation method of identifying body waves from a distant seismic source.) *Izv. AN SSSR, Fizika Zemli*, No. 8, 90–8.

Kubotera, A. 1974. Volcanic tremors at Aso volcano. *Physical volcanology*, 29–47.

Omori, F. 1911. The Usu-san eruption and earthquakes and elevation phenomena. *Bull. Imp. Earthqu. Invest. Comm.*, **5**(1), 1–38.

Rykunov, L. N. 1967. *Mikroseysmy. Eksperimental'nyye kharakteristiki yestestvennykh mikrovibratsiy grunta v diapazone periodov 0,07–8 sek.* (*Microseisms. Experimental characteristics of natural microvibrations of the ground in the period range 0.07–8 s.*) M., Nauka, 88 pp.

Sassa, K. 1935. Volcanic micro-tremor and eruption earthquakes. *Mem. Coll. Sci.*, Kyoto Univ., **18**, 255–93.

Savarenskiy, Ye. F. 1972. *Seysmicheskiye volny.* (*Seismic waves.*) M., Nedra, 294 pp.

Shimozoru, D., Kamo, K. & Kinoshita, W. T. 1966. Volcanic tremor of Kilauea volcano, Hawaii, during July–December, 1963. *Bull. Earthqu. Res. Inst.*, Univ. Tokyo, **44**, Pt. 3, 1093–1133.

Tanaka, Y. 1968. A seismometrical study of Izu-Oshima (I). *Pap. Met. Geophys.*, **19**(4), 627–50.

Tanaka, Y. 1969. A seismometrical study of Izu-Oshima (II). *Pap. Met. Geophys.*, **20**(4), 385–416.

Tokarev, P. I. 1963. O tipakh seysmicheskikh voln pri vulkanicheskom drozhanii i fazovykh skorostyakh ikh rasprostraneniya. (On types of seismic waves during volcanic tremors, and the phase velocities of their propagation.) In: *Sovremennyy vulkanizm severo-vostoka Sibiri. (Present-day volcanism of North-East Siberia.)* M., Nauka, pp. 103–9.

The influence of the 19-year tidal cycle on large-scale eruptions and earthquakes in Kamchatka, and their long-term prediction

V. A. Shirokov

The present work is a continuation of the author's research into the influence of astronomical factors on the occurrence of severe earthquakes and volcanic eruptions (Abdurakhmanov *et al.*, 1971, 1976; Fedotov *et al.*, 1976b; Shirokov, 1973, 1974; Fedotov *et al.*, 1977). This paper examines the connection between the tidal period of 18.6 years and the time when the most severe Kamchatka earthquakes and volcanic eruptions occur. On the basis of a study of this link, a prediction is made for severe instances of this kind for the next 25 years. It is shown that the Great Tolbachik Fissure Eruption (GTFE), like all the preceding eruptions of volcanoes on Kamchatka with a volume of erupted material of $V \geqslant 0.5$ km³, occurred as a result of the influence of the 18.6-year tidal period on the volcanic process.

Basic data

In this paper we take large-scale events to mean volcanic eruptions with a volume of erupted material $V \geqslant 0.5$ km³ and earthquakes with a magnitude $M \geqslant 7.5$.

Eruptions. Seven dated volcanic eruptions with $V \geqslant 0.5$ km³ are known in Kamchatka: Shiveluch in 1854 and 1964 (Gorshkov & Dubik, 1969; Piyp & Markhinin, 1965), Ksudach in 1907 (Piyp, 1941), Klyuchi in 1937–38 and 1945 (Menyaylov, 1947; Naboko, 1947; Piyp, 1956), Bezymyannyy in 1955–56 (Gorshkov & Bogoyavlenskaya, 1965), and the GTFE in 1975–76 (Fedotov *et al.*, 1976a). The total volume of products of these eruptions is equal to 12 km³ (Table 1), which approximates to $\frac{4}{5}$ of the volume of all the material erupted from the Kamchatka volcanoes since the middle of the last century.

A comparison of the large-scale eruptions of the Kamchatka volcanoes with those of the world (Abdurakhmanov *et al.*, 1976) indicates that in the twentieth century volcanic activity on Kamchatka has considerably exceeded the level of activity in other regions of the earth. Together with Japan and the Kuril Islands, Kamchatka is also the region with the highest seismicity on our planet (Gutenberg & Richter, 1954). It may thus be considered potentially one of the most rewarding geodynamic polygons for the study of the volcanic process and its relation to seismicity.

We shall give some brief information about the seven large eruptions, and about the activity of the Kamchatka volcanoes both during the years when these eruptions took place and in the periods immediately adjoining them in time.

1853–1855. On the night of 17/18 February 1854 a catastrophic eruption of the volcano Shiveluch occurred – the first to be historically recorded. Altogether, the eruption of 1854 was more powerful than that of 1964 and was comparable with the eruption of Bezymyannyy in 1956, or maybe even exceeded it (Gorshkov & Dubik, 1969). From 1853 to 1855 there were eruptions of the volcanoes Opala, Mutnovskiy, Gorelyy, Avachinskiy and Koryakskiy (Vlodavets & Piyp, 1957).

1907–1909. After the Kamchatka volcanoes had been quiet for over two years, on 28/29 March 1907 Ksudach underwent a powerful explosive eruption which in scale and character may be ranked with the catastrophic explosions of Bezymyannyy (1956) and Shiveluch (1964). It was

Table 1. *Large eruptions of Kamchatkan volcanoes with volume of eruption material* $V \geqslant 0.5\ km^3$

Name of volcano	Time of main eruption paroxysm	Volume of erupted material (km³)
Shiveluch	Feb. 1854	1–3, (Gorshkov & Dubik, 1969)
Ksudach	March 1907	3 (Piyp, 141)
Klyuchi	Feb. 1938	0.5 (Menyaylov, 1947; Naboko, 1947)
Klyuchi	Jan. 1945	0.8 (Piyp, 1956)
Bezymyannyy	March 1956	3 (Gorshkov & Bogoyavlenskaya, 1965)
Shiveluch	Nov. 1964	1.2 (Piyp & Markhinin, 1965)
Tolbachik	Dec. 1975	2 (Fedotov *et al.*, 1965*a*)

accompanied by the destruction of the monolithic extrusive part of the central cone of Ksudach's caldera. The volume of ejected resurgent material was estimated at 0.5–1 km³ (Dubik & Menyaylov, 1971); an enormous quantity of juvenile material was dispersed over the whole of Kamchatka in the form of pumice and ash (Piyp, 1941). In August 1907 a glow was observed above Klyuchevskaya Sopka and lava seen to be pouring from its eastern slope, and in 1909 there was a moderate eruption of ash in the summit crater. In 1908 Karymskiy erupted, and in 1909 an eruption of Avachinskiy was accompanied by effusion of lava (Vlodavets & Piyp, 1957).

1937–1938. After a few years of low activity, Ploskiy Tolbachik became active in February 1937. Two months later one of the most interesting eruptions of Klyuchevskaya Sopka began, lasting two years (Menyaylov, 1947). The culmination of the eruption was the breakthrough of the subordinate vent Bilyukay in February 1938, when 0.1 km³ of lava issued in the first ten days alone, the lava fountain reaching 250 m in height (Naboko, 1947). As a result of the eruption nine subordinate vents were formed on a strip 15 km long and at heights of between 900 and 4800 m (Menyaylov, 1947; Naboko, 1947). In 1937 and 1938 intensive explosive activity was observed in Avachinskiy and Karymskiy. In 1938 Mutnovskiy erupted. Over the next six years only the Ploskiy Tolbachik eruption, with the breakthrough of a lateral vent in 1941, can be included among the more severe events (Vlodavets & Piyp, 1957).

1944–1946. From November 1944 onwards, over a period of ten months, six volcanoes on Kamchatka began to erupt (Vlodavets & Piyp, 1957). The first in this series of events was the eruption of Shiveluch, on which a new summit extrusion formed between 1946 and 1948. On 1 January 1945 a paroxysmal explosion of Klyuchevskaya Sopka occurred, which ejected 0.6 km³ of bombs, lapilli and ash in a few hours, followed by lava flowing from the summit crater for twenty days. Two lava flows appeared on the slopes of Klyuchevskaya Sopka during subsidiary eruptions of the craters of the Yubileynaya group and Apakhonchich in June 1945 and October 1946. The most violent eruption of Avachinskaya Sopka this century, during which a layer of ash 45 cm thick fell on the sea coast 25 km from the volcano, took place on 28 February 1945. In June 1945 ash explosions of Mutnovskiy were recorded. From 1945 to 1947 Karymskiy was in a state of severe eruption. During 1945 and 1946 ash explosions of the volcano

Malyy Semyachik were observed. Gorelyy was also active in 1947 (Vlodavets & Piyp, 1957). Thus in the course of two years all the most active volcanoes erupted, whereas over the few years before and after this period volcanic activity was very subdued.

1955–1957. On 22 October 1955 the first eruption of Bezymyannaya Sopka in historical times began. A tremendous explosion on 30 March 1956 destroyed the summit of the volcano. In the crater that was formed, measuring 1.7×2.8 km, a summit extrusion quickly grew, reaching a height of 400 m in two years. Its rate of growth was at its greatest in 1956–57 (Gorshkov & Bogoyavlenskaya, 1965). In July–August 1956 a subsidiary explosion of Klyuchevskaya Sopka occurred, forming the Vernadskiy and Kryzhanovskiy craters and causing two lava flows. In 1956 and 1957 Zhupanovskiy, Karymskiy and Koryakskiy erupted (Vlodavets & Piyp, 1957).

1964–1966. The seismic preparation of the gigantic Shiveluch eruption in November 1964 lasted almost 10.5 months (Tokarev, 1967). The volcano showed no signs of activity until a mighty directed explosion took place (Gorshkov & Bogoyavlenskaya, 1965), which destroyed the dome of the central summit and several other domes. In their place a new crater formed, consisting of two conjoined funnels, each of which was as much as 2 km in diameter. Four months after this explosion the greatest eruption of Bezymyannyy for the past fifteen years took place (Dubik & Menyaylov, 1969). In 1965 an ash eruption was recorded in the summit crater of Klyuchevskaya Sopka, and in October 1966 the subordinate Piyp Craters broke through in its north-east slope (Kirsanov, 1968).* Karymskiy was also active during these years (Tokarev *et al.*, 1969).

1974–1977. In August 1974 the Fourth VVS subordinate eruption occurred,† against a background of moderate eruption in Klyuchevskaya Sopka's summit crater. On 6 July 1975 the GTFE began (Fedotov *et al.*, 1976a), and lasted more than seventeen months. During it, Karymskiy, which had been dormant for the past two years, became active. The record of the period we are considering ends with the eruption of Bezymyannyy in March 1977, the most violent eruption of this volcano for the past ten years.

* Named after B. I. Piyp.
† Named after the Fourth All-Union Volcanological Conference (VVS).

Thus, during the years closest to those when the large-scale eruptions with $V \geqslant 0.5$ km³ took place, other volcanoes of Kamchatka were also active. This tendency was pointed out by B. I. Piyp (1956), who proposed that the causes of the various disturbances in the equilibrium of the magma were 'as likely to be cosmic, such as maxima of lunar/solar attraction...or sun-spot maxima...as they are to be intratelluric, i.e. of the kind that effect gigantic movements of the masses in the deeper parts of the lithosphere, and earthquakes of corresponding energy' (Piyp, 1956, p. 136).

Earthquakes. Let us examine the distribution in time from 1904 of earthquakes in Kamchatka with a magnitude $M \geqslant 7.5$ and depth of focus 0–100 km. There is a continuous record of earthquakes of this force from that year (Fedotov & Bagdasarova, 1974). Altogether 12 earthquakes have occurred; 7 were recorded in 1904–1923, 5 in the years 1952–1971. The almost thirty-year pause from 1923 to 1952 indicates an uneven distribution of events in time. This phenomenon is probably connected with processes on a planetary scale, since a noticeably lower level of seismic activity is recorded for the Earth as a whole during this period (Ben'of, 1961). When events of $M \geqslant 7.5$ occurred there would be a conspicuous increase in seismic activity on the level $M \geqslant 7$. These are the corresponding time intervals and the number of earthquakes of $M \geqslant 7$ within them: 1904–05 ($N = 5$), 1915–17 ($N = 2$), 1922–23 ($N = 4$), 1952–53 ($N = 5$), 1959–60 ($N = 3$), 1969–71 ($N = 3$). Since 1904, 38 earthquakes of $M \geqslant 7$ have been recorded, of which 22 took place within the intervals given, which in turn make up less than 10% of the time of the whole period of observations. The density of the earthquakes of $M \geqslant 7$ is six times higher during these time intervals than the average for the whole period.

To summarise, it can be seen that, firstly, during the years closest to large-scale eruptions and earthquakes the density of events of lesser strength is noticeably greater than the average; secondly, intensive bursts of seismic and volcanic activity continue for approximately two to three years; and thirdly, periods of increased seismic activity do not coincide with periods of increased volcanic activity.

On the influence of the 19-year tidal cycle on the occurrence of large-scale eruptions and earthquakes

Some scientists are coming to the conclusion that the 18.6-year tidal cycle influences the occurrence of earthquakes (Lamakin, 1966) and volcanic eruptions (Hamilton, 1973; Mank, 1973). It certainly has a considerable influence on the occurrence of severe earthquakes and eruptions in Kamchatka (Shirokov, 1974). This enabled the present author to make a long-term forecast of earthquakes in Kamchatka of $M \geqslant 7$ (Fedotov *et al.*, 1976b; Fedotov *et al.*, 1977) and to predict the eruption and effusion of lava from Klyuchi volcano in 1974 (Shirokov, 1973).

It is well known that the declination of the moon is not constant but varies from 18° 19′ to 28° 35′ over a period of 18.61 years. As a result, the decay in the potential of the moon's tide-forming force contains a fairly considerable component of this cycle proportional to the value $(1-3 \sin^2 \phi \cos \psi$, where ϕ is the latitude of the place and ψ the longitude of the ascending node of the moon's orbit). The 19-year tidal cycle is comparable in magnitude with the well-known monthly tidal period (Doodson, 1921). It is safe to assume that this tidal cycle, which plays an important part in the long-term variability of the dynamics of the ocean, will have its effect upon the solid earth as well.

Let us attempt to evaluate the influence of the tidal cycle on eruptions of volume $V \geqslant 0.5$ km³. For dates of eruptions let us take the time of the main paroxysm (see Table 1). It is noticeable that the difference in time between the Bezymyannyy and Tolbachik pair of events is 19.3 years, and between Klyuchi (1938) and Bezymyannyy 18 years; i.e. these eruptions are confined to one and the same phase of the 18.6 tidal period with a deviation of less than 4% from the average value of the phase. Four other eruptions are confined to another phase of the tidal period. The deviation from the average value of the phase is also less than 4%. The distribution of all eruptions according to the phase of the 18.6 tidal period is shown on Fig. 1(*a*), where the phase equal to 0 corresponds to the time of maximum declination δ_{max} of the moon. In the current cycle, the preceding time δ_{max} corresponds to March 1969. The hypothesis that distribution of eruptions according to the 18.6-year tidal phase is not fortuitous is correct by Pirson's criterion with a level of reliability of more than 0.99. The probability of the distribution of the eruptions being fortuitous is $\approx 10^{-6}$.

A similar distribution of events within combined periods of 18.6 years can also be observed for earthquakes of magnitude $M \geqslant 7.5$ (see Fig. 1(*b*)). The events are confined to two phases of the cycle as in the case of the eruptions.

A comparison of distribution of large-scale eruptions and earthquakes according to phases of the lunar tide shows that the maxima of seismic and volcanic activity do not coincide in time. However, there is a clear phase correlation between the distribution of earthquakes and eruptions.

The correlation coefficient is at a maximum ($r = 0.58$) for a shift of 3 years within the phase, eruptions lagging behind earthquakes.

Thus the lunar tide exerts a considerable influence on the most important occurrences of seismic and volcanic activity. Most scientists regard tidal phenomena as merely a trigger mechanism of earthquakes and volcanic eruptions (Lamakin, 1966; Mank, 1973). In our view, it is perfectly admissible that the tides affect the process of build-up and release of tensions in the Earth's crust and upper mantle. Apparently, the protracted, directed action of the tides actually determines the scale of this phenomenon.

Long-term prediction of earthquakes and eruptions

The fact that events are clearly restricted to narrow intervals in the 18.6-year tidal phases enables us to predict future violent eruptions and earthquakes. 'Disturbed' phases, which are marked in the figure as black rectangles, comprise no more than $\frac{1}{4}$ the length of the period, which makes

Fig. 1. Distribution (of eruptions and earthquakes) within lunar tidal periods of 18.6 years duration. (*a*) Distribution of large Kamchatkan eruptions with a volume of erupted material $V \geqslant 0.5$ km³ during the period 1850–1976. (*b*) As (*a*), for Kamchatkan earthquakes between 1964 and 1976; with magnitude $M \geqslant 7.5$, depth of focus 0–100 km. Below is given a time scale for the current period. δ_{max}, epochs of maximum declination of the moon in 19-year cycles. Circles indicate the number of eruptions and earthquakes, black bands indicate the duration of eruptions and earthquakes.

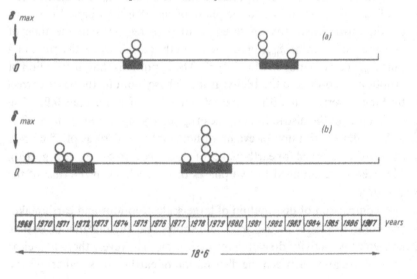

it possible to count on an effective forecast. Predictions have been made for the next 25 years (Table 2). Disturbed phases have been calculated to the nearest half a year, and their length has been taken as equivalent to 2 years. For instance, in the next disturbed phase, the most probable time for earthquakes corresponds to November 1978; rounding it off to half a year gives the beginning of 1979. The disturbed phase therefore corresponds to the years 1978–79, and so on.

This table shows the disturbed phase to which the beginning of the GTFE corresponds. This phase had been identified by us earlier (Shirokov, 1973), although no indication was given that it would be a large-scale eruption.

One can make several suggestions about the possible locations of future events. A forecast of earthquakes with $M \geqslant 7\frac{3}{4}$ has in fact been made by S. A. Fedotov (Fedotov *et al.*, 1976b, Fedotov *et al.*, 1977). In attempting to assess the probability of the various volcanoes erupting, one should take into consideration that six out of the seven large-scale eruptions occurred in the region of the Northern Group of volcanoes in Kamchatka. Of these, violent eruptions occur most frequently in Klyuchevskaya Sopka (Piyp, 1956), suggesting it is more likely that the next large-scale eruption will take place in this volcano than in any of the other volcanoes of Kamchatka. P. I. Tokarev (1971) thinks a violent eruption of the Avachinskiy volcano highly probable.

Conclusions

1. An analysis of the distribution in time of the large-scale eruptions in Kamchatka with a volume of ejected material $V \geqslant 0.5 \text{ km}^3$, and of

Table 2. *Long-term prognosis of the time of initiation of large Kamchatkan eruptions and earthquakes*

Probable dates of eruptions with $V \geqslant 0.5 \text{ km}^3$ ($p \approx 0.6$)	Probable dates of earthquakes of $M \geqslant 7.5$, $H = 0$–100 km ($p \approx 0.6$)
July 1974–July 1976	1978–1979
1982–1983	1990–1991
1993–1994	July 1966–July 1998

The magnitude p corresponds to the probability of the initiation of one or more events in each of the intervals listed.

240   *Shirokov*

earthquakes with a magnitude $M \geqslant 7.5$, shows that the principal events of seismic and volcanic activity recorded in the present century in Kamchatka were confined to narrow intervals of the phases (approximately 2 years) of the 18.6-year tidal cycle. The probability of such a distribution of events being fortuitous is negligibly small: $\approx 10^{-5}$–10^{-6}. This suggests that the events are closely bound up with the long lunar tidal cycle.

2. Within the 18.6-year period, one can identify two active phases both for earthquakes and for eruptions, the eruption phases lagging 3 years behind the earthquake phases and neither overlapping the other. This indicates a possible connection between seismic and volcanic processes.

3. Like all the previous large-scale eruptions of the Kamchatka volcanoes, the Tolbachik eruption of 1975–76 occurred as a result of the influence on the volcanic process of the 19-year component in the tide-forming force of the moon.

4. On the basis of the patterns identified, predictions have been made for eruptions of $V \geqslant 0.5$ km³ and earthquakes of $M \geqslant 7.5$ over the next 25 years. Disturbed periods comprise 22% of the time.

References

Abdurakhmanov, A. I., Firstov, P. P. & Shirokov, V. A. 1974. *Vozmozhnaya svyaz' vulkanicheskikh izverzheniy s tsiklichnost'yu solnechnoy aktivnosti. XV General'naya assambleya MGGS, tezisy. Simpozium: Vulkanizm i zemletryaseniya verkhney mantii. (A possible link between volcanic eruptions and cycles of solar activity. 15th General Assembly of MGGS, theses. Symposium on volcanism and earthquakes in the upper mantle.)* M., pp. 3–4.
Abdurakhmanov, A. I., Firstov, P. P. & Shirokov, V. A. 1976. Vozmozhnaya svyaz' vulkanicheskikh izverzheniy s odinnadtsatiletney tsiklichnost'yu solnechnoy aktivnosti. (A possible link between volcanic eruptions and the eleven-year cycle of solar activity.) *Byul. vulkanol. stantsiy*, No. 52, 3–10.
Ben'of, G. 1961. *Nakopleniye i vysvobozhdeniye deformatsiy po nablyudeniyam sil'nykh zemletryaseniy. Slabyye zemletryaseniya. (The build-up and release of deformations according to observations of severe earthquakes. Small earthquakes.)* M., IL, pp. 199–211.
Doodson, A. T. 1921. The harmonic development of the tide-generating potential. *Proc. Royal Society.* S.A.T., **100**, NA-704.
Dubik, Yu. M. & Menyaylov, I. A. 1969. Novyy etap eruptivnoy deyatel'nosti vulkana Bezymyannogo. (A new stage in the eruptive activity of Bezymyannyy volcano.) In: *Vulkany i izverzheniya. (Volcanoes and eruptions.)* M., Nauka, pp. 38–77.
Dubik, Yu. M. & Menyaylov, I. A. 1971. Gazogidrotermal'naya deyatel'nost' kal'dery Ksudach. (Gaseous and hydrothermal activity of the Ksudach caldera.) *Byul. vulkanol. stantsiy*, No. 47, 40–3.
Fedotov, S. A. & Bagdasarova, A. M. 1974. Seysmichnost' Kamchatki i Komandorskikh ostrovov v 1897–1961 gg. po dannym instrumental'nykh nablyudeniy. (The seismicity of Kamchatka and the Komandor Islands 1897–1961 from instrument data.) In: *Seysmichnost' i seysmicheskiy prognoz, svoystva verkhney mantii i ikh svyaz' s vulkanizmom na Kamchatke. (Seismicity and seismic prediction, the properties of the upper mantle and their connection with volcanism in Kamchatka.)* Novosibirsk, Nauka, pp. 7–34.

Fedotov, S. A., Khrenov, A. P. & Chirkov, A. M. 1976a. Bol'shoye treshchinnoye Tolbachinskoye izverzheniye 1975 g., Kamchatka. (The Great Tolbachik Fissure Eruption of 1975 in Kamchatka.) *Dokl. AN SSSR*, **228**(5), 1193–6.

Fedotov, S. A., Sobolev, G. A. *et al.* 1976b. Dolgosrochnyy i probnyy kratkosrochnyy prognoz sil'nykh kamchatskikh zemletryaseniy. (Long-term and experimental short-term prediction of severe Kamchatka earthquakes.) In: *Poiski predvestnikov zemletryaseniy. (Searching for precursors to earthquakes.)* Tashkent, Izd-vo Filiala AN, pp. 49–61.

Fedotov, S. A. *et al.* 1977. Long- and short-term earthquake prediction in Kamchatka. *Tectonophysics*, **37**, 305–21.

Gorshkov, G. S. & Bogoyavlenskaya, G. Ye. 1965. *Vulkan Bezymyannyy i osobennosti yego poslednego izverzheniya (1955–1963 gg.). (Bezymyannyy volcano and features of its latest eruption (1955–1963).)* M., Nauka, 171 pp.

Gorshkov, G. S. & Dubik, Yu. M. 1969. Napravlennyy vzryv na vulkane Shiveluch. (A directed explosion on Shiveluch volcano.) In: *Vulkany i izverzheniya. (Volcanoes and eruptions.)* M., Nauka, pp. 3–37.

Gutenberg, B. & Richter, C. F. 1954. *Seismicity of the Earth and associated phenomena.* Princeton Univ. Press, 306 pp.

Hamilton, W. L. 1973. Tidal cycles of volcanic eruptions: fortnightly to 19 yearly periods. *J. Geophys. Res.*, **78**(17), 3363–75.

Kirsanov, I. T. 1968. Izverzheniya Klyuchevskogo vulkana v 1966 g. s proryvom pobochnykh kraterov imeni B. I. Piypa. (The eruption of the Klyuchi volcano in 1966 and the breakthrough of the Piyp subordinate vents.) *Byul. vulkanol. stantsiy*, No. 44, 11–29.

Lamakin, V. V. 1966. Periodichnost' baykal'skikh zemletryaseniy. (The periodicity of the Baykal earthquakes.) *Dokl. AN SSSR*, **170**(2), 410–13.

Mank, F. J. & Johnson, M. I. S. 1973. On the triggering of volcanic eruptions by earth tides. *J. Geophys. Res.*, **78**(17), 3356–62.

Menyaylov, A. A. 1947. *Dinamika i mekhanizm izverzheniy Klyuchevskogo vulkana v 1937–1938 gg. (The dynamics and eruptive mechanism of the Klyuchi volcano in 1937–1938.)* Trudy Labor. vulkanol. i Kamchatskoy vulkanol. stantsii, Issue 4, pp. 3–91.

Naboko, S. I. 1947. *Izverzheniye Bilyukaya, pobochnogo kratera Klyuchevskogo vulkana v 1938 g. (The eruption of Bilyukay, a subordinate vent of the Klyuchi volcano, in 1938.)* Trudy Labor. vulkanol. i Kamchatskoy vulkanol. stantsii, Issue 5, pp. 3–134.

Piyp, B. I. 1941. O sile izverzheniya vulkana Ksudach v marte 1907 g. (The force of the eruption of Ksudach volcano in March 1907.) *Byul. vulkanol. stantsii*, No. 10, 23–9.

Piyp, B. I. 1956. *Klyuchevskaya sopka i yeye izverzheniya v 1944–1945 gg. i v proshlom. (Klyuchevskaya Sopka and its eruptions of 1944–1945 and the past.)* Trudy Labor. vulkanol., Issue 11, 309 pp.

Piyp, B. I. & Markhinin, Ye. K. 1965. Gigantskoye izverzheniye vulkana Shiveluch 12 noyabrya 1964 g. (The gigantic eruption of Shiveluch volcano on 12 November 1964.) *Byul. vulkanol. stantsiy*, No. 39, 28–34.

Shirokov, V. A. 1973. Kosmos i vulkany. (The cosmos and volcanoes.) In: *Chelovek i stikhiya. (Man and the elements.)* L., Gidrometeoizdat, pp. 26–8.

Shirokov, V. A. 1974. Vliyaniye kosmicheskikh faktorov na geodinamicheskuyu obstanovku i yeye dolgosrochnyy prognoz dlya severo-zapadnoy chasti Tikhookeanskoy tektonicheskoy zony. (The influence of cosmic factors on a geodynamic situation, and its long-term prediction for the northwestern part of the Pacific Ocean tectonic zone.) In: *Geodinamika vulkanizma i gidrotermal'nogo protsessa. Kratkiye tezisy k IV Vses. vulkanologicheskomu soveshchaniyu. (The geodynamics of volcanism and the hydrothermal process. Short theses for the IVth All-Union Volcanological Conference.)* Petropavlovsk-Kamchatskiy, 49 pp.

Tokarev, P. I. 1967. Gigantskoye izverzheniye vulkana Shiveluch 12 noyabrya 1964 i yego predvestniki. (The gigantic eruption of Shiveluch volcano on 12 November 1964, and phenomena preceding it.) *Izv. AN SSSR, Fizika Zemli*, No. 9, 11–22.

Tokarev, P. I. 1971. O dolgosrochnom prognoze izverzheniy Avachinskogo vulkana. (Long-term prediction of eruptions of Avachinskiy volcano.) *Byul. vulkanol. stantsiy*, No. 47, 33–6.

Tokarev, P. I., Firstov, P. P. & Lemzikov, V. K. 1969. Seysmologicheskiye issledovaniya na vulkane Karymskom v 1966 godu. (Seismological studies of Karymskiy volcano in 1966.) *Byul. vulkanol. stantsiy*, No. 45, 21–31.

Vlodavets, V. I. & Piyp, B. I. 1957. Katalog deystvuyushchikh vulkanov Kamchatki. (A catalogue of the active volcanoes of Kamchatka.) *Byul. vulkanol. stantsiy*, No. 25, 5–95.

The focal mechanism and dynamic parameters of volcanic earthquakes preceding the Great Tolbachik Fissure Eruption of 1975

V. M. Zobin

Introduction. The Great Tolbachik Fissure Eruption (GTFE) began on 6 July 1975 and continued until the end of December 1976. The eruption was preceded by an intensive swarm of volcanic earthquakes, which began on 27 June 1975. A general description of the eruption and the seismic activity connected with it is given by Fedotov *et al.* (1976). The course of seismic preparation was characteristic of a swarm of volcanic earthquakes preceding eruptions of subsidiary vents (Gorel'chik & Zobin, 1971). Evidently, in terms of the physics of its flow processes, the seismic preparation of fissure eruptions is close to that of subsidiary vents. During the eruption, seven cinder cones appeared along a newly formed eruptive fissure up to 3 km long.

This paper discusses the focal mechanism and dynamic parameters of the volcanic earthquakes that preceded the eruption of Cone I; the characteristics of the foci of 20 volcanic earthquakes with $K_{S1.2}^{\phi 68} \geqslant 9.5$ are analysed. A catalogue of the earthquakes examined is given in Table 1. The epicentres of the earthquakes are shown in Fig. 1.

According to data on the distribution of the epicentres of earthquakes with $K_{S1.2}^{\phi 68} \geqslant 9.5$, two stages can be identified in the development of the swarm, the turning point being the occurrence of two earthquakes with magnitude $M = 5.0$. The epicentres of the first stage are positioned fairly randomly, but in the second stage they are concentrated within the narrow linear zone of tectonic disruptions that are associated with areal volcanism. It is interesting to note that Cone I grew on the epicentre of a severe earthquake with $M = 5.0$.

The focus mechanism of volcanic earthquakes. Research by the author (Zobin, 1970 and other papers) has shown that Vvedenskaya's method

243

Table 1. *Data on focus mechanism of earthquakes preceding the Great Tolbachik Fissure Eruption*

Date, 1975	Time at focus	Coordinates of epicentre		H (km)	K	Compression stress	
		Lat.	Long.			Azimuth(°)	e°
27 June	16–08	55° 41'	160° 13'	0–5	9.5	180	30
28 June	11–14	55° 41'	160° 14'	0–5	9.8	353	42
28 June	11–51	55° 39'	160° 08'	5–10	9.6	342	10
28 June	12–45	55° 41'	160° 13'	0–5	10.7		
28 June	13–50	55° 42'	160° 10'	15–30	10.3		
28 June	16–15	55° 42'	160° 13'	10–15	9.9		
29 June	15–08	55° 38'	160° 14'	10–15	10.8	330	36
29 June	17–02	55° 43'	160° 12'	5	9.6	186	2
29 June	22–28	55° 42'	160° 17'	5–10	10.6	130	8
30 June	14–31	55° 42'	160° 13'	0–5	10.2	350	24
1 July	04–07	55° 42'	160° 15'	0	9.6	175	32
2 July	01–49	55° 40'	160° 10'	0	9.8	5	6
2 July	07–10	55° 39'	160° 16'	10	11.0	187	20
2 July	07–34	55° 41'	160° 15'	10–20	11.5	166	4
2 July	07–43	55° 38'	160° 15'	0–5	10.6	162	33
2 July	13–00	55° 41'	160° 14'	5–10	9.5	214	32
3 July	21–52	55° 38'	160° 14'	0–5	10.5	303	35
4 July	09–53	55° 43'	160° 16'	0	10.1	130	4
4 July	21–36	55° 40'	160° 14'	0	9.5	100	3
6 July	23–13	55° 41'	160° 16'	0	9.5	300	60

Table 1. (cont.)

Date 1975	Time at focus	Tension stress Azimuth(°)	e°	Intermediate stress Azimuth(°)	e°	Rupture plane I Azimuth(°)	e°	Components of movement along dip	Components of movement along strike
27 June	16–08	84	30	310	60	40	60	0.0	−1.0
28 June	11–14	216	42	106	24	14	24	0.0	−1.0
28 June	11–51	80	30	228	60	32	80	+0.44	−0.90
28 June	12–45								
28 June	13–50								
28 June	16–15	96	36	210	30	32	90	0.87	−0.50
29 June	15–08	95	42	276	48	58	62	+0.53	−0.85
29 June	17–02	40	8	268	80	86	90	0.17	+0.98
29 June	22–28	90	40	224	40	36	80	+0.77	−0.54
30 June	14–31	70	22	316	50	30	50	−0.21	−0.98
1 July	04–07	270	36	98	54	52	60	+0.40	−0.92
2 July	01–49	287	20	54	60	54	90	0.50	−0.87
2 July	07–10	76	16	280	76	32	76	+0.10	−0.99
2 July	07–34	72	45	254	45	35	58	+0.71	−0.71
2 July	07–43								
2 July	13–00	332	32	94	40	2	40	0.0	+1.0
3 July	21–52	207	6	108	55	78	70	−0.50	+0.87
4 July	09–53	32	50	224	40	4	54	+0.62	−0.79
4 July	21–36	190	18	360	70	50	75	+0.21	+0.98
6 July	23–13	36	8	133	30	94	46	−0.67	+0.74

Table 1. (cont.)

Date 1975	Time at focus	Rupture plane II				Precision Class
		Azimuth(°)	e°	Components of movement along dip	Components of movement along strike	
27 June	16-08	130	90	0.50	+0.87	B
28 June	11-14	106	90	0.91	+0.40	B
28 June	11-51	116	64	+0.24	+0.97	B
28 June	12-45					
28 June	13-50					
28 June	16-15	120	30	0.0	+1.0	A
29 June	15-08	132	64	+0.53	+0.85	A
29 June	17-02	176	80	0.0	-1.0	B
29 June	22-28	108	42	+0.26	+0.97	A
30 June	14-31	126	80	-0.54	+0.77	A
1 July	04-07	132	70	+0.56	+0.83	A
2 July	01-49	145	60	0.0	+1.0	B
2 July	07-10	120	84	+0.24	+0.97	A
2 July	07-34	106	60	+0.57	+0.82	A
2 July	07-43					
2 July	13-00	92	90	0.77	-0.54	B
3 July	21-52	158	60	-0.37	-0.93	B
4 July	09-53	70	60	+0.70	-0.72	A
4 July	21-36	146	80	+0.24	-0.97	A
6 July	23-13	156	60	-0.74	-0.67	A

(1969), which is based on a dislocation model of the focus, may be used to study the focus mechanism of volcanic earthquakes. Nodal lines were drawn mainly from the first arrivals of P waves, taken from seismograms of the Kamchatka regional network, and using data on the signs of the first arrivals of S waves. Fig. 2 gives examples of nodal lines for the foci of volcanic earthquakes (stereographic projection, upper hemisphere).

Table 1 summarises determinations of the focus mechanism for seventeen earthquakes. We were not able to construct nodal lines for the earthquakes.

Fig. 1. The Great Tolbachik Fissure Eruption 1975. Stages in the development of the epicentre field for volcanic earthquakes of energy class 9.5 and higher.
1, Cone I (active); 2, epicentres of the volcanic earthquakes (the number alongside corresponds to the number in Table 1); 3, epicentre of the most powerful earthquake swarm; 4, zone of areal volcanism. On the left is a stereogram with nodal lines for the most powerful earthquake of the swarm.

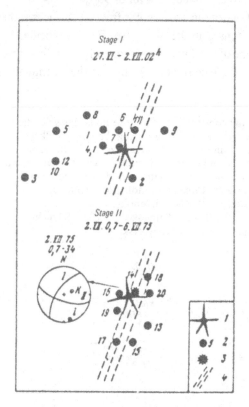

248   *Zobin*

In Table 1 the precision of the nodal lines is given: precision class A, ± 5–10°, precision class B, ± 10–15°.

All the earthquakes examined have a predominant horizontal displacement component in their foci, the majority also having an upthrust component. The axes of compression and tension stresses are close to horizontal. The orientation of the stress axes in the earthquake foci varied throughout the swarm.

Fig. 3 shows the azimuth variation of the axis of compressional stress in relation to time. One can see that from 27 June to 2 July the orientation of this stress axis varied within the narrow range of 150–180°. After the severe earthquakes of 2 July a reorientation of the stress system can be observed, which did not survive to the beginning of the eruption. Let us estimate the statistical significance of the hypothesis that the stress system was reorientated. Taking the criterion 2×2 from the tables for conjugate features (Bol'shev & Smirnov, 1968), we calculate that our hypothesis is statistically significant to a confidence level of 95%.

We should note that one of the two possible fracture surfaces at the focus of the most severe earthquake in the swarm (see Fig. 1) coincided in strike with the dislocation zone of areal volcanism. Possibly, therefore, this nodal surface corresponds to the true fracture surface at the earthquake focus.

Fig. 2. Samples of the construction of P wave nodal lines for the volcanic earthquakes in the swarm preceding the Great Tolbachik Fissure Eruption 1975 and P wave Fourier spectra for the same earthquakes. On the stereograms black circles mark the stations which recorded compression waves, white circles mark dilatation waves. I–II, potential rupture surfaces. Circles with dot, principal axes.
Abbreviations for stations: KRN = Kronoki;
APKh = Apakhonchich; BRG = Beregovo; KLCh = Klyuchi;
KRB = Krutoberegovo; KZR = Kozyrevsk.

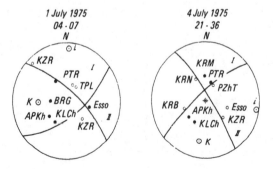

Fig. 2 (*cont.*)

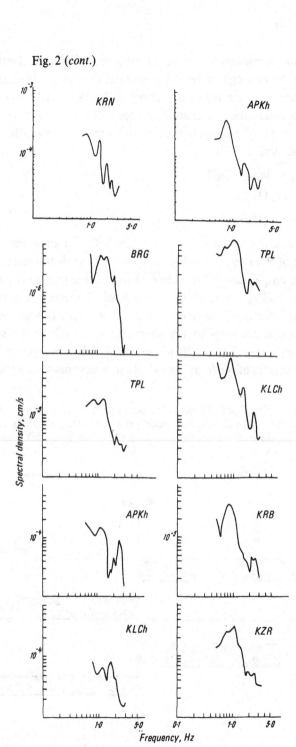

250   Zobin

The dynamic parameters of the foci of volcanic earthquakes. Twenty-four
Fourier spectra of longitudinal waves were constructed for 14 earthquakes
from recordings of the regional network (VEGIK, SVK). This made it
possible to determine the seismic moment M_0, the radius of circular
dislocation r_0, and the values of the upthrust stresses $\Delta\sigma$ using the methods
of Hanks & Wyss (1972).

$$M_0 = 4\pi\rho V_p S_0 d/R(P), \tag{1}$$

$$r_0 = 0.35 V_P/f_0, \tag{2}$$

$$\Delta\sigma = \tfrac{7}{16} M_0/r_0^3 \tag{3}$$

In these calculations density $\rho = 2.7$ g/cm^3, $V_p = 4$ km/s, the radiation
orientation of P waves $R(P) = 0.4$, and d is the hypocentral distance (km).
The double amplitude of displacements near a free surface is taken into
account. S_0 and f_0 are respectively the level of spectral density of the
long-period branch of the spectrum and the frequency attending the
transition from the long- to the short-period branch of the spectrum
(angular frequency). The results of the calculations are given in Table 2.

For the majority of earthquakes, calculations were made using data from

Fig. 3. Variations in time of the orientation of the compression axes
and the magnitudes of the faulting stresses in the foci of the volcanic
earthquakes of the swarm preceding the Great Tolbachik Fissure
Eruption 1975.

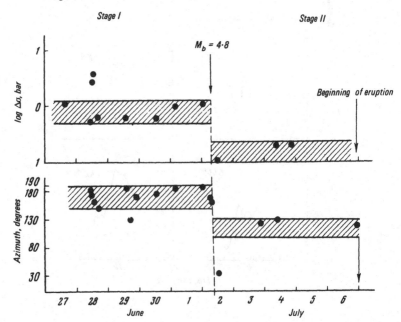

one or two stations. Dynamic parameters were calculated from data of five stations for only two earthquakes (see Fig. 2). This enabled us to calculate standard deviations from the magnitudes examined: $\sigma(M_0) = 0.6$–0.8×10^{21} dyne cm (26–37%). $\sigma(f_0) = 0.02$–0.08 Hz (2%).

One can assume that on the whole the error in calculating the dynamic parameters from one station does not exceed 2σ.

The majority of spectra were constructed using recordings made at Klyuchi (epicentral distance 75–80 km) and Kronoki (epicentral distance 110 km) stations. The results were compared with data for tectonic earthquakes both in the literature (Thatcher & Hanks, 1973; Riznichenko, 1976) and from four local Kamchatka earthquakes chosen for comparison. The epicentres of the tectonic earthquakes and the P-waves spectra are shown in Fig. 4. The main data on these earthquakes are given in Table 3.

Seismic moment M_0 and energy class $K^{\phi\,68}_{S1.2}$. Fig. 5 gives determinations of the seismic moment M_0 according to data from Klyuchi and Kronoki stations as a function of $K^{\phi\,68}_{S1.2}$. The same values for four tectonic earthquakes and an averaged curve of the relation $M_0 = f(K)$ taken from Yu. V. Riznichenko's work (1976) are also given here. The limits of the 70% confidence level of the given curve are indicated. One can see that the values M_0 of volcanic earthquakes as a function of energy class fall within the limits of the 70% confidence level, the majority of points lying above the averaged curve. The M_0 values for tectonic earthquakes lie below the averaged curve. From Fig. 5 one can see that for one and the same energy class $K^{\phi\,68}_{S1.2}$ the values of seismic moment M_0 are higher for volcanic earthquakes than for tectonic ones by almost half an order, which indicates the relative predominance of the long-period component in the radiation of the foci of volcanic earthquakes.

Focus size. The dimensions of foci were determined on the assumption that the focus can be modelled by a circular dislocation of radius r_0, which is perfectly admissible for weak earthquakes (Riznichenko, 1976). Calculations of r_0 were made from formula (2) and are given in Table 2.

In the calculations the wave velocity V_p in the focal zone was taken to be equal to 4.0 km/s. Then $r = 1.5/f_0$.

Since, broadly speaking, an evaluation of the velocity of longitudinal waves in the focal zone can vary with the depth of the focus, we analysed the f_0 values direct, and not the r_0 values calculated from them.

Table 2. *Dynamic characteristics of foci of the volcanic earthquakes preceding the Great Tolbachik Fissure Eruption from P-wave spectra (1975)*

Date	Time of focus	$K_{S_{1,2}}^{\Phi_{obs}}$	Seismic station	f_0 (Hz)	S_0 (cm/s)	$M_0 \times 10^{21}$ (dyne/cm)	$\bar{M}_0 \times 21^{21}$ (dyne/cm)	r_0 (km)	$\Delta\sigma$ (bar)	$\Delta\bar\sigma$ (bar)
27 June	16–08	9.5	Krn	1.8	5.0×10^{-5}	1.45	1.45	0.8	1.1	1.1
28 June	11–14	9.8	Krn	1.6	4.6×10^{-5}	1.33	1.3	0.9	0.71	0.5
28 June	11–51	9.6	Klch	1.1	6.0×10^{-5}	1.26		1.4	0.22	
28 June	12–45	10.7	Krn	1.6	1.8×10^{-4}	5.2	5.2	0.9	2.8	2.8
28 June	13–50	10.3	Krn	1.5	3.0×10^{-4}	8.7	8.7	1.0	3.8	3.8
28 June	16–15	9.9	Klch	1.2	1.2×10^{-4}	2.5	2.6	1.3	0.5	0.6
29 June	15–08	10.8	Krn	1.3	9.0×10^{-5}	2.6		1.2	0.76	
29 June	17–02	9.6	Klch	1.0	2.3×10^{-4}	4.8	4.8	1.5	0.62	0.6
29 June	22–28	10.6	Klch	0.8	4.0×10^{-4}	8.4	8.4	1.9	0.55	0.6
30 June	14–31	10.2	Klch	1.8	1.0×10^{-5}	1.3		0.8	1.0	
1 July	04–07	9.6	Krn	1.6	2.1×10^{-4}	4.0	1.6	0.9	0.44	0.7
			Tlb	1.4	1.6×10^{-5}	1.4		1.1	0.5	
			Brg	1.6	3.0×10^{-6}	0.3		0.9	0.8	
			Apkh	1.4	1.2×10^{-4}	1.5		1.1	0.5	

Date	Time at focus	$K^{\Phi_{68}}_{S1.2}$		Seismic station								
2 July	01–49	9.8		Krn	1.7	5.8×10^{-5}	1.7	1.7	0.9	1.1	1.1
2 July	07–10	11.0		Ptr	0.36	2.2×10^{-3}	200	200	4.2	1.2	1.2
2 July	07–34	11.5		Ptr	0.36	2.4×10^{-3}	220	220	4.2	1.3	1.3
2 July	07–43	10.6									
2 July	13–00	9.5		Klch	0.6	2.1×10^{-4}	4.4	4.4	2.7	0.1	0.1
3 July	21–52	10.5									
4 July	09–53	10.1		Klch	1.0	6.6×10^{-5}	1.4	1.4	1.6	0.15	0.2
				Klch	1.0	5.4×10^{-5}	1.1		1.5	0.14	
4 July	21–36	9.5		Krb	1.1	2.0×10^{-5}	0.8		1.4	0.13	
				Kzr	1.2	2.2×10^{-4}	4.5	2.0	1.3	0.10	0.2
				Apkh	1.0	2.6×10^{-4}	3.1		1.5	0.4	
				Tlb	1.3	6.0×10^{-6}	0.5		1.2	0.15	
6 July	23–13	9.5									

Table 3. *Dynamic parameters for the foci of weak tectonic earthquakes (Kamchatka) in 1975*

Date	Time at focus	$K^{\Phi_{68}}_{S1.2}$	Depth of hypocentre (km)	Seismic station	Epicentric distance (km)	S_0, (cm/s)	f_0 (Hz)	M_0 (dyne/cm)	r_0 (km)	$\Delta\sigma$ (bar)
10 Jan.	17–05	9.0	10–20	Klyuchi	102	6.0×10^{-5}	3.8	1.7×10^{20}	0.5	1.1
14 March	03–27	9.4	10	Kronoki	185	1.1×10^{-5}	2.3	5.5×10^{20}	0.65	0.9
15 May	09–29	9.6	30	Kronoki	80	2.0×10^{-5}	2.6	4.3×10^{20}	0.58	1.0
1 Jan.	23–31	9.1	10–20	Klyuchi	105	$3.5–10^{-6}$	3.6	1×10^{20}	0.42	0.6

Fig. 4. Epicentres of tectonic earthquakes and P-wave spectra of these earthquakes.
1, seismic stations; 2, epicentres of earthquakes.

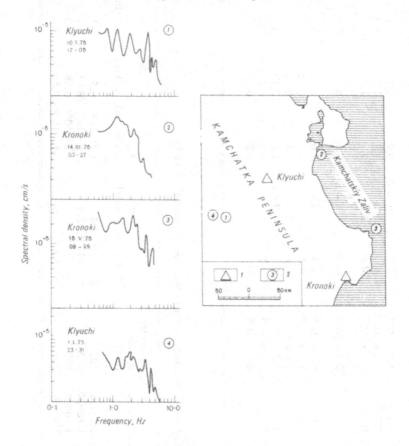

Fig. 5. Dependence of seismic moment M_0 on energy class K.
1, data for volcanic earthquakes; 2, data for tectonic earthquakes.

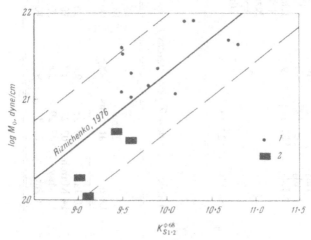

Fig. 6 gives the values of f_0 according to data from Klyuchi station as a dependence of the energy class $K^{\phi 68}_{S1.2}$. The values f_0 for tectonic earthquakes are also given here. Obviously our conclusion that the predominating P periods have greater values for volcanic earthquakes than for tectonic earthquakes of the same energy class is confirmed.

Magnitudes of fault stresses $\Delta\sigma$ were calculated from formula (3) for a focus in the form of a circular dislocation with radius r_0. According to Thatcher & Hanks (1973), the normal values $\Delta\sigma$ for tectonic earthquakes lie between 1.0–10 bar. Most of our calculations of $\Delta\sigma$ for volcanic earthquakes lie between 0.1–1.0 bar, i.e. in the zone of small fault stresses. It is interesting to note that the calculations of $\Delta\sigma$ for the tectonic earthquakes of Kamchatka also lie in this zone.

Fig. 3 shows the changes in the value of $\Delta\sigma$ for volcanic earthquakes in the course of the swarm. It is curious that at the same time as the elastic stresses in the volcanic earthquake foci were being reorientated, the values of $\Delta\sigma$ fell sharply. If between 27 June and 2 July the values of $\Delta\sigma$ were oscillating between 0.5–3.8 bar, then after 2 July they equal 0.1–0.2 bar.

Applying the method used to evaluate the hypothesis concerning the reorientation of the principal stress axes, we assessed the hypothesis of a decrease in value of $\Delta\sigma$ before the beginning of the eruption of the New Tolbachik Volcanoes and found it significant at a confidence level of 95%.

Fig. 6. Dependence of angular frequency f_0 of longitudinal waves on the energy class K.
1, data from tectonic earthquakes; 2, 3, data from volcanic earthquakes (2, before the strong earthquake of 2 July 1975, 3, after it).

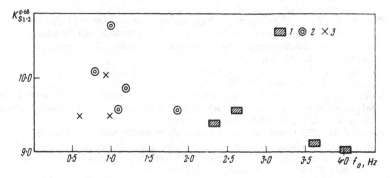

256   *Zobin*

Conclusions

1. The foci of the volcanic earthquakes that preceded the GTFE had a greater seismic moment M_0 and a lower frequency of radiation than those of tectonic earthquakes of the region in the same energy class and with the same depth of focus.

2. During the development of the volcanic earthquake swarm preceding the eruption, the following phenomena were observed immediately before the eruption:

(*a*) a reorientation of the stress system in the foci of the volcanic earthquakes;

(*b*) an increase in the predominating periods of P waves, and as a result smaller fault stresses. Physical parameters in the volcanic earthquake foci started to vary immediately after the most severe earthquakes of the swarm.

3. The rupture plane in the focus of the severest earthquake in the precursor swarm had a strike close to the orientation of the eruptive fissure.

References

Bol'shev, L. N. & Smirnov, N. V. 1968. *Tablitsy matematicheskoy statistiki.* (*Mathematical statistics tables.*) M., Izd-vo Vychislitel'nogo tsentra AN SSSR, 352 pp.

Fedotov, S. A., Gorel'chik, V. I. & Stepanov, V. V. 1976. Seysmologicheskiye dannyye o magmaticheskikh ochagakh, mekhanizme i razvitii bazal'tovogo treshchinnogo Tolbachinskogo izverzheniya v 1975 godu na Kamchatke. (Seismological data on the magmatic foci, mechanism and development of the Tolbachik basalt fissure eruption of 1975 in Kamchatka.) *Doklady AN SSSR* **228**(6), 1407–10.

Gorel'chik, V. I. & Zobin, V. M. 1971. O kharaktere razvitiya royev zemletryaseniy v oblastyakh aktivnogo vulkanizma na Kamchatke. (On the nature of earthquake swarms in areas of active volcanism in Kamchatka.) In: *Vulkanizm i glubiny Zemli.* (*Volcanism and the Earth's depths.*) M., Nauka, pp. 118–19.

Hanks, T. S. & Wyss, M. 1972. The use of body-wave spectra in the determination of seismic source parameters. *Bull. Seismol. Soc. Am.,* **62**, 561–89.

Riznichenko, Yu. V. 1976. Razmery ochaga korovogo zemletryaseniya i seysmicheskiy moment. (The dimensions of the focus of a crustal earthquake, and the seismic moment.) In: *Issledovaniya po fizike zemletryaseniy.* (*Studies of the physics of earthquakes.*) M., Nauka, pp. 9–26.

Thatcher, W. & Hanks, T. C. 1973. Source parameters of Southern California earthquakes. *J. Geophys. Res.,* **78**(35), 8547–76.

Vvedenskaya, A. V. 1969. *Issledovaniya napryazheniy i razryvov v ochagakh zemletryaseniy pri pomoshchi teorii dislokatsiy.* (*Studying stresses and ruptures in earthquake foci using the theory of dislocations.*) M., Nauka, 136 pp.

Zobin, V. M. 1970. O mekhanizme vulkanicheskikh zemletryaseniy, svyazannykh s izverzheniyem vulkana Shiveluch v noyabre 1964 goda. (On the mechanism of volcanic earthquakes connected with the eruption of Shiveluch volcano in November 1964.) *Izvestiya AN SSSR,* Fizika Zemli, No. 3, 31–6.

Atmospheric shock waves accompanying the Great Tolbachik Fissure Eruption (September, 1975)

P. P. Firstov, V. V. Adushkin and A. V. Storcheus

Atmospheric waves occurring during catastrophic eruptions have long attracted the attention of scientists. The first solution to the problem of the propagation of disturbances in the atmosphere in the case of the Earth's actual profile was found by Pekeris (1939), on the basis of a study of atmospheric waves produced during the eruption of the volcano Krakatoa in 1883. The atmospheric wave during this eruption was so great that it was recorded by barographs of low sensitivity in meteorological stations throughout the world. The TNT equivalent of an explosion in the atmosphere capable of causing such an atmospheric wave would be 100–150 megatons (Press & Harkrider, 1966).

The atmospheric waves from the catastrophic eruption of Bezymyannyy volcano on 30 March 1956 have been studied in a number of works (Gorshkov, 1957; Pasechnik, 1958; Murayama, 1969). The atmospheric wave produced by this eruption was not as great as that of Krakatoa; it travelled little more than once round the globe and its TNT equivalent is estimated at 8 megatons.

A number of scientists have devoted papers to the study of weak atmospheric waves accompanying volcanic eruptions (Stewart, 1959; Tokarev, 1964, 1967; Minakami et al., 1970; Tanaka, 1967; Tanaka et al., 1974). In them microbarographic phenomena are regarded as one of the parameters that characterise eruptions. The authors attempted to estimate the depths of occurrence of volcanic explosions (ejections) using kinematic parameters of seismic and atmospheric waves. Thus a depth of 180 m was obtained for the volcano Asama (Minakami et al., 1970) and 50 m for Akita – Komaga – Take (Tanaka et al., 1974). This indicates that, in the first approximation, individual ejections can be equated with ejection by explosion. In these publications there is no mention of the form of the pulse

257

of an atmospheric wave recorded in a nearby zone, despite the fact that this and its dynamic parameters provide information about the nature of the source. In their turn, the source and the forces operating during the ejections give an idea of the physics of the explosive process. It was with this aim in mind that atmospheric waves at the eruption of Cone II of the Great Tolbachik Fissure Eruption (GTFE) were recorded in September 1975.

Microbarographic phenomena were registered using a wide-band EDMB-VI microbarograph (Pasechnik & Fedoseyenko, 1958), linked to a GB-IV galvanometer with a natural frequency of $f_0 = 15$ Hz and damping of $D_0 = 3$.

The amplitude–frequency characteristics of the microbarograph were

Fig. 1. Sensitivity curve for microbarograph.

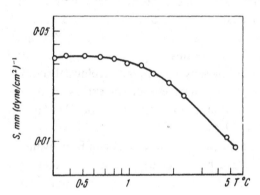

Fig. 2. Forms of atmospheric shock waves, constructed from mean statistical parameters.

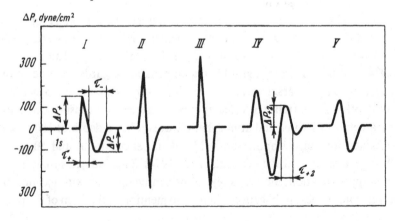

determined by a calibrating device consisting of a hermetic chamber and a supercharger system. The latter consisted of a pump with a volume of 15 cm^3 and a reducer giving a pressure oscillation frequency of 0.2–2 Hz. Fig. 1 shows the sensitivity curve of the microbarograph.

An OSB-1MP oscillograph was used as a recording system. The roll past speed was 150 mm/min. Observations were made at Vodopadnaya station, 9 km away from Cone II.

In the middle of September 1975 the eruption was of an intermittent nature (Fedotov *et al.*, 1977). On 10 September the most liquid lava of this stage of the eruption began to flow from the northern and southern boccas of Cone II. Explosive activity increased somewhat although there were periods of absolute quiet. On 14 September the explosive–effusive activity of Cone II ceased.

From the form of the recording one can provisionally identify five types of atmospheric waves generated during the GTFE (Fig. 2).

Type I is the classic form of weak shock wave at a considerable distance from the source (Sadovskiy, 1945). It has a sharp front of positive pulse and a more prolonged negative pulse of smaller amplitude.

Type II is a form of recording with a sharp forward front of positive and negative pulses. The rear front of the negative pulse is close to the exponent.

Type III differs from the second type in having a sharp rear front of the negative pulse.

Type IV has oscillations consisting of clearly expressed pulses – two positive and one negative.

Type V has an almost quasi-sinusoidal form with blunt pulse fronts of both signs.

The following parameters of atmospheric waves were examined by analogy with shock waves produced by explosions of high explosive (HE) in the air (Sadovskiy, 1945): (1) the maximum excess pressure in the compression phase ΔP_+; (2) the duration of activity of the positive pulse τ_+, by which is meant the time interval between the forward front and the boundary where excess pressure is equal to zero; (3) the size of the pulse as calculated from the formula

$$I_+ = \int_0^{\tau_+} \Delta P(t)\, \mathrm{d}t;$$

and (4) parameters characterising the release phase ΔP_-, τ_-, I_-.

The qualitative division of atmospheric waves into types is supported by statistical average data (Table 1) obtained by averaging the pulse parameters recorded at Vodopadnaya station (distance = 9 km) on 14 September 1975 from 08.00 to 12.00. Only pulses which can be identified with certainty as being one or another were chosen. The distributions of excess pressure and the magnitudes of pulse are close to the log. normal, although the asymmetry and excess are not zero. The distribution of the period of pulse activity is close to normal.

The positive pulse of type III has the greatest excess pressure $\overline{\Delta P}_+ = 324$ dyne/cm$^2\pm0.16$ unit log., and the shortest period of activity $\bar{\tau}_+ = 0.39\pm0.09$ s. The fifth type of atmospheric wave is the weakest, with $\overline{\Delta P}_+ = 123$ dyne/cm$^2\pm0.11$ unit log., and a period $\bar{\tau}_+ = 0.57\pm0.09$ s.

The parameters $\alpha = \bar{\tau}_+/\bar{\tau}_-; \beta = \overline{\Delta P}_+/\overline{\Delta P}_-$ differ fundamentally for each type. The negative pulse in types II and IV is greater than the positive one: $\beta_{II} = 0.89; \beta_{IV} = 0.72$, and the pulses of types III and V are almost equal: $\beta_{III} = 1.03; \beta_V = 0.97$, whilst type I has a characteristically large positive pulse $\beta_I = 1.27$. The period of activity of the positive pulse is less than that of the negative pulse for all types. The smallest value is $\alpha_I = 0.54$, the greatest $\alpha_{III} = 0.74$.

Table 1. *Mean statistical parameters for atmospheric shock waves*

Type of wave	No.	log ΔP_+		τ_+		log I_+	
		m	σ	m	σ	m	σ
I	30	2.19	0.11	0.44	0.09	1.52	0.18
II	38	2.38	0.22	0.40	0.10	1.66	0.29
III	20	2.51	0.16	0.33	0.09	1.77	0.21
IV	24	2.24	0.17	0.59	0.12	1.62	0.35
V	45	2.09	0.11	0.57	0.09	1.56	0.16

Table 1. *(cont.)*

Type of wave	No.	log ΔP_-		τ_-		log I_-	
		m	σ	m	σ	m	σ
I	30	2.06	0.18	0.82	0.28	1.59	0.38
II	38	2.43	0.21	0.66	0.20	1.94	0.32
III	20	2.52	0.18	0.53	0.31	1.93	0.21
IV	24	2.38	0.11	0.90	0.15	2.02	0.15
V	45	2.12	0.12	0.80	0.18	1.68	0.19

The relationship between the period of activity of the positive pulse τ_+ and the size of the charge of solid HE (C) is expressed by the following formula: $\tau_+ = 1.4\ R^{\frac{1}{2}}C^{\frac{1}{2}}$, and that between the maximum excess pressure and the weight of the HE charge (Sadovskiy, 1945):

$$\Delta P = 0.84C^{\frac{1}{3}}/R + 2.7C^{\frac{2}{3}}/R^2 + 6.95C/R^3.$$

In our case ($P = 60$–600 dyne/cm²; $R = 9$ km) we can disregard the second and third terms in the latter formula since they are as small as $0 = 0(C^{\frac{2}{3}})$. Then by means of some transformations we can obtain $\tau_+ = 0.014\Delta P^{\frac{1}{2}}$, which gives grounds for searching for a parabolic relation $\tau_+ = \tau_+(\Delta P_+)$ for the experimental values. Fig. 3 shows the correlation

Fig. 3. Correlation fields for the various types of the dependencies $\tau_+ = \tau_+(\Delta P_+)$.

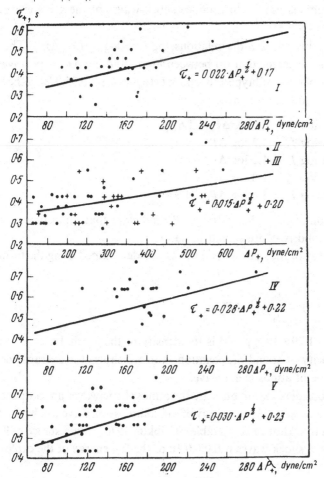

field $\tau_+ = \tau_+(\Delta P_+)$ for each type of atmospheric wave. Using the method of least squares we arrived at the following relations:

$$\text{I}\quad \tau_+ = 0.022\Delta P_+^{\frac{1}{2}} + 0.17;$$

$$\left.\begin{matrix}\text{II}\\\text{III}\end{matrix}\right\}\tau_+ = 0.015\Delta P_+^{\frac{1}{2}} + 0.20;$$

$$\text{IV}\quad \tau_+ = 0.028\Delta P_+^{\frac{1}{2}} + 0.22;$$

$$\text{V}\quad \tau_+ = 0.030\Delta P_+^{\frac{1}{2}} + 0.23.$$

From a comparison of experimental and calculated relations it follows that the graphs closest to a parabolic relation are those of types II and III. The excess of 0.2 s throughout over the calculated value for τ_+ may be explained by the fact that Sadovskiy's formulae (1945) were obtained using solid HE, whereas in the present case the HE may consist of explosive volcanic gases (Fenner, 1950; Gushchenko, 1965), of which a considerably larger volume is needed to produce shock-waves of the same energy as those from solid HE.

Of great interest are the relations $\log I_+ = \log I_+ (\log \Delta P_+)$. The coefficient of the correlation (ρ) between $\log I_+$ and $\log \Delta P_+$ is positive for all samples of every subtype and amounts to $\rho = 0.84$–0.95. Using the least squares method we obtained:

$$\text{I}\quad \log I_+ = 1.4 \log \Delta P_+ - 1.6;$$
$$\text{II}\quad \log I_+ = 1.2 \log \Delta P_+ - 1.3;$$
$$\text{III}\quad \log I_+ = 1.1 \log \Delta P_+ - 1.0;$$
$$\text{IV}\quad \log I_+ = 1.8 \log \Delta P_+ - 2.3;$$
$$\text{V}\quad \log I_+ = 1.6 \log \Delta P_+ - 1.8;$$

The relation $\log I_+ = \log I_+ (\log \Delta P_+)$ indicates that atmospheric wave types II–III and IV–V are similar in nature.

The energy of the atmospheric waves was determined using the formula (Gubkin, 1970):

$$E'(R) = \frac{4\pi R^2}{\rho_0 C_0} \int_\tau \Delta P^2 \, dt,$$

where $\rho_0 = 1.29 \times 10^{-3}$ g/cm^3 is the density of the air and $C_0 = 334$ m/s is the velocity of sound propagation. Table 2 shows the average energy of each type of atmospheric wave.

If one considers that in our case the atmospheric waves are propagated in semi-space, then the true energy of the compression wave will be double. According to Okhotsimskiy's tables (Gubkin, 1970), at a distance of 9 km the energy of shock waves is 0.08–0.10 of the full energy of the explosion

E_0 for $\Delta P = 10^2$–10^3 dyne/cm^2; not taking into account the energy lost in sound vibrations.

The TNT equivalent of the given energy falls within 300–1200 kg, the specific heat of the TNT being taken as 1000 kcal/kg (Lovlya *et al.*, 1976).

Let us examine the special features of each type of air wave. The relations $\tau_+ = \tau_+ (\Delta P_+)$, log I_+ = log I_+ (log ΔP_+), as well as the ratios $\Delta P_+/\Delta P_-$; τ_+/τ_- and the form of the recording show that the division into types is justified and that the mechanism producing each type is different.

In terms of the form of its recording, the first type of atmospheric wave is a weak shock wave with all the latter's characteristic features: a jump in pressure $\Delta P_+ > \Delta P$; $\tau_+ < \tau_-$; $E_+ \approx E_-$. The nature of the origin of shock waves of this type is problematical.

For atmospheric waves of both the second and the third type there is apparently a pressure jump, but the frequency characteristics of the apparatus and the inadequate roll-past speed prevent it from being recorded. A special feature of type II is $\Delta P_+ < \Delta P_-$; $I_+ < I_-$, i.e. the full

Table 2. *Mean energy of atmospheric waves of types I–V*

Energy	I	II	III	IV	V
$E' \times 10^{15}$ (erg)	0.6	1.7	2.5	1.1	0.6
$E' \times 10^{16}$ (erg)	1.3	3.4	5.0	2.2	1.1
G' (kg)	310	800	1200	520	260

Fig. 4. Form of atmospheric wave from an explosion to ejection at different depths of charge emplacement (Reed, 1970).

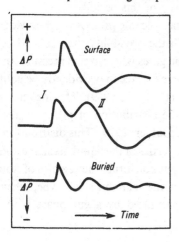

pulse is not equal to 0. At the same time the rear negative pulse is close to its exponent. It is possible that this type of shock wave occurs when a mixture of ash and gas escapes at supersonic speed. The exponent of the rear front of the negative pulse is explained by the sudden condensation of the erupted matter.

The third type of atmospheric wave is close to the shock wave in its dynamic features: $\Delta P_+ > \Delta P_-$; $\tau_+ < \tau_-$. This type of shock wave may occur as a result of the burning and detonation of explosive volcanic gases. There are, for instance, calculations (Fenner, 1950; Gushchenko, 1965), which show that it is in principle possible for chain reactions to take place in volcanic gases.

In the form of its recording, the fourth type is reminiscent of atmospheric waves accompanying explosions causing ejections. Fig. 4 shows atmospheric waves recorded during explosions with ejections and with the charge at various depths (Reed, 1970). Two positive pulses can be identified. The first is connected with the development of a cavern and the formation of a dome as a result of the ground being forced out towards a free surface, and the second with an escape of gases into the atmosphere; the velocity of the dispersal of the ground in the epicentral area being close to the rate of movement of the free surface (Dokuchayev *et al.*, 1963).

In this case the pressure being built up by the dome is related to the velocity (v) of the elevation of the free surface: $\Delta P_0 = v\rho_0$ s. If one considers that atmospheric waves dampen proportionally to distance, then $\Delta P(R) = r \times \Delta P_0/R$, where r is the radius of the dome. Taking $r = 5\text{--}10$ m, then for the maximum excess pressure of atmospheric waves of type IV $\Delta P_{\max} = 200$ dyne/cm^2 the velocity will be $v = 140\text{--}70$ m/s, which agrees entirely with data from a film survey of this period.

The important role of gases in the volcanic process is acknowledged by many volcanologists, together with the possibility that isolated volumes exist ('bubbles'), filled with volcanic gases where excess pressure can build up. The movement of lava near a free surface may evidently be controlled by the laws of two-phase mixing (Droznin, 1969). The movement of a two-phase flow is determined by the distribution of phases in the space occupied by the flow (Khaddard & Dakler, 1970). This distribution varies with the velocity of the flow, the properties of the liquid, its shape and other factors. These numerous distributions constitute the regimes of flow. One of the regimes of two-phase flow is the 'cannon shell' type, when areas form during movement that are saturated by a gas phase with excess pressure.

The fifth type of atmospheric wave has much in common with the fourth and can be regarded as an extreme case where there is an escape of gases only. It should be noted that variation in the depth of the ejection accounts for the variety of forms of the transitional type (from IV to V) of atmospheric waves.

Five types of atmospheric shock waves, then, have been identified among those recorded whilst Cone II of the GTFE was active. They differ in the form of their recordings and their mean statistical parameters. There is a distinct difference between the source of atmospheric waves of types II and III and that of types IV and V. Shock waves of types II and III can occur when there is a supersonic discharge of a mixture of ash and gas, and when explosive gases are detonated. Atmospheric waves of types IV and V are evidently connected with processes taking place deep down in the magma conduit. Information about atmospheric shock waves produced during eruptions is useful from the point of view of studying the physics of the explosive process. Since the causes of atmospheric shock waves are so various, fundamental studies of them are needed, using filming and seismic techniques.

References

Dokuchayev, M. M., Rodionov, V. N. & Romashov, A. N. 1963. *Vzryvy na vybros.* (*Ejection explosions.*) M., Nauka.

Droznin, V. A. 1969. O prirode vulkanicheskikh izverzheniy. (On the nature of volcanic eruptions.) In: *Trudy 1-oy gorodskoy konferentsii uchenykh i spetsialistov.* (*Transactions of the 1st city conference of scientists and specialists.*) Petropavlovsk-Kamchatskiy, pp. 4–6.

Fedotov, S. A., Chirkov, A. M., Andreyev, V. N. *et al.* 1977. Kratkoye opisaniye khoda treshchinnogo izverzheniya v 1965 godu. (A short description of the course of the Tolbachik fissure eruption in 1965.) *Byul. vulkanol. stantsiy,* No. 53, 3–15.

Fenner, N. 1950. The chemical kinetics of the Katmai eruption. *Am. J. of Sci.,* p. 248.

Gorshkov, G. S. 1957. Izverzheniye sopki Bezymyannoy (predvaritel'noye soobshcheniye). (The eruption of Bezymyannaya Sopka (preliminary description).) *Byul. vulkanol. stantsii,* No. 26, 69–70.

Gubkin, K. Ye. 1970. Rasprostraneniye vzryvnykh voln. (The propagation of explosion waves.) In: *Mekhanika v SSSR za 50 let.* (*50 years of mechanics in the USSR.*) M., vol. 2, pp. 115–16.

Gushchenko, I. I. 1965. *Peply severnoy Kamchatki i usloviya ikh obrazovaniya.* (*Ashes of northern Kamchatka and the conditions in which they were formed.*) M., Nauka, pp. 122–5.

Khaddard, N. D. & Dakler, A. D. 1970. Rezhimy techeniya gorizontal'nogo dvukhfaznogo potoka. (Flow regimes of a horizontal two-phase stream.) In: *Dostizheniya v oblasti teploobmena.* (*Achievements in the field of heat exchange.*) M., Mir, pp. 7–29.

Lovlya, S. A., Kaplan, B. L., Mayorov, V. V. *et al.* 1976. *Vzryvnoye delo.* (*Blasting.*) M., Nedra, 90 pp.

Minakami, T., Utibori, S., Hiraga, S. *et al.* 1970. Seismometrical studies of volcano Asama (Part I) – Seismic and volcanic activities of Asama during 1934–1969. *Bull. Earthqu. Res. Inst. Tokyo Univ.,* **48**(2), 235–301.

266   Firstov et al.

Murayama, N. 1969. Propagation of atmospheric pressure waves produced by the explosion of volcano Bezymianny of March 30, 1956 and transport of the volcanic ashes. *Quart. J. Seism.*, **33**(1), 1–11.

Pasechnik, I. P. 1958. Seysmicheskiye i vozdushnyye volny, voznikshiye pri izverzhenii vulkana Bezymyannyy 30 marta 1956 g. (Seismic and atmospheric waves during the eruption of Bezymyannyy volcano on 30 March 1956.) *Izv. AN SSSR*, Ser. geofiz., No. 9, 1121–6.

Pasechnik, I. P. & Fedoseyenko, N. Ye. 1958. Elektrodinamicheskiy mikrobarograf s gal'vanometricheskoy registratsiyey. (An electrodynamic microbarograph with galvanometric recording.) *Izv. AN SSSR*, Ser. geofiz., No. 1, 121–31.

Pekeris, F. 1939. The propagation of a pulse in the atmosphere. *Proc. Roy. Soc. Am.*, **171**, 131–49.

Press, F. & Harkrider, D. 1966. Air–sea waves from the explosion of Krakatoa. *Seismic*, **154**(3754), 1325–7.

Reed, F. W. 1970. *Air blast effects. Symp. on Engineering with nuclear explosives; January 14–16*. Las Vegas, Nevada, vol. 2, pp. 1070–91.

Sadovskiy, M. L. 1945. Opytnyye issledovaniya mekhanicheskogo deystviya udarnoy volny vzryva. (*Experimental studies of the mechanical action of a shock wave from an explosion*.) Trudy Seysmol. In-ta, No. 116, 114 pp.

Stewart, K. N. 1959. Air waves from a volcanic explosion. *Meteor. Mag.*, **88**, 1–3.

Tanaka, J. 1967. On the explosion-earthquake at the volcano Sakurazima. *Bull. Volcanol. Soc. Japan*, **12**(1), 4–26.

Tanaka, K., Kashara, M. & Hori, S. 1974. Research on Akita–Komage–Take (11) Focal depth of explosion earthquake. *Sci. Rep. Tokyo Univ.*, Ser. V, Geophys., **22**(1), 9–18.

Tokarev, P. I. 1964. Registratsiya vzryvov Klyuchevskogo vulkana v 1962 g. (Recording explosions of the Klyuchi volcano in 1962.) *Byul. vulkanol. stantsiy*, No. 37, 52–9.

Tokarev, P. I. 1967. Gigantskoye izverzheniye vulkana Shiveluch 12 noyabrya 1964 g. i yego predvestniki. (The gigantic explosion of Shiveluch volcano on 12 November 1964, and phenomena preceding it.) *Izv. AN SSSR*, Fizika Zemli, No. 9, 11–22.

Deformations of the Earth's surface in the vicinity of the New Tolbachik Volcanoes (1975-1976)

S. A. Fedotov, V. B. Enman, M. A. Magus'kin,
V. Ye. Levin, N. A. Zharinov and S. V. Enman

Geodetic work in the vicinity of the Great Tolbachik Fissure Eruption (GTFE) began on 13 July 1975, a week after Cone I became active. The programme of research included measuring horizontal and vertical movements. Horizontal displacements were examined using optical range-finders at various distances from the new volcanoes of 1975. Elevations and subsidences were measured by trigonometric and geometric levelling. These methods were used in combination with phototheodolite stereo-surveys to record the growth in height and volume of the new cones and the changes in their shape. To study the tilt of the Earth's surface, automatic tiltmeters installed in underground lava tubes were employed.

During the first two months studies were made within a radius of 4 km around the erupting cones, where the maximum deformations were occurring. Then it became possible to widen the area of research to 10 km (Fig. 1). Geodetic measurements in the vicinity of the Southern Break-through (Cone VIII) began in October 1975 and tiltmeter measurements in August 1976.

This paper presents the main results of the work at the Northern Breakthrough and tiltmeter data obtained in a lava cave at the Southern Breakthrough.

Vertical displacements in the area of the Northern Breakthrough during the eruption

On 28 and 29 July severe deformations of the north-west slope of Gora 1004 began (Fig. 2 (*b*)). In 12 hours (from 22.00 on 28 July to 10.00 on 29 July) the slope rose 30 m. Then a fissure formed between this rise and Gora 1004, from which the first lava began to flow at 10.30 hours.

Theodolite readings taken from 22 July to 11 August showed a general

267

Fig. 1. Sketch map of the location of geodetic points in the vicinity of the Northern Breakthrough.
1, optical range-finder lines; 2, levelling track; 3, tilt-measuring stations; 4, fissures formed during the eruption: *a*, eruptive fissures; *b*, visible at surface; *c*, probable position of deep-seated feeder dyke; 5, sources of lava flows; 6, cinder cones of 1975; 7, lava flows; 8, Holocene cinder cones; 9, Holocene tension fissure.

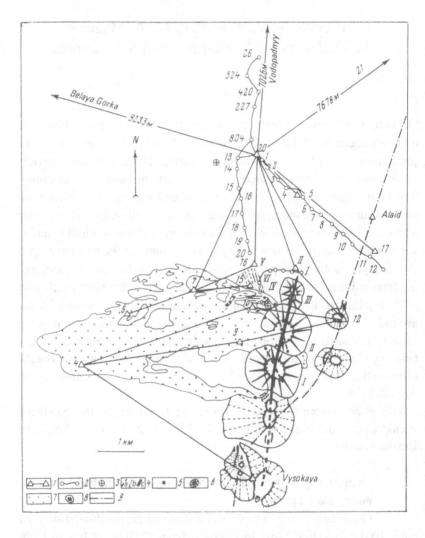

bulging of the region of the Northern breakthroughs from 22 to 29 July (see Fig. 2), the Point 12 region experiencing the greatest uplift (2.9 m) in relation to Point 4. The Gora Vysokaya and Gora Alaid points were raised by 1.4 and 1.8 m respectively. After 29 July, i.e. after the breakthrough of the southern bocca, the area began to subside, and continued to do so until 7 August. In 9 days the maximum subsidence of Point 12 was 4.5 m, during which time the Gora Vysokaya and Gora Alaid points subsided by 2.2 and 2.6 m respectively. After this the region underwent intensive elevation, probably as a result of preparation for the breakthrough of Cone II.

Fig. 2. (*a*) Alteration in height of points: Gora Vysokaya, Point 12, and Gora Alaid relative to Point 4 (for the period 22 July to 11 August 1975). (*b*) Uplift of a block on the northern slope of Gora 1004 before lava broke through the southern bocca.
1, Gora Vysokaya; 2, Point 12; 3, Gora Alaid, moment of formation: sh, of sharra; sb, of southern bocca; nb, of northern bocca. The solid line below the main figure marks the time during which Cones I and II were active.

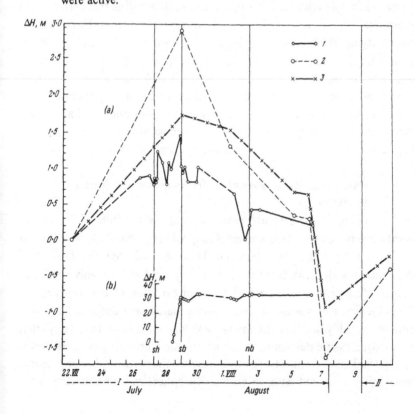

Trigonometric levelling was taken from Point 4 to Gora Vysokaya and from Point 12 to Gora Alaid and Gora Vysokaya. The graphs in Fig. 2 have been compiled from measurements, on the following assumptions. The level of 22 July (when recordings began) was provisionally taken as the starting point for measurements of changes in the height of all points. Readings were taken with second theodolites in one direction, sometimes every hour. In unfavourable weather conditions and with lines of up to 4 km, overmeasurements of the order of 0.5 m are possible. Nevertheless, the main trend of displacements can be taken to be accurate. The amplitude of subsidence of the points from 29 July to 7 August may be 20–25% less than that indicated, amounting to approximately 2.5, 2.0 and 1.5 m for Points 12, Gora Alaid and Gora Vysokaya respectively.

According to these data, from the beginning of the eruption on 6 July until the moment when lava appeared at the surface on 27–29 July, during the period when only a powerful ash–gas stream and pyroclastics were being ejected from the crater of Cone I, magma was accumulating at a comparatively shallow depth. The decrease in heights from 29 July until Cone I ceased activity on 7–9 August, shows that during this period when there were big outflows of lava from the southern and northern boccas, the cavity in which lava was accumulating from 6 to 29 July apparently emptied. It is difficult to determine the precise form, dimensions, position and depth of this cavity from the single line, Vysokaya – Point 12 – Alaid, along which relative heights were measured. One can only assume that it was that of a sill up to several kilometres long and not less than several metres thick (possibly 10–15 m). Its depth must have been less than that of the powerful release of gas from the magma, i.e. some few kilometres.

Data on the dimensions of feeder fissures of the Northern Breakthrough

From 26 to 29 July measurements were taken of horizontal displacements using a Geodimeter-8 range-finder from Point 4 on Points 9, 10 and Vysokaya (see Fig. 1). They showed that the horizontal deformations along all three lines were synphasal and roughly similar in magnitude (0.5–4 cm with a measurement precision of 0.5 cm) (Fig. 3). Thus the first new volcano (Cone I) was situated on a deep feeder fissure running roughly north-south and not less than 4 km long. In the days that followed, a fissure developed to the north where on 9 August Cone II was formed and on 17 August Cone III. At this time the horizontal displacements were being measured with an EOK-2000 optical range-finder across the

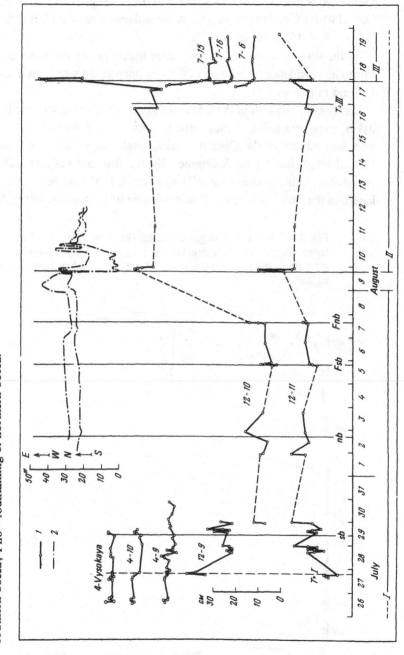

in line length; 2, tilts. Moments of formation of visible fissures: T-I – during the activity of Cone I; T-III – before the breakthrough of Cone III. Moments of formation of: sb – southern bocca; nb – northern bocca; Fsb – fountaining of southern bocca; Fnb – fountaining of northern bocca.

fissures on which the cones had appeared (lines 12–10 and 12–11; see Fig. 1), and with a Geodimeter-8 on the western flank of the deep feeder fissure (lines 7–6, 7–15, 7–16; see Fig. 3).

At the surface, above this main feeder fissure of the Northern Breakthrough, a sequence en échelon of fissures opened, on which grew Cones I, II and III (see Fig. 1).

Data on the breadth of these fissures, i.e. on the thickness of the feeder dykes, were obtained as Cones II and III formed. On 9 August 1975 in the very first minute of the Cone II breakthrough, the visible length of the breakthrough fissure, as determined by the distance between extreme eruptive ejections, was measured from Point 6. In the second minute the length of the fissure was 80 m; 20 minutes later it had reached 450 m. After

Fig. 4. (*a*) Increase in length of the breakthrough fissure of Cone II. (*b*) Increase in width of the breakthrough fissure of Cone III. (*c*) Magma pressure in dyke; 1, when $E = 4 \times 10^4$ kg/cm^2; 2, when $E = 1 \times 10^5$ kg/cm^2.

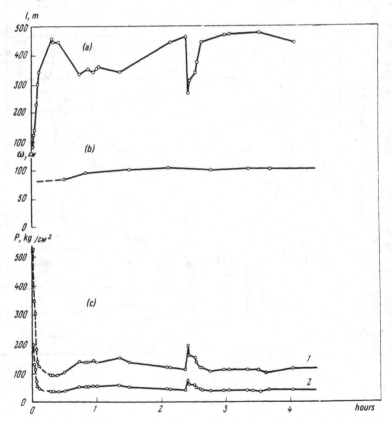

that, in the course of 2.5 hours, the length of the eruptive fissure varied, with a minimum of 270 m, and when it had reached its maximum – 480 m – three hours after the breakthrough, Cone II began to form (Fig. 4(*a*)). On 17 August, a few hours before the breakthrough of Cone III, measurements began to be made with the geodimeter of the lines 7–6, 7–15 and 7–16; they were being subjected to compression that had apparently begun ten hours before the breakthrough. The compression of lines 6–7 and 7–15 amounted naturally to 8 and 20 cm. The damping of horizontal deformations may be approximated by an empirical relation of the type $\Delta L = 4.1(3.5 - r)^2$, where ΔL is the displacement of the points and r their distance from the fissure. The displacement of a point on the rim of the fissure proves to amount to 52 cm, i.e. the width of the fissure slightly exceeds 1 m.

From the moment when the fissure was formed its breadth apparently changed analogously to changes in the lines 6–7 and 7–15. This made it possible to calculate variations in the width of the fissure (Fig. 4(*b*)). During the first minute of the eruption it was about 80 cm (its maximum breadth being 1 m). A fracture in the Earth's surface evidently occurred a few hours earlier and the fissure in the bedrock was much longer than that which could be seen on the surface of the pile of friable pyroclastics.

The dynamics of the injection of magma from the Northern Breakthrough along feeder fissures

If one regards the rapid deformations that occur within the first few minutes and hours as elastic, one can work out the excess pressure of the magma as it breaks through. Given a density for the upper units of lavas and pyroclastics of $\rho = 2$ g/cm³ and a velocity of longitudinal waves in the surface layer of $V_1 = 1400$ m/s, Young's modulus, E, is $2(1+\sigma)\rho V_S^2 = 4 \times 10^4$ kg/cm², where σ is Poisson's ratio and V_S is the velocity of the transverse waves.

The internal pressure of viscous liquids in vertical fissures is $P = \omega_0 E/4$ $(1 - \sigma^2) l$ (Zheltov, 1975), where ω_0 is the full breadth of the fissure, and l is half its length.

The excess pressure of the magma and gases in the feeder dykes proved to be modest, with a maximum of about 100 kg/cm² (Fig. 4 (*c*)).

When $E = 1 \times 10^5$ kg/cm², which is obtained when the velocity of longitudinal waves $V_1 = 2$–2.5 km/h for a thickness of 1 km (the same value for E is given in reference books for welded tuffs), the curve of excess pressure lies higher and the maximum pressure is approximately 250 kg/cm².

The vertical displacement of the area of the Northern
Breakthrough from levelling data

With the help of precise levelling a number of the features of vertical deformations in the area of the Northern Breakthrough have been established. The first levelling profile (bench marks I–IV), with a length of 1.1 km, was set up on 5 August 1975 two kilometres north of Cone I. Four levellings were made along this course on 7, 11, 13 and 15 August. By the end of August a more extended network had been set up. Three profiles: 1–12; 843–20; 804–26 with lengths of 3.0; 1.8; 1.2 km converge on one nodal point. Regular, repeated measurements on these profiles began on 28 August.

Repeated levellings of the first profile, I–VI, showed that the bench marks rose during the formation of Cone II. Between 7 and 11 August bench mark VI was displaced by 11.3 cm (bench mark I being taken as the point of origin). This corresponds to a tilt of about 40″ to the north-east. No vertical displacements were observed after 11 August.

The repeated levelling of section (V–VI) on 17 August recorded a

Fig. 5. (*a*) Results of levelling 20–9 for the period from 3 Sept. 1975 to 27 June 1976. (*b*) Displacements of markers for line 766–12 from 17 June to 12 August 1976.

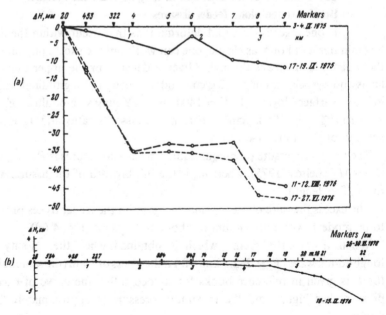

displacement of 20 mm on a base of 300 m (a tilt of 15″) to the south-west, caused by preparation for the breakthrough of Cone III on 17 August.

Results from repeated readings of the levelling network established that the area of the northern cones subsided gradually from September 1975 onwards.

One portion of the line (20–9) was levelled in 1975 and 1976 (Fig. 5 (*a*)). The settling of the bench marks relative to Point 20 did not exceed 5 cm for the period 3 September 1975 to 27 June 1976. Furthermore, towards the middle of August 1976 a tendency to uplift was observed for bench marks 11 and 12, which are closest to the fault line running along the watershed of the dome to Ploskiy Tolbachik (Fig. 5 (*b*)).

Another levelling line (26–20) was measured three times from 20 September 1975 to 15 September 1976. Here the rate of subsidence of the bench marks is higher; it reaches 35 cm/year relative to Point 843 and exceeds 40 cm/year relative to Point 26 (Fig. 6).

The vector of tilt over a year, constructed at Point 20 from levelling data, exceeds 40″ and is directed towards the centre of the Northern Breakthrough

Fig. 6. Sinking of levelling marks relative to marker 26 from 20 September 1975 to 15 September 1976.

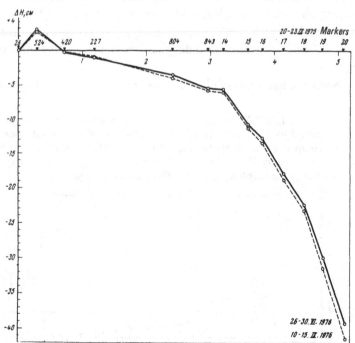

(Cone II). It is considerably greater than the tilt over the same period at the Severnaya tiltmeter station. This is explained by the rapid increase in rate of subsidence of bench marks with proximity to the new volcanoes. It is also possible that most of the line 843–20 is located in the fissure zone and the high rates of subsidence are caused by local factors.

Tiltmeter studies

Observations of deformations of the ground surface were made using tiltmeters at two stations situated in lava tubes in the vicinity of the Northern Breakthrough. High-stability automatic levelled UN-P tiltmeters were used with a sensitivity of about 0.1 seconds of arc and with a calculation discretion of 1 or 12 hours (Dobrokhotov, 1972).

The first station operated 0.8 km to the north of Cone I from 28 July to 12 August 1975. No noticeable tilt was observed during the eruption of Cone I (see Fig. 3). Considerable tilts of variable signs (totalling more than 40 seconds of arc) preceded the appearance of Cone II.

The second station began work on 10 October 1975. It was situated further from the Northern Breakthrough, 3 km to the north-west of the nearest cone, III (766, see Fig. 1). The apparatus was set up in a lava tube over 100 m long at about 20 m underground. The tilt curves from both components showed a substantial decrease in tilt from March 1976, especially in the north-south vector (Fig. 7 (1), (2)). The general orientation of the tilt is shown on a vertical diagram constructed for the same period (Fig. 8).

In one and a half years of observations the ground surface was recorded

Fig. 7. 1, 2, Tilts along two components in northern cave for the period from 10 October 1975 to 5 February 1977; 3, 4, tilts along two components in southern cave for the period from 20 August 1976 to 15 February 1977. The vertical dashed lines mark the beginning and end of the eruption at the Southern Breakthrough.

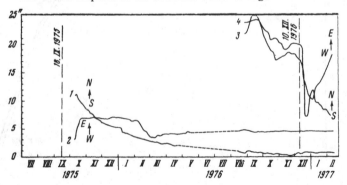

as subsiding in a south-south-easterly direction. The full tilt vector reaches 10″, which corresponds to a change in excess of 5 cm on a line 1 km in length.

The results of geodetic levelling and tiltmeter observations in the area of Point 20 from 10 October 1975 to 15 February 1977 confirm and complement each of the methods. The levelling revealed the differential character of movements in the area. The rates of displacement of the bench marks after Cones I, II and III ceased activity increase the closer they are to the volcanoes. Tiltmeter observations have established the non-stationary nature of movements in the time following the activity of these cones. According to geodetic and tiltmeter data, in the area where the tiltmeter station is situated the direction of movement of the surface and the sign of the displacements coincide.

The rapid tilt of 4″ to the east from 10 to 20 October 1975 probably indicates that stresses dropped and subsidences occurred along faults

Fig. 8. Vector diagram of tilt in northern cave for the period from 10 October 1975 to 20 November 1976.

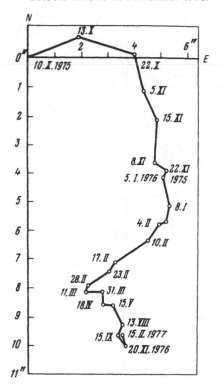

parallel to the main one along the watershed, on the axis of the cinder cone zone. Subsequently, the area of the northern adit inclines primarily to the south, and only in January–February 1976 does the tilt assume a south-westerly direction.

The Yuzhnaya tiltmeter station equipped with two levelled UN-P meters was set up in a lava tube not far from the surface, two kilometres north-north-west of Cone VIII. The ceiling of this cave is no more than 2–3 m below the surface and its large dimensions made it impossible to control its temperature. Consequently the seasonal range of the air temperature here in 1976–77 was about 8 °C. Apart from serious difficulties in using the tiltmeters (in autumn and spring frequent control of the size of the levelling bubbles is required), this causes appreciable seasonal tilts to the pediment.

However, inclinations of volcanic origin during the eruption undoubtedly exceeded any possible interference by many times. The graphs in Fig. 7

Fig. 9. Vector diagram of tilt in the southern cave for the period from 30 August 1976 to 11 February 1977.

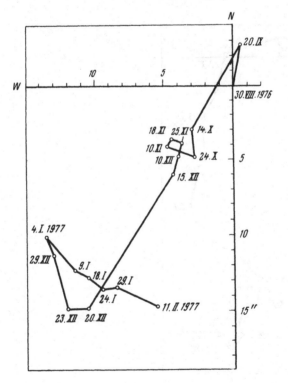

(3), (4) show the course of tilt in the southern cave over the period from August 1976 to February 1977. What is interesting here is the conspicuous change in direction of tilt from the end of October until the end of the eruption, and the conspicuous acceleration of tilt along both components immediately after the eruption. The diagram in Fig. 9 shows the changes in the tilt vector.

During the last two months before the eruption ceased (from 14 October 1976), the direction of tilt changed radically. Possibly this was connected with the last activisation of the volcano, during which (on 7 and 23 November) two active vents formed in succession on the north-eastern slope of the cone.

The fact that there was a south-west tilt from 10 to 20 December indicates the presence of an active fissure with azimuth 300–320°. All three vents of Cone VIII are positioned in the same direction. Evidently this fracture was the breakthrough fissure of Cone VIII. Its position is further confirmed by other geodetic data.

It is possible that from 20 December 1976 to 4 January 1977 a new breakthrough (or a continuation of the eruption) was under preparation. The north-westerly direction of tilt was a sign that stresses were building up along the main north-easterly fissure extending along the watershed of the zone to Ploskiy Tolbachik. However, despite the fact that the activisation was accompanied by a severe swarm of local earthquakes* and noticeable elongations of the lines around the cone, no continuation of the eruption occurred. Whether this was a hidden intrusion of magma and why the stresses were dissipated remains unknown, but from 4 January 1977 the tilt was consistently to the south-east. The most likely cause was subsidence of the sides of the north-north-east fissure.

Horizontal deformations of the area of the Northern Breakthrough after the eruption

From September 1975 to February 1977 measurements were made of lines 1–10 km long in the area of the Northern Breakthrough using optical range-finders – principally Geodimeter Model 8 (see Fig. 1). The point of origin for the measurements was Point 20. The nature of the changes in the length of the lines is shown in Fig. 10. Measurements made in September 1975 showed that appreciable deformations occurred over several days during the preparation and breakthrough of Cone VIII. The

---

* See Fedotov *et al*. The development of the Great Tolbachik Fissure Eruption in 1975 from seismological data (this volume).

9.2 km line to Belaya Gorka changed by 7 cm, which was several times larger than synchronous changes in the shorter lines. One may postulate the existence of an active fault of north-easterly strike west of Point 20 and forming the western limit of the horst-like uplift of the zone of volcanic activity. According to the data, strain of a roughly north–south direction in the zone affected by the eruption exceeds 10–12 km, and possibly even 25–30 km if one takes into account the simultaneous major deformations of the Ploskiy Tolbachik crater.

Until March 1976 the lines to the north of Point 20 tended to increase in length, while the short ones to the south tended to contract. The

Fig. 10. Alteration of length of lines for the period from September 1975 to February 1977.

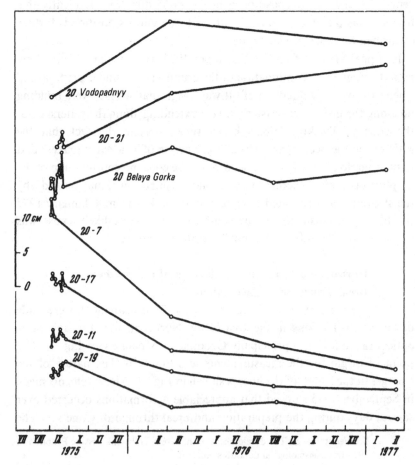

substantial drop in stresses in the area of the Northern Breakthrough during this period was evidently connected with the intensive discharge at the Southern Breakthrough. Later, until February 1977, the multitonne compression of the shorter southern lines and the elongation of the northern ones gradually dies down. This may be partly explained by the subsidence of the Northern Breakthrough area, which was at its greatest at points in the vicinity of the new volcanoes (see Figs. 5–7), and is possibly connected with the collapse of underground chambers and the creation of a subsidence funnel after the cones – which weigh more than 1.5 milliard tonnes – had formed.

Conclusions

1. Hypothetically between the beginning of the eruption and the breakthrough of the southern bocca of Cone I, a sill-like injection of magma with a thickness between a few metres and a few tens of metres was formed not far beneath the Northern Breakthrough. This reservoir was then drained before the eruption of Cone I ceased.

2. An evaluation of the dimensions of the feeder dyke of the Northern Breakthrough on the basis of horizontal deformations measured on the western flank of the breakthrough fissure of Cone II showed that during the formation of each of the first three cones the feeder dyke was little more than 1 m thick.

3. The magnitude of excess pressure of magma during the breakthrough to the surface was modest: 100–250 kg/cm^2.

4. The local vertical deformations of 28–29 July in the area of the southern bocca were connected with the pressure of the magma forcing its way to the surface and were as much as 30 m.

5. Subsidence in the area of the Northern Breakthrough cones began immediately after the termination of the eruption and proceeded differentially at rates of 5–40 cm/year.

6. The fact that horizontal deformations and inclinations in the vicinity of the Northern Breakthrough were dying out by the beginning of 1977, indicated a drop in the stresses caused by eruptive and post-eruptive processes.

7. Tilt measurements in the vicinity of the Southern Breakthrough showed the presence of two main fractures: a submeridional fault which can be traced to Ploskiy Tolbachik, and a 'feathering' fault which passes along an azimuth of 300–320° through Cone VIII (obviously coinciding with the direction of the breakthrough fissure of this cone).

References

Dobrokhotov, Yu. S. 1972. Izmereniye naklonov zemnoy poverkhnosti vysokochuvstvitel'nymi urovnyami. (Measuring tilt of the Earth's surface using highly sensitive levels.) In: *Medlennyye dvizheniya zemnoy kory.* (*Slow movements of the Earth's crust.*) M., Nauka, 229 pp.
Zheltov, Yu. P. 1975. *Mekhanika neftegazonosnogo plasta.* (*The mechanics of an oil- and gas-bearing layer.*) M., Nedra, 450 pp.

Analysis of the methods and results of geodetic studies in the Tolbachik geodynamic polygon, 1975–1976

A. P. Kiriyenko, V. A. Shul'man and
Yu. P. Nikitenko

Introduction

At the present time questions relating to contemporary tectonic processes are at the centre of attention not only of specialised scientific organisations, but also of industrial ones, among which the Central Directorate of Geodesy and Cartography (GUGK) occupies one of the leading positions.

Studies by Bulanzhe & Meshcheryakov (1962), Kashin (1973), Fedotov & Enman (1973) and a number of others, have indicated that geodetic methods can contribute much to solving the problems of contemporary geotectonics.

In recent years this has led to intensive studies being undertaken in the USSR's highly seismic region on the Kamchatka Peninsula, studies which have proved to be extremely apposite.

On 6 July 1975 the Great Tolbachik Fissure Eruption (GTFE) began. A description of the dynamics of the eruption has been given in a number of papers and communications (Fedotov *et al.*, 1976). This eruption is considered the most powerful in historical times in the Kuril–Kamchatka volcanic belt. From fissures that had formed in the Earth's crust, liquid lava of basaltic type emerged at the surface with a volume of around 1.5 km^3.

As a result of the eruption, which lasted until 10 December 1976, eight volcanic cones were formed on the surface. The biggest of them are Cones I, II and III of the Northern Breakthrough, and Cone VIII of the Southern Breakthrough.

The study of the GTFE was carried out from the moment of its inception by the exertions of the Institute of Volcanology of the DVNTs AN SSSR (Fedotov *et al.*, 1976), which enlisted representatives of the leading Earth sciences – geophysicists, geodesists and geologists.

From July 1976 onwards team No. 2 of GUGK carried out geodetic work over the eruption zone comprising about 4.5 thousand km².

The purpose of this work was to discover the nature of the areal distribution of deformations of the surface, to study the processes of accumulation and stabilisation of stresses in the crust in the region of the eruption, to determine the volumes of the products of the eruption (lava, bombs, ash, etc.), and to identify the changes in the topography of the locality. The results obtained from repeated geodetic measurements will help us to establish the nature of the tectonic processes in the crust and discover the ways in which the processes of preparation of eruptions and earthquakes are related to changes in the Earth's surface, with a view to predicting catastrophic eruptions and earthquakes. The results of these geodetic measurements enable us to assess the reliability of the various hypotheses about the block structure of the crust, the existence of trans-crustal faults, and their dynamics.

It is well known that the efforts of GUGK over many years culminated in the creation of a continuous triangulation and levelling network over the entire territory of the USSR. Thus it was decided to adopt as a starting point the network that had been set up in the Tolbachik area between 1971 and 1973.

Between July and October 1976 team No. 2 carried out the following work over the Tolbachik geodynamic polygon: (1) measurement of horizontal and vertical angles according to the class II programme; (2) measurements of the 36 sides of the triangulation according to the class I programme of base measurements; (3) class II geometrical levelling; (4) a stereotopographic survey on a scale of 1:25000.

The present paper examines the methods and results of measurements in the established geodetic network (triangulation surveys, linear measurements).

The existence of the state geodetic network of classes II and III, constructed in the region of the eruption in 1971–73, enabled us to determine the nature of the areal distribution of the horizontal and (to a certain extent) vertical deformations of the Earth's surface. With this aim the mathematical processing of the measurements was carried out in the following order: (1) preliminary processing, evaluation of accuracy; (2) calculation of the differences in horizontal directions; (3) equalisation of the triangulation networks of 1971 and 1976; (4) calculation of the vectors of horizontal displacements of the triangulation points, and an assessment of their accuracy; (5) calculation of the differences in the equalised sides of the triangulation; (6) calculation of the differences in the excesses and equalised bench marks of the points, as determined by geodetic levelling.

An analysis of the results of the mathematical processing has shown that in the region of the GTFE tectonic movement occurred in the Earth's surface of the order of 30 cm between 1971 and 1976. The nature of the areal distribution of the deformations of the surface has been determined, and the existence and position of a trans-crustal fault confirmed from geophysical data (Anosov *et al.*, 1974).

The results of measurements

The geodetic network on the Tolbachik geodynamic polygon consists of 25 points (Fig. 1). Of these, 24 coincide with the points of the state network, and only Point 19 has been constructed afresh.

From 2 August 1970 to 16 September 1973 on a sector of the network coinciding with the new network of the geodynamic polygon, measurements of angles and directions according to the class II programme (direction weighting 24–35) were carried out at Point 21, and according to the

Fig. 1. Triangulation network of the Tolbachik polygon. Change in direction for 1971–75.

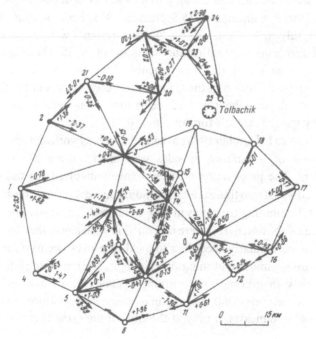

class III programme (direction weighting 18) at three points. The instruments and methods of measurement, the maximum discrepancies of directions and angles in the programme, the percentage of repeated measurements, were as stipulated by the *Instruktsiya...(Instruction...)*, (1966).

An analysis of the accuracy of the free components of the figures showed that the standard deviation for the measured angles of class II is equal to $\pm 1.02''$. The distribution of the closure errors of the triangles is near normal.

According to a number of authors (Tokarev *et al.*, 1974; Fedotov & Kondratenko, 1975; Fedotov *et al.*, 1974), between August and September 1973 seismic activity in the Klyuchi group of volcanoes, to which Ploskiy Tolbachik belongs, was low. In 1971, when most of the measurements were carried out, a single earthquake at a depth of more than 150 km ($K_{S\,1.2}^{\phi\,68} = 10$) was recorded in the Tolbachik region.

The results of the measurements carried out on the sector of the state geodetic network, aligned with the new network of the geodynamic polygon, may be taken as the starting point for the analysis of the horizontal movements of the triangulation points.

The measurements of 1976 were made at a time when the eruption at the Southern Breakthrough was drawing to a close (the eruption at the Northern Breakthrough finished on 15 September 1975; Fedotov *et al.*, 1976).

From 31 July to 15 October class II measurements were made of the angles and directions, with a direction weighting of 24–35. The results and the methods used satisfy the requirements of the Instruction. Analysis established the distribution of the values of free components of the figures as near normal, and the standard deviation of the measured angle, obtained by using Ferrero's formula, as $\pm 0.73''$.

From 28 July to 11 October 1976 on the Tolbachik network high-precision measurements of a number of sides were made, using SG-3 visual range-finders. The programme of measurements involved the use of 12 procedures, divided into four series. Depending on weather conditions, two methods of distributing the series in time were used. Thirty-six lines were measured in all, of which 61% were measured in such a way that the series were distributed regularly according to local time over twenty-four hour periods: morning, midday, evening and midnight. The remaining lines were measured only in evening and morning visibility, with two series of measurements (morning and evening) in one day. Temperature, humidity and pressure were measured only at the ends of lines, the first two using standard Assman aspirational psychrometers, the last using aneroid barometers. The time of the measurements of atmospheric parameters at the points where the reflectors were set up was agreed with the leader of the field party, and measurements were made at intervals of 10 mins. from the beginning to the end of the observations. The psychrometers were set up at the height at which the signal was propagated. The reduction elements were determined graphically. To evaluate the stability of the constant corrections to the receiver–transmitters, measurements were made before and after the linear readings, at a control base situated in the Petropavlovsk–Kamchatskiy urban district. This control base, extending 2990.317 m, was measured with a wire with a relative error of 1:670000. Constant adjustments to the instrumentation were introduced into the results of the measurements. These adjustments were determined in the factory, since they are close in size to the adjustments obtained at the control base. At the stations the estimate of the accuracy of the line measurements was made according to the deviations from the average value obtained by the various methods. The relative standard deviation of the measurements of the inclined distances varies from 1:16000000 to 1:2500000, a fact that indicates the care with which the time of the measurements was chosen, the favourable conditions for the passage of the signal and the steady operation of the instruments during measurement.

Mathematical processing: analysis of results

Preliminary processing of the measurement data. The aim of the preliminary processing of the results of the geodetic measurements is to reduce the values obtained to a particular mathematical surface, to estimate the precision of the results of measurements and to analyse the results in order to discover possible miscalculations.

In reducing measured directions to a plane surface, the same formulae were used as for observations in both 1971 and 1976, thus preserving identical reduced values in an unmodified gravitational field.

Into the measured direction, reduced to the centres of the markers, were introduced corrections for the curvature of the geodesic line on the plane surface, and these were calculated according to the formula: $\delta_{i,j} = ((x_i - x_j)/6R^2)(2y_i + y_j)$, where x and y are the coefficients of points i and j, and R is the average radius of the curvature of the ellipsoid.

The precision of the measured values was calculated according to accepted formulae. In the triangulation survey the standard deviation of the measured angle was calculated according to Ferrero's formula

$$m_\beta = \pm \sqrt{\left(\frac{[WW]}{3n}\right)},$$

where W is the misclosures of the triangles, and n is the number of triangles. Also, the quality of the angular measurements was assessed from the values of free components of the polar conditions.

The measured inclined distances were calculated according to the deviations from the mean in the readings obtained by each procedure.

$$M_D = \pm \sqrt{\left(\frac{[VV]}{k(k-1)}\right)},$$

where V are the deviations from the mean, and k is the number of procedures.

Then the relative standard deviations of the inclined distances were calculated.

In reducing the inclined distances to a plane surface the following adjustments were introduced:

(1) for the inclination of the line

$$\Delta D_\alpha = -\frac{h^2}{2D} - \frac{h^4}{(2D)^3},$$

where h is the excess of the ends of the line and D the measured inclined distance reduced to the centres of the points;

(2) for the transition to the surface of the ellipsoid

$$\Delta D_{el} = -\frac{H_m + l_{P_m}}{R'} D_0 + \frac{(H_m + l_{P_m})^2}{R'^2} D_0,$$

where H_m is the mean height of the line, l_{P_m} is the mean height of the quasi-geoid above the ellipsoid, D_0 is the distance between the stations reduced to the horizon, and R' is the radius of the curvature of the normal section reduced according to the azimuth of the measured line;

(3) for the transition from the ellipsoid to the plane

$$\Delta D_{pl} = D_{el}\left(\frac{y_m^2}{2R^2} + \frac{\Delta y^2}{24R^2} + \frac{y_m^4}{24R^4}\right),$$

where y_m is the mean ordinate of the line, Δy is the difference of the ordinates of the ends of the line, and R the mean radius of curvature.

To evaluate the precision of the linear measurements as a whole throughout the network in the triangles, the angles were calculated from the sides, reduced to a plane surface. From the differences of the measured and calculated angles, the standard deviation of the differences is calculated:

$$m_d = \pm \sqrt{\left(\frac{[dd]}{n}\right)},$$

where d is the differences of the angles, and n the number of differences. Let us use the relation

$$m_d^2 = m_{\beta\tau}^2 + m_{\beta1}^2,$$

where $m_{\beta\tau}$ and $m_{\beta1}$ are the standard deviations of the measured and calculated angles respectively, when obtaining a value for $m_{\beta1}$.

The relative standard deviation of the reduced line is calculated from the formula

$$\frac{m_D}{D} = \frac{m_{\beta1}}{\sqrt{(2\rho'')}}.$$

When $m_d = \pm 0.98''$, and $m_{\beta\tau} = \pm 0.68''$, the value of $m_D \times D = 1 : 430\,000$.

Equalisation of the triangulation networks

In equalising the triangulation networks of 1971 and 1976, the following conditions were adopted:

(1) the networks were equalised as being free, since there are no sufficiently clear confirmations of the block structure of the crust; thus from the many points a pair was chosen that is situated at a maximum distance from the active volcanoes and is presumably located on a single block;

(2) for the triangulation networks of 1971 and 1976 the same initial data were taken: the length and directional angle of one side, and the coordinates of the starting point;

(3) the length of the base side was calculated from measurements made in 1976 (inclined distance reduced to a plane);

(4) the directional angle of the base side was calculated according to the coordinates determined from measurements made in 1971;

(5) to determine the reliability of the equalised values, a comparison was made between the measured and equalised sides of the triangulation (maintaining the scale of the network). The equalisation was carried out in several variants (each of which had its own initial geodetic data).

Networks have now been completely equalised according to the first variant (initial points 1 and 2) and second variant (initial points 16 and 17).

The equalisation of the networks was carried out according to A. I. Balashov's programme (1971), which specifies an iteration method for solving normal equations.

The equalisation of the networks of the triangulation survey of 1971 and 1976 has a number of distinguishing features. In the region of the Ploskiy and Ostryy Tolbachik volcanoes there is a 'window' in the continuous triangulation network. In the 1976 network this 'window' is slightly reduced in size as a result of the addition of Point 19, a fact that has reduced the error in the position of Point 25 relative to the initial positions by 53% compared with the network of 1971.

Given the particular features of the network and of the programme employed (accumulation of errors in the position of points as the latter increase in number), the triangulation network of 1971 has been equalised in the following manner.

First the triangulation of class II was equalised without Points 22, 23, 24 and 25; the coordinates of these points were then calculated from cotangent formulae for the angles of the triangles. The coordinates of the class III triangulation points were determined from equations of the systems which they form with the class II points.

The triangulation network of 1976 was equalised in two variants with the same initial points (1 and 2), with a view to choosing the best one. In the first variant the network was equalised using triangular measurements alone. In the second variant linear measurements were used. Since the programme we employed was only designed for equalisation of a triangulation, the task was to 'restore' the complete triangulation network using linear measurements. For this purpose we followed P. A. Gaydayev's (1970) proposal that in the class II triangulation network some directions need not be measured, but instead measurements should be made of two sides of the triangle. In this way it is possible to calculate the two angles in the triangle which have not been measured, from the two sides that have and the angle between them. Using this proposition, we 'restored' the complete triangulation network.

A comparison of the measured and equalised sides of the triangulation of 1976, obtained in the first and second variants, showed (Table 1) that the scale of the network is least distorted in the course of equalising the continuous network.

In equalising the triangulation networks of 1971 and 1976 from the second variant (initial points 16 and 17), the methods of determining the coordinates of the 1971 epoch remained unchanged, while the coordinates of the 1976 points were determined by equalising the entire network.

Table 1. *Varieties of measured and equalised sides of the 1976 triangulation*

		Variant I		Variant II	
Lines	S_{meas} (m)	S_1 (m)	$d_1 = S_1 - S_{\text{meas}}$ (cm)	S_2 (m)	$d_2 = S_2 - S_{\text{meas}}$ (cm)
3–20	14729.964	0.135	+17.1	9.949	−1.5
3–19	16232.642	2.718	+7.6	2.618	−2.4
5–15	11994.405	4.543	+13.8	4.465	+6.0
3–14	12793.001	3.146	+14.5	3.064	+6.3
3–1	22890.696	0.783	+8.7	0.581	−11.5
3–2	17423.388	3.538	+17.0	3.318	−7.0
3–21	17291.569	1.735	+16.6	1.519	−5.0
13–7	25260.832	0.926	+9.4	0.814	−1.8
13–11	14973.300	3.387	+8.7	3.268	−3.2
13–12	9921.780	1.882	+10.2	1.777	−0.3
13–16	13912.521	2.676	+15.5	2.508	−1.3
13–18	21532.594	2.785	+19.1	2.605	+1.1
13–15	14153.609	3.721	+11.2	3.661	+5.2
7–10	9448.253	8.299	+4.6	8.241	−1.2
7–14	16105.375	5.480	+10.5	5.374	−0.1
7–8	15825.065	5.201	+13.6	5.101	+3.6
8–9	7102.047	2.046	−0.1	2.001	−4.6
8–4	21721.526	1.489	−3.7	1.530	+0.4
8–5	19483.468	3.462	−0.2	3.328	−14.0
14–13	10940.972	1.044	+7.2	0.992	+2.0
14–15	6278.585	8.586	+0.1	8.554	−3.1
14–10	6671.682	1.722	+4.0	1.673	−0.9
12–11	11767.914	8.013	+9.9	7.892	−2.2
12–16	8949.568	9.546	−2.2	9.474	−9.4
2–21	11860.454	0.636	+18.2	0.386	−6.8
24–25	16644.008	4.268	+26.0	3.923	−8.5
24–23	8891.817	1.774	−4.3	1.777	−4.0
20–21	15537.282	7.458	+17.6	7.246	−3.6
20–23	9751.254	1.267	+1.3	1.191	−6.3
6–5	12301.793	1.811	+1.8	1.734	−5.9
8–5	15079.563	9.598	+3.5	9.493	−7.0
19–15	10034.362	4.396	+3.4	4.345	−1.7
23–25	11801.474	1.658	+18.4	1.411	−6.3
16–17	14420.415	0.484	+6.9	0.452	+3.7

Analysis of the differences in the measured and equalised values

Following L. A. Kashin's (1973) proposal, a comparison was made of the measured directions of 1971 and 1976. This gave the first information concerning the areal distribution of deformations of the earth's surface (see Fig. 1).

The directions of 1971 are shown here by continuous lines. The initial directions are marked by arrows on the lines. The remaining arrows indicate the directions and size (in seconds of arc) of the change in the directions of 1976 relative to those of 1971.

The points at which differences exceed twice the error are shown by blacked-in circles. They are obviously situated in a zone of considerable movement, which includes the Southern and Northern Breakthroughs as well as extending north of them. This indicates the possibility of comparing directly measured values in order to obtain an initial qualitative picture of the changes effected in the locality.

The next source of information about the deformations of the surface is the vectors of displacement of the triangulation points.

The modulus of the vector of displacement of a point was calculated from the formula:

$$r = \sqrt{[(x_{II}-x_I)^2 + (y_{II}-y_I)^2]},$$

where x_I, y_I, x_{II}, y_{II} are the coordinates of points corresponding to the epochs of 1971 and 1976 respectively.

The standard deviation of the vectors was determined from the formula:

$$m_r = \pm \sqrt{(M_{x_I}^2 + M_{x_{II}}^2 + M_{y_I}^2 + M_{y_{II}}^2)},$$

where M_x and M_y are the standard deviations of the coordinates of a point relative to its origin and are obtained from the results of the equalisation.

The directional angle θ of the direction of the vector r was determined from the following relation

$$\tan \theta = (y_{II}-y_I)/(x_{II}-x_I).$$

The standard deviation of the directional angle of the vector was determined from the approximated formula

$$m_\theta = \rho^0(M/r)\sqrt{2},$$

where M is the error in the position of the point relative to its initial coordinates.

The results of the calculation of the displacement vectors of points and an assessment of their accuracy are given in Tables 2 and 3. Fig. 2 shows each vector with its moduli above and the errors in the moduli (in centimetres) below.

If we compare the directions and values of the displacement vectors of the triangulation points, we can conclude that in the region of the Northern and Southern Breakthroughs significant deformations of the Earth's surface have been recorded. The direction of the displacement vectors indicates that the zone expanded as a whole. Over the area bounded by Points 2, 3, 20 and 21 a significant compression of the surface occurred.

A comparison of the data of Fig. 2 (*a*) and (*b*) shows that in determining the displacement vectors from the results of equalising free networks it is necessary to use a few variants of the initial geodetic data. None of the variants on its own gives a reliable picture of an areal deformation of the surface. This is one of the drawbacks of representing movement in the form of vectors.

Table 2. *Vectors of point-displacements (variant 1 of levelling)*

No. of point	Vector module (cm)	Standard deviation of vector	Direction angle θ	Standard deviation of $m\theta$
3	18.3	±8.6	346° 43′	32.4°
4	10.0	10.0	218° 56′	69.9°
5	13.5	12.3	108° 34′	61.4°
6	19.2	16.5	120° 18′	57.6°
7	20.8	12.6	85° 35′	34.8°
8	7.9	8.8	4° 20′	69.0°
9	15.2	11.7	61° 14′	47.1°
10	31.2	13.3	78° 09′	24.5°
11	37.6	19.0	71° 22′	31.6°
12	40.9	21.1	44° 39′	30.0°
13	35.4	17.1	57° 55′	29.4°
14	32.4	12.7	63° 39′	25.3°
15	33.6	12.9	60° 09′	25.2°
16	35.8	24.0	39° 06′	41.6°
17	29.7	25.8	15° 13′	58.0°
18	33.3	21.5	7° 09′	44.0°
20	13.6	12.7	194° 27′	66.4°
21	10.9	8.1	185° 49′	44.0°
22	33.0	27.5	211° 37′	54.2°
23	30.5	23.6	230° 12′	58.7°
24	46.2	27.8	236° 33′	40.5°
25	29.3	18.1	306° 24′	69.4°

The next stage in the analysis of areal deformations of the surface was to calculate the differences in the sides of the 1971 and 1976 triangulations as obtained by equalising the networks according to the first and second variants.

In Fig. 3 the zone of expansion of the Earth's surface (differences of sides more than 30 cm) is clearly visible. It is bordered in the west and east by a zone of compression (compensation). The whole northern sector of the network underwent compression during the period 1971–76.

It is necessary to bear in mind that the coordinates of Points 22, 23, 24 and 25 of the 1971 epoch were obtained from the cotangent formulae (Points 20 and 21 were taken as origins); for this reason the possibility cannot be excluded that these particular points were displaced as a result of the displacement of Points 20 and 21.

From the available data it is possible to construct a diagram showing the approximate position of the faults in the crust (see Fig. 3 (*a*)). The faults

Table 3. *Vectors of point displacements* (*variant 2 of levelling*)

No. of point	Vector module (cm)	Standard deviation of vector	Direction angle θ	Standard deviation of $m\theta$
1	19.6	±24.9	20° 17′	66.5°
2	16.6	24.3	61° 17′	76.5°
3	16.6	15.8	336° 00′	70.4°
4	29.6	23.8	297° 47′	57.5°
5	16.5	20.5	307° 51′	89.0°
6	17.9	17.3	261° 58′	69.3°
7	6.6	13.4	276° 55′	145.8°
8	13.8	15.5	310° 52′	80.6°
9	8.7	12.0	326° 18′	123.0°
10	11.4	13.7	102° 42′	73.3°
11	2.5	11.5	2° 20′	329.4°
12	5.6	6.4	56° 35′	81.4°
13	10.0	7.3	62° 40′	59.3°
14	17.2	10.9	69° 34′	46.1°
15	20.8	13.3	97° 00′	38.2°
18	4.3	8.5	277° 57′	121.4°
20	27.2	17.2	173° 02′	43.6°
21	20.2	23.2	135° 12′	80.3°
22	45.5	27.3	180° 15′	34.5°
23	43.1	23.0	197° 41′	30.7°
24	59.6	27.2	204° 30′	26.3°
25	24.6	17.0	256° 35′	39.6°

Fig. 2. Vectors of horizontal displacements of triangulation points. (*a*)
Variant 1 (line 1–2 taken as fixed). (*b*) Variant 2 (line 16–17 taken as
fixed). 1, base line; 2, new cones; 3, main tension fissure.

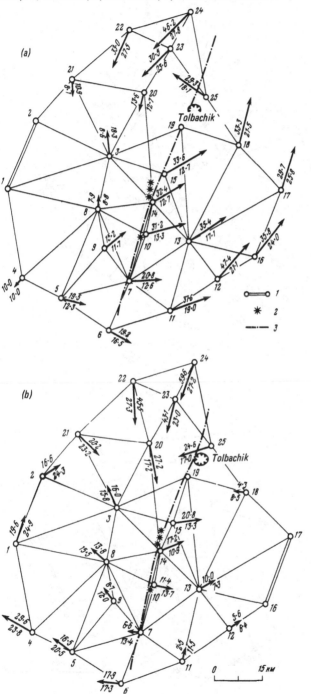

Fig. 3. Vector for alteration of triangulation sides and heights for 1971–76. (*a*) Variant 1. (*b*) Variant 2.

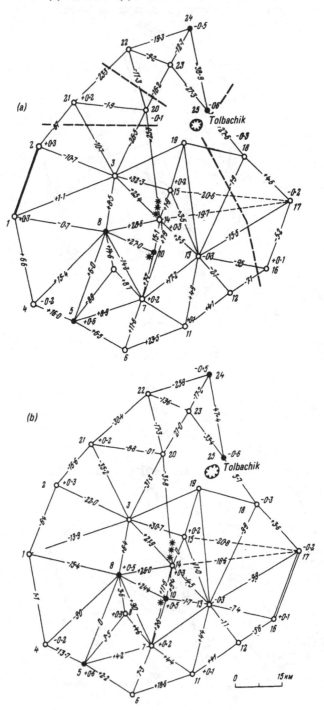

that we have traced explain the distribution of the differences in the sides and of the directions of the displacement vectors. The geodetic data confirm the presence of a trans-crustal fault passing through Ploskiy Tolbachik volcano (Anosov *et al.*, 1974). The radial arrangement of the faults with Ploskiy Tolbachik at their centre, is interesting. Equally interesting is the fact that during the eruption of the Northern Breakthrough a gigantic pit 1700 m across and 700 m deep appeared on the volcano's summit.

The precise position of these faults in the locality can be determined (1) by geodetic methods, using a condensation of the networks in plan and in elevation, and (2) by indirect methods using geophysical, geological and geomorphological studies.

The last stage in the analysis of the results of measurements in the 1971 and 1976 epochs was to compare the direct and inverse excesses obtained from geodetic levelling. The differences in the one-sided excesses fluctuate within $+0.88$ and -1.42 m. The maximum average difference of the direct and inverse excesses of 1971 and 1976 is 0.75. Thus no catastrophic vertical deformations have been revealed on the geodynamic polygon.

An assessment of the accuracy of the differences in the double measurements of average direct and inverse excesses obtained from geodetic levelling showed that in 1971 and 1976 the excesses were measured with the same accuracy, ± 3 cm to 1 km.

According to the results of equalising, the maximum standard deviation of the height of a point is ± 10 cm.

Fig. 3(*a*) and (*b*) show the changes in the heights of points obtained from the formula:

$$\Delta h = H_{1976} - H_{1971},$$

where H is the equalised heights of the points.

In equalising the geodetic levelling of 1971 and 1976, the heights of Points 6 and 12 (obtained from geometrical levelling) are taken as the starting points.

It should be noted that the greatest positive differences in heights coincide with the zone of expansion, and the maximum negative differences with the zone of contraction.

It would appear that the expansion of the Earth's surface in the region of the new breakthroughs of the order of 30 cm is accompanied by an elevation of 50–60 cm. The contraction of the Earth's surface is accompanied by subsidence that is also of 50–60 cm.

Conclusions

1. On the Tolbachik geodynamic polygon geodetic methods have been used to determine deformations of the Earth's surface caused by the Great Tolbachik Fissure Eruption of 1975–76 and have obtained numerical values for them.

2. According to the geophysical data, the area affected by these deformations is between 1000 and 1500 km². Our results compel us to enlarge this area considerably.

Evidently, the zone of deformations extends northwards and north-eastwards and takes in Bezymyannyy and Klyuchevskaya Sopka volcanoes. In the south the zone is likewise not limited to the territory of the polygon.

3. The results obtained permit a 'shear' interpretation of the process, and moreover enable us to mark the position of hypothetical radial faults in the region of Ploskiy Tolbachik volcano.

However, another interpretation of the process is possible. For this reason reliable conclusions about the primary causes of the changes that occurred may be reached only by involving the whole complex of methods that were used to study this eruption.

4. The results and experience of our work on the Tolbachik geodynamic polygon have not only scientific but methodological significance:

(*a*) we have ascertained that the state geodetic networks of class II and III triangulation can be effectively used to determine deformations of the Earth's surface;

(*b*) it has been shown that in projecting new geodynamic polygons it is necessary to retain the centres of the existing networks although the scheme of the network can be altered in exceptional cases, since it is necessary to secure the maximum number of coinciding directions that can be used to compare the directly measured values;

(*c*) it has been shown that useful information can be extracted from the results of comparing directly measured angular elements;

(*d*) recommendations have been made as to the choice of initial geodetic data for equalising the two periods of observations, and this is extremely important; if the scale of the network is set by means of direct measurement, then the orientation of the network without azimuthal measurement remains arbitrary, a fact that affects the size of the transverse shift of the triangulation points;

(*e*) it has been shown that the equalisation of free networks necessitates the use of programmes based on a rigorous method of solving normal equations;

(*f*) it has been confirmed that to control the equalisation of the triangulation it is possible and advantageous to use directly measured sides, thus increasing the significance of linear measurements.

Since volcanic regions are distinguished by a dissected relief, considerable fluctuations in heights, and difficult accessibility, measures should be taken to devise methods and apparatus that will increase the accuracy of geodetic levelling.

This would also make it possible to determine with the requisite accuracy changes in height, and to perform a reduction of the measured lengths of lines with an accuracy that would enable us to effect combined equalisation of the linear–angular network.

It would be advantageous to include in future work on geodetic polygons the Laplace azimuths situated on various sides of the active zone. This would make it possible to calculate more fully the probable rotations of the individual blocks of the Earth's surface, and would give an idea of how deviations from the vertical vary with time.

An analysis of the deformations of the Earth's surface must include a comparison of directly measured values (directions, sides), calculation of the displacement vectors of points, an assessment of their accuracy, and a calculation of the differences in equalised values (of the sides and their directional angles).

References

Anosov, G. I., Balesta, S. T. *et al.* 1974. Osnovnyye cherty tektonicheskogo stroyeniya Klyuchevskoy gruppy vulkanov (Kamchatka) v svyazi s yeye glubinnoy strukturnoy. (Basic features of the tectonic structure of the Klyuchi group of volcanoes (Kamchatka) in connection with its deep structure.) *Dokl. AN SSSR*, **219**(5), 1192–5.

Balashov, A. I. 1971. Uravnivaniye trigonometricheskikh setey na ETsVM Minsk 22. (Equalising trigonometric nets on a Minsk 22 ETs computer.) In: *Uravnivaniye geodezicheskikh setey na ETsVM Minsk 22. (Equalising geodetic nets on a Minsk 22 ETs computer.)* Gomel', pp. 55–70.

Bulanzhe, Yu. D. & Meshcheryakov, Yu. A. 1962. Izucheniye sovremennykh dvizheniy zemnoy kory. (The study of contemporary movements of the earth's crust.) *Geofiz. byul.*, No. 12.

Fedotov, S. A. & Enman, V. B. 1973. Programma geodezicheskikh rabot po izucheniyu sovremennykh dvizheniy zemnoy kory na Kamchatke. (A programme of geodetic work for the study of contemporary movements of the earth's crust in Kamchatka.) In: *Sovremennyye dvizheniya zemnoy kory. (Contemporary movements of the earth's crust.)* Tartu, No. 5, pp. 267–73.

Fedotov, S. A., Enman, V. B. *et al.* 1976. Vnedreniye bazal'tov i obrazovaniye pitayushchikh treshchin Bol'shogo Tolbachinskogo izverzheniya 1975 g. po geodezicheskim dannym. (The injection of basalts and the formation of feeder fissures in the Great Tolbachik eruption of 1975 from geodetic data.) *Dokl. AN SSSR.* **229**(1), 170–3.

Fedotov, S. A. & Kondratenko, A. M. 1975. Zemletryaseniya Kamchatki i Komandorskikh ostrovov. (Earthquakes of Kamchatka and the Komandor Islands.) In: *Zemletryaseniya v SSSR. (Earthquakes in the USSR.)* M., Nauka, pp. 150–62.

Fedotov, S. A., Tokarev, P. I. *et al.* 1974. Detal'nyye dannyye o seysmichnosti Kamchatki i Komandorskikh ostrovov (1965–1968 gg.). (Detailed data on the seismicity of Kamchatka and the Komandor Islands (1965–1968).) In: *Seysmichnost' i seysmicheskiy prognoz, svoystva mantii i ikh svyaz' s vulkanizmom na Kamchatka.* (*Seismicity and seismic prediction, the properties of the upper mantle and their connection with volcanism in Kamchatka.*) Novosibirsk, Nauka, pp. 35–46.

Gaydayev, P. A. 1970. O proyektirovanii optimal'noy geodezicheskoy seti 2-ogo klassa. (Projecting an optimal geodetic net of the 2nd class.) *Geodeziya i kartografiya,* No. 1, 7–11.

Instruktsiya o postroyenii gosudarstvennoy geodezicheskoy seti SSSr. (*Instruction on setting up the State Geodetic Network of the USSR.*) 1966. M., pp. 29–48.

Kashin, L. A. 1973. O postanovke izucheniya sovremennykh dvizheniy zemnoy kory geodezicheskimi metodami. (Organising the study of contemporary movements of the earth's crust by geodetic methods.) In: *Sovremennyye dvizheniya zemnoy kory.* (*Contemporary movements of the earth's crust.*) Tartu, No. 5, pp. 341–7.

Tokarev, P. I., Shirokov, V. A. & Zobin, V. M. 1974. Seysmichnost' rayona severnoy gruppy vulkanov Kamchatki. (The seismicity of the region of the northern group of volcanoes in Kamchatka.) In: *Seysmichnost' i seysmicheskiy prognoz, svoystva verkhney mantii i ikh svyaz' s vulkanizmom na Kamchatke.* (*Seismicity and seismic prediction, the properties of the upper mantle and their connection with volcanism in Kamchatka.*) Novosibirsk, Nauka, pp. 46–52.

22 Tilt of the Earth's surface during the formation of Cone II of the Great Tolbachik Fissure Eruption in 1975

N. A. Zharinov, Yu. S. Dobrokhotov,
M. A. Magus'kin and S. V. Enman

Among the studies carried out during the Great Tolbachik Fissure Eruption (GTFE) were measurements of the tilt and vertical displacement of the Earth's surface using equalised UN-P tiltmeters (Dobrokhotov, 1972) and levelling. In this paper we examine the results of measurements carried out from 28 July to 15 August 1975, which reflected deformations of the Earth's surface preceding and accompanying the formation of Cone II.

Two tiltmeters were set up on 28 July 0.8 km to the north of the active Cone I in the lava tube of an old flow at a depth of 6 m and about 6 m from the entrance to the tube (Fig. 1). One instrument was set up in a near east–west direction (azimuth 78°), the other in a near north–south direction (azimuth 345°). They were erected on a firm base of lava, in which small platforms were hollowed out for the step-bearings. To prevent the apparatus moving during earthquakes and volcanic tremor, these bearings were cemented to the base.

The sensitive elements of the UN-P tiltmeters are cylindrical levels, prepared for astronomical instruments, with a scale value of about one second (where the distance between calibrations is 2 mm). Photo-optical recording of the movements of the bubbles in the levels was carried out automatically every hour. The sensitivity of the apparatus to tilt is approximately 0.2″/km of the recording. In each apparatus three levels are arranged parallel to each other. During exposure on film a group of six luminous points is photographed – these are the positions of the ends of the bubbles in the levels. There is an error of ±0.1–0.2″ in the determination of tilt of the apparatus.

Readings on the instruments were recorded from 28 July to 12 August. From 28 July to 2 August the tilt along each component was near zero.

302   *Zharinov* et al.

Fig. 2 shows the course of the tilts from 2 August to 12 August along the east–west and north–south components. This graph reflects events connected with the termination of the operation of Cone I and the commencement of that of Cone II (Fedotov *et al.*, 1976c; Fedotov *et al.*, 1976a). In the night of 6/7 August Cone I became quiet and lava stopped flowing from the northern bocca. At this time the tiltmeters registered an increase in the tilt of the Earth's surface south-westwards of up to 5.4″; the surface subsiding in a south-westerly direction. Explosive activity began again at

Fig. 1. Sketch map showing situation of fissures, levelling signs, levelling points and tiltmeter point.
1, trigonometrical levelling lines; 2, levelling course; 3, tiltmeter point; 4, fissures formed during the eruption: *a*, eruptive fissures; *b*, probable position of feeder dyke; 5, source of lava flow; 6, cinder cones, 1975; 7, lava flows; 8, Holocene cinder cones; 9, Holocene tension fissures; the position of the fissures is shown as they were on 15 August 1975; 10, precipitous rims of 1975 craters and boccas; 11, precipitous rims of Holocene cones.

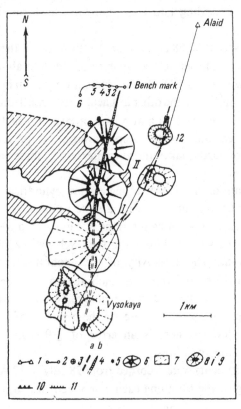

Breakthrough I at 06.00 on 7 August. On 8 August at about 18.00 hours Cone I began to eject a pale-coloured ash, and after continuing like this for about 11 hours, finally quietened. The termination of Cone I's activity was accompanied by an intense tilt to the north-east, numerically equivalent to 18″ (6″ to the north and 17″ to the east).

On 9 August at 18.50 hours 0.8 km north of Cone I, which by that time had ceased its activity, a fissure opened on a relatively level site and Cone II began to form. In the first hours of the eruption 12–15 vents were functioning, in a line more than 350 m long and with an azimuth of about 345°. Subsequently the eruption was concentrated in the central vents, and 200 m from the observation point of the tilts (measured from the nearest vent) Cone II began to grow. Recordings show (see Fig. 2) that before Cone II broke through, the tilts reached 45″ (mostly westerly) along a more or less east–west direction at right angles to the supply fissure. After 18.00 hours on 9 August immediately before Cone II broke through, and on 10 August after 12.00 hours, large jumps in the tilt were observed, the bubbles in the levels went off the scale, and the sizes of the tilts were estimated by the number of turns of the lifting screw. These approximate values are shown in Fig. 2 by dashed lines. Cone II went on operating fairly quietly until 14.30 hours on 10 August, after which a five-hour interval ensued in its activity. The stable regime of Cone II's activity was characterised by tilts of variable sign with an amplitude of up to 4–5″ and a slight east–west trend; tilts in the north–south direction for this period were small. During the interval in Cone II's activity the surface tilted to the north-east. Recordings continued until 12 August when the instruments had to be taken out of the lava tube to avoid advancing lava.

Fig. 2. Progress of tilt from north–south and east–west components (2–12 August 1975). 1, progress of east–west tilting; 2, progress of north–south tilting. Stars mark a break in operation of Cone I, arrows mark a break in operation of Cone II.

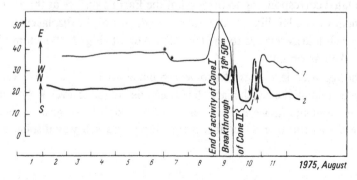

The great rate of change in tilt recorded during the opening of the fissure and the formation of Cone II, as well as the change in sign, enable us to propose that developing deformations were elastic in character. The difference in the east–west and north–south directed tilts preceding the breakthrough of Cone II indicates that the injection of the magma occurred along a fissure of some length. And indeed, as we have already said, the fissure was over 350 m long. The unresolved vector of the tilt is directed to the source of the deformations. For the tilts preceding the breakthrough of Cone II, the vector coincides with an azimuth of 70°. Consequently, it was in this direction from the observation station that the source was situated. Considering that the foci of the earthquakes before the breakthrough of Cone II were situated at a depth of 0–5 km (Fedotov *et al.*, 1976b) and the time from the appearance of the large deformations until the breakthrough was 26 hours, it is not difficult to work out the average speed of the rise of the magma. It was approximately 95 m/h, with a maximum of 190 m/h. A similar speed was obtained for the breakthrough of Cone I (Fedotov *et al.*, 1976b).

The vertical displacements and tilt over a large base have been determined by levelling data. The vertical deformations connected with the eruptions of Cone II were registered by the movements of bench marks on profile 1–6, which had been laid on 5 August 2 km north of Cone I and fixed by marks on the rock (see Fig. 1). The direction of the 1–5 portion of the profile is east-west, while that of 5–6 is oriented to the south-west. Its total length is 1.1 km. The first levelling of the profile was carried out on 7 August, the others on 11, 13 and 15 August, i.e. after the formation of Cone II.

The levelling was carried out using the class II method, an Ni-007 Zeiss (East Germany) leveller and Invar staves made by the EOMZ factory, and with footpieces. It took 2–2.5 hours to measure the profile in one direction.

From the results of four levellings, the heights of bench marks were calculated corresponding to the level of the Earth's surface at the average moment of one levelling, and graphs were drawn of the displacements of the bench marks relative to the level of 7 August (Fig. 3). Bench mark 1 was taken as origin.

The repeated levellings of profile 1–6 made on 11, 13 and 15 August revealed that all the bench marks had risen relative to the level of 7 August. The greatest uplifts were observed in the period 7–11 August, i.e. when Cone II was forming. Over this period bench mark 6 was displaced by 12.3 cm.

The change in the direction of the profile at point 5 enabled us to determine from levelling data the size and direction of the overall vector of tilt in the period when Cone II was forming. Between 7 and 11 August the tilt amounted to 40″, in a north-easterly direction (Fig. 4).

The measurements carried out on 13 and 15 August showed that the tendency of the bench marks to rise remained, that bench mark 5 underwent greatest vertical displacement and the direction of tilt changed to south-east and south.

The results of trigonometric levellings carried out along the sides: Point 12 – Gora Alaid, Point 12 – Gora Vysokaya, show that between 6 and 11 August Point 12 dropped 13 cm relative to Gora Alaid and rose 3 cm relative to Gora Vysokaya (see Fig. 1).

From the results of the tiltmeter and levelling studies from 2 to 15 August we can draw the following conclusions.

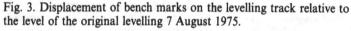

Fig. 3. Displacement of bench marks on the levelling track relative to the level of the original levelling 7 August 1975.

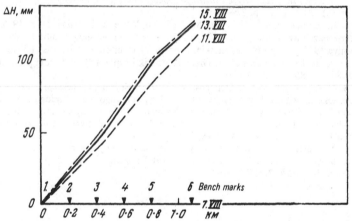

Fig. 4. Vector diagram of tilt from levelling data (7–15 August 1975).

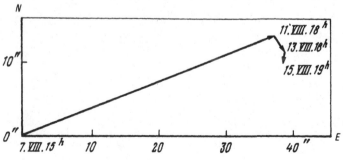

1. According to tiltmeter readings over a period of 13 hours preceding the breakthrough of Cone II, a tilt to the west formed, in a direction at right angles to the feeder fissure, exceeding 45″.

2. According to the levelling data, the total tilt in the period when Cone II was forming was 40″ directed to the north-east.

3. The unavoidable breaks in recording the tilt mean that we cannot compare the size of the general tilt of the Earth's surface, and its direction, with the data obtained by levelling.

4. When the operation of Cones I and II was established, no tilt greater than 2–4″ was observed, despite the fact that the tiltmeters were situated in immediate proximity to the active vents of the volcano.

5. The vertical movements affect a region stretching more than 6 km from north to south.

References

Dobrokhotov, Yu. S. 1972. Izmereniye naklonov zemnoy poverkhnosti vysokochuvstvitel'nymi urovnyami. (Measuring tilt of the earth's surface with highly sensitive levelling.) In: *Medlennyye dvizheniya zemnoy kory*. (*Slow movements of the earth's crust*.) M., Nauka, pp. 229–38.

Fedotov, S. A., Enman, B. V., Magus'kin, M. A., Levin, V. Ye. & Zharinov, N. A. 1976a. Vnedreniye bazal'tov i obrazovaniye pitayushchikh treshchin bol'shogo Tolbachinskogo izverzheniya 1975 goda po geodezicheskim dannym. (The injection of basalts and the formation of the feeder fissures of the Great Tolbachik Eruption of 1975 from geodetic data.) *Dokl. AN SSSR*, **229**(1), 170–3.

Fedotov, S. A., Gorel'chik, V. I. & Stepanov, V. V. 1976b. Seysmologicheskiye dannyye o magmaticheskikh ochagakh, mekhanizme i razvitii bazal'tovogo treshchinnogo Tolbachinskogo izverzheniya v 1975 godu na Kamchatke. (Seismological data on the magma chambers, mechanism and development of the Tolbachik basalt fissure eruption of 1975 in Kamchatka.) *Dokl. AN SSSR*, **228**(6), 1407–10.

Fedotov, S. A., Khrenov, A. P. & Chirkov, A. M. 1976c. Bol'shoye treshchinnoye Tolbachinskoye izverzheniye 1975 goda. (The Great Tolbachik Fissure Eruption of 1975.) *Dokl. AN SSSR*, **228**(5), 1193–6.

Changes in height, volume and shape of the New Tolbachik Volcanoes of the Northern Breakthrough

M. A. Magus'kin, V. B. Enman and
V. S. Tselishchev

During the Great Tolbachik Fissure Eruption (GTFE) of 1975 the changes in the height, volume and shape of the nascent volcanoes of the Northern Breakthrough were monitored by regular geodetic observations and a phototheodolite survey from points and bases not more than 1.5 km from the sites of the breakthroughs (Fig. 1).

The observations of the first active cone, begun on the sixth day of the eruption, consisted of using Theo-010A and T2 theodolites to measure the absolute height of the cone, its width in its upper part and at the base, and the average height of the trajectory and dispersal of bombs. Phototheodolite surveys were begun on the tenth day of the eruption using the Photheo 19/1318 equipment. Their aim was to determine the deformations of the cone itself, the thickness of the lava flows, and their velocity, using methods of stereophotogrammetry. The laboratory processing of the photo-theodolite photographs was carried out on a stereocomparator, the coordinates and heights of the points being subsequently calculated by an Odra computer.

Cone I. The formation of Cone I can be said to have begun with the fissure eruption of the morning of 6 July 1975 in the southern zone of the Tolbachik cinder cones 880 m above sea level and at a point with coordinates 55° 41′ N and 160° 13′ E. During the first few days the eruption was concentrated at one vent, and a cinder cone formed about 50 m high and with an irregular base drawn out in a north-easterly direction (Fig. 2).

No notable deformations in the growth of the cone were observed before 27 July. However, the north-western projection at the base of the cone is

307

Fig. 1. Sketch map showing position of cones and geodetic observations.
1, base of phototheodolite survey; 2, lava flows; 3, sources of lava flows; 4, the 1975 cinder cones; 5, Holocene cinder cones; 6, precipitous rims of 1975 craters and boccas; 7, precipitous rims of Holocene craters.

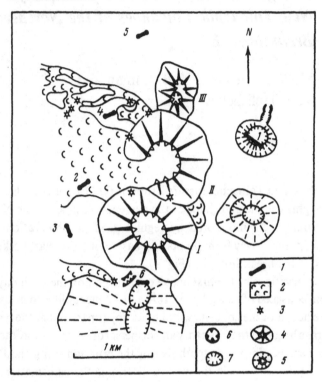

Fig. 2. Alteration of Cone ɪ.

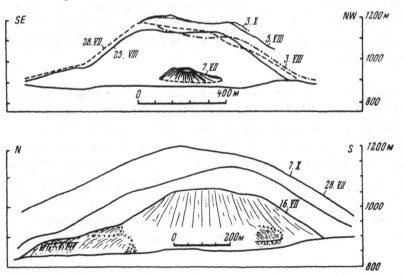

interesting. From the survey data, this projection moved 80 m between 16 and 28 July. There was a similar projection, smaller in size and less mobile, in the south-western portion of the base of the cone. Evidently these projections were portions of very viscous lava flows concealed beneath the constant fall of coarse pyroclastics and bombs; for this reason their existence and the fact that they were moving were difficult to establish visually.

The phototheodolite survey carried out on 28 and 29 July from bases 1 and 4 showed subsidence of the western slope of Cone I by approximately 30 m and the formation of a pit on its north-western slope (see Figs. 2 and 3(*a*)). The lower boundary of the pit was situated at a height of 95 m from the foot of the cone, while the upper boundary could be traced to the summit. The pit had a maximum width of 160 m, a depth of 22 m, and was generally elongated along an azimuth of 340°. The volume of the material that had caved in was 0.0012 km³. These phenomena occurred simultaneously with the formation of the southern bocca and the emergence of the first lava flow, which set off westwards at a speed of 100 m/h.

In the night of 29 July the ejection of pyroclastic material from the crater increased sharply and this high degree of explosive activity was maintained until 9 August. A second survey carried out on 1 August from base 4 showed that the pit that appeared on 29 July had disappeared, and the western edge of the summit of the cone had become approximately 7 m

Fig. 2 (*cont.*)

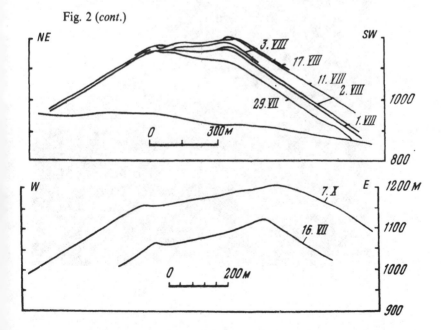

higher than the eastern (see Fig. 2). This could have been caused by the wind changing from east to west and carrying the pyroclastics in one direction. From 29 July to 8 August the average growth in volume of the cone in twenty-four hours was 6.5×10^6 m³.

Fig. 3(*b*) shows a stereoscopic photograph from the phototheodolite survey taken on 2 August at 16.50 hours, 30 minutes after the formation of the northern bocca, and a topographic plan drawn from the double images. Subsequently the development of the bocca and the movement of the lava flow were monitored by a regular survey from base 4. The mouth of the lava river is at a height of 960 m and is situated on the cone approximately 30 m above its foot. The upper part of the bocca coincides with the lower boundary of the pit which formed on 29 July. The lava flow in the first minutes and hours of its existence was 3.2 m thick and moved at a speed of 100–130 m/h with an incline of 5.7°. Fig. 3(*c*) illustrates the shapes of the cone and the bocca on 11 August, the second day after they had ceased activity.

In the course of its eruption Cone I attained an absolute height of 1212 m. Its relative height was 332 m. However, the process of settlement

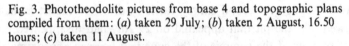

Fig. 3. Phototheodolite pictures from base 4 and topographic plans compiled from them: (*a*) taken 29 July; (*b*) taken 2 August, 16.50 hours; (*c*) taken 11 August.

and destruction of the summit portion of the crater had begun even as Cones I and II were forming. By the time the Northern Breakthrough had ceased activity (15 September), the absolute height of Cone I was 1204 m and approximately six months later it had shrunk another 2.5 m. Cone I has a volume of 0.165 km³. In shape it presents an almost perfect cone with slopes of not more than 30°.

Cone II. As with Cone I, the formation of Cone II began with a fissure eruption, when the activity of a number of conduits was concentrated in one conduit. In approximately 48 hours from the moment of breakthrough, the height of the cone attained 130 m. The average increase in the volume of the cone over this time constituted 5.6 × 10⁶ m³/day. The way in which it grew can be seen in Fig. 4. It is characteristic that whilst Cone III was being formed and was active, the increase in the height and volume of Cone II slowed down greatly. The approximate increase in the volume of the cone from 18 to 25 July was 0.6 × 10⁶ m³/day, i.e. 10 times less than before the

Fig. 3(*a*) (*cont.*)

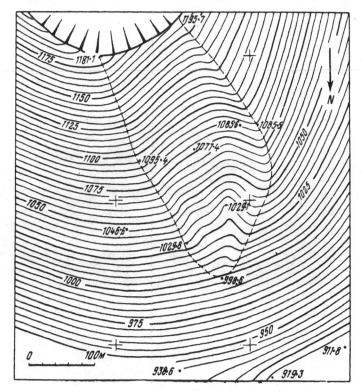

Fig. 3*b* and *c*

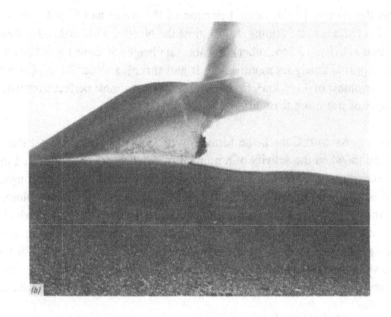

Fig. 3b and c

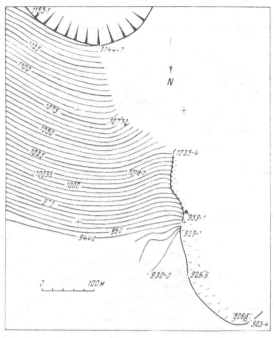

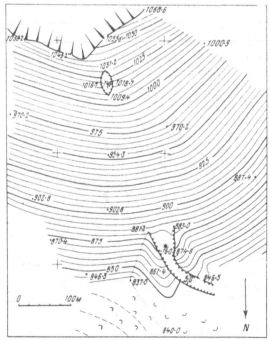

Fig. 4. Change in profile of Cone II.

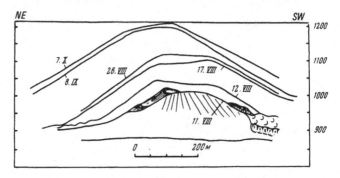

Fig. 5. Change in profile of Cone III.

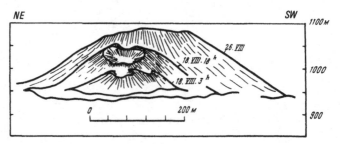

Fig. 6. Increase in height and volume of the cones.
1, change in volume; 2, change in height.

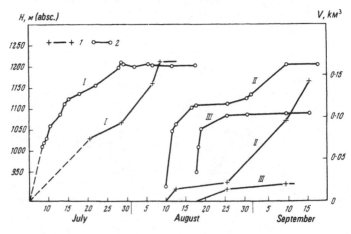

appearance of Cone III. After Cone III had ceased activity on 25 August, the eruption was concentrated in the crater of Cone II, and its volume and height began to increase sharply. From 25 August to 15 September the average increase in volume of Cone II was 5.7×10^6 m³/day.

On 11 August as Cone II was forming, a lava flow 15 m thick poured from its central crater. Subsequently new flows poured out onto this flow and their total thickness attained 35 m at 100–150 m from the foot of the cone.

After the conclusion of the eruption, Cone II had an absolute height of 1206 m, its relative height was 291 m, and the absolute height of its foot was 915 m. The volume of the cone amounted to 0.142 km³. About six months later Cone II had settled by 5 m. The cone is horseshoe-shaped, as its south-western portion was borne forward by the lava flow.

Cone III. Cone III began to form in roughly the same way as the first two, but for a fairly long time there were two craters on its summit. The cone began to grow from a height of 950 m at its foot, and its formation in the first few hours proceeded extremely intensively (Fig. 5). In a period of 6 hours after breakthrough, a cone formed with a height of 50 m and in the next 15 hours it grew another 60 m. The average increase in its volume during the period of activity (17–25 August) was 1.8×10^6 m³/day. In approximately eight days of activity, the third breach formed a cone 141 m high and with a volume of 0.020 km³. The absolute height of the cone at the end of the eruption was 1091 m and in approximately half a year it decreased by 4.0 m. Cone III has a double crater, its base is stretched out from south to north, and it has slopes of less than 30°.

To summarise, the Northern Breakthrough of the GTFE was active for a total of 72 days and formed three large cinder cones: the first was 332 m high, the second 291 m, and the third 141 m; their absolute heights are 1212, 1206 and 1091 m respectively. The total volume of the cones is 0.327 km³ (Fig. 6).

References

Fedotov, S. A., Enman, B. V., Magus'kin, M. A., Levin, V. Ye. & Zharinov, N. A. 1976a. Vnedreniye bazal'tov i obrazovaniye pitayushchikh treshchin Bol'shogo Tolbachinskogo izverzheniya 1975 g. po geodezicheskim dannym. (The injection of basalts and the formation of the feeder fissures of the Great Tolbachik eruption of 1975 from geodetic data.) *Dokl. AN SSSR*, **229**(1), 170–3.

Fedotov, S. A., Khrenov, A. P. & Chirkov, A. M. 1976b. Bol'shoye treshchinnoye Tolbachinskoye izverzheniye 1975 g. Kamchatka. (The Great Tolbachik Fissure Eruption of 1975 in Kamchatka.) *Dokl. AN SSSR*, **228**(5), 1193–6.

The structure of the New Tolbachik Volcanoes from seismic data

S. T. Balesta, A. A. Kargapol'tsev and
G. B. Grigoryan

The aim of seismic research in the region of the New Tolbachik Volcanoes was to study the structural position of the Tolbachik Valley zone of areal volcanism, to identify the conduits of the newly-formed volcanoes, and to determine the location of the chambers feeding these volcanoes and their characteristics. Work was carried out while these volcanoes were active in 1975–76. The seismic vibrations were recorded by seismometers of the Poisk type and Tayga tele-controlled seismographs. S-205 seismogeophones with a natural frequency of 5 Hz were used. The space between observations was 100 m. Extensive seismic data were obtained, characterising the deep structure of the New Tolbachik Volcanoes.

In 1975 the basic method of study of the deep structure of the New Tolbachik Volcanoes was seismic shooting. The actual explosion and detection points of the stations were situated on various sides of the volcanoes in such a way as to provide information about the passage of seismic waves directly beneath the active cones of the Northern Breakthrough. During this seismic study, within the crystalline basement that rests in this region at a depth of 7–8 km, an anomalous zone of attenuation of seismic waves was discovered directly beneath the cones. Furthermore, this zone was revealed in two seismic profiles, which gave grounds for speaking of the existence of an inhomogeneity within the crystalline basement beneath the Northern Breakthrough, and suggesting that it is connected with the magma-feeding zone. We should note, moreover, that the southern termination of the zone in the direction of the Southern Breakthrough could not be established from seismic data in 1975, since the profile lay within this shadow zone.

The system of observations implemented was, of course, designed to expose deep-seated inhomogeneities beneath the new volcanoes, and so did

non-longitudinal explosion points; 4, position of upper inhomogeneity; 5, most probable position of lower inhomogeneity; 6, Cones I, II, III of the Northern Breakthrough; 7, zones of seismic shadow: *a, b,* from explosion point 1 on profile of 1/1975, *c, d,* from explosion point 7 on profile of 1/1976, *e, f,* from explosion point 8 on profile of 1/1976. The letters for the shadows correspond to those on the seismogram mounting.

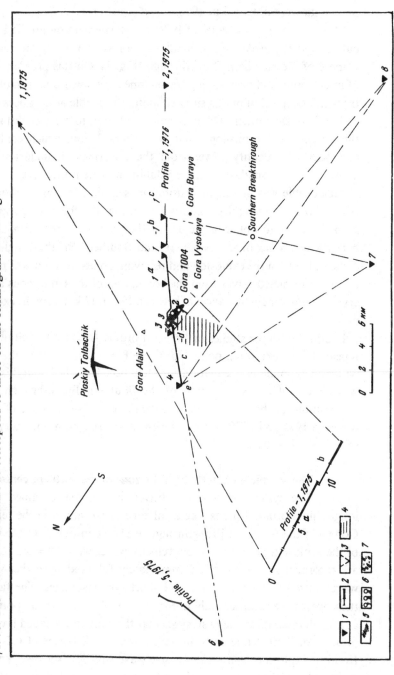

not permit us to construct a detailed seismic section across the areal zone. Such data were obtained in 1976.

Most of the seismic studies of 1976 were carried out on profile 1, which cuts across the strike of the areal zone of volcanism in the region of Cone II of the Northern Breakthrough (Fig. 1). On this profile a system of intercepting and overlapping travel–time curves was obtained of waves from eight explosion points situated along the profile at distances of 2 to 25 km from each other. The maximum explosion to receiver distance on this profile from explosion point No. 6 was 38 km, which enabled us confidently to identify waves from the uppermost boundaries of the division to the boundary of the crystalline basement inclusive.

This profile was investigated using the seismic shooting method from two explosion points (No. 7 and No. 8 on Fig. 1), which were selected in such a way as to obtain a picture of the passage of the seismic waves beneath the cones of the Northern and Southern Breakthroughs. This distance between the explosion and receiving points meant it was possible to record refracted waves from the boundaries of the Cretaceous and the crystalline basement (distances of between 19 and 32 km from the explosion points).

In addition to the studies of 1975, on profile 1/1975 observations were repeated from explosion point No. 1 (see Fig. 1) on a part of the profile where an anomalous attenuation zone had been identified, and this profile was continued in the direction of the Southern Breakthrough. These observations enabled us to compare the picture of the passage of seismic waves in 1975 and 1976, as well as identify the southern boundary of the seismic shadow discovered in 1975.

The seismic section through the zone of the northern cones

The system of observed travel–time curves obtained on the longitudinal profile 1 across the areal zone of volcanism in the region of Cone II of the Northern Breakthrough, is shown in Fig. 2. In the vicinity of the explosion points a few low velocity waves with $V^* = 1.5$–2.0 km/s are recorded, which are often immediately followed by higher-velocity waves with $V^* = 3.5$–3.8 km/s and short synphasal axes. The fact that these waves are recorded only in the vicinity of the explosion points and at a small interval, strongly suggests that there are interbedded lava flows present, of limited thickness and extension. At distances of 0.5–2.0 km from the explosion points an intensive wave with $V^* = 3.8$–4.5 km/s can be traced, which in its kinematic and dynamic parameters belongs to the refracted class. This is the wave for which the most detailed system of intercepting and overlapping travel–time curves has been obtained. The boundary velocities (4.0–4.4 km/s) and characteristic shape of the recording of this wave attribute it to the top of the Neogene sedimentary–volcanogenic deposits.

From the more distant explosion points (2 and 6) a system of intercepting travel–time curves of waves was obtained, refracted at the boundaries of the Cretaceous and crystalline basements. The attribution of these waves to the said discontinuities is based on values obtained for the boundary velocities, as well as on the kinematic and dynamic characteristics of the waves as revealed by both the present studies and by previous GSZ investigations in the region of the Klyuchi group of volcanoes (Utnasin *et al.*, 1974). Although the identification of these seismic discontinuities with stratigraphic complexes is somewhat provisional, we shall use these terms in the analysis that follows. This, of course, does not preclude the

Fig. 2. Travel–time curves for waves observed. EP = explosion point.

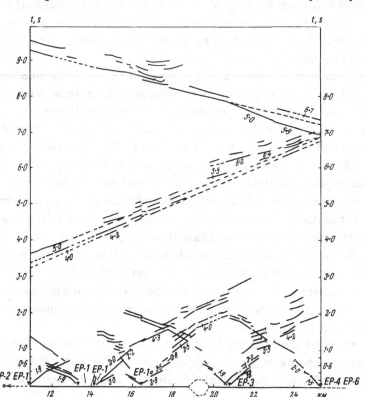

possibility of attributing some of the seismic discontinuities to horizons within the stratigraphic subdivisions.

We should note one important feature of the travel–time curves obtained from explosion points 2 and 6 (see Fig. 2). At distances of 28–34 km from explosion point 6 and 24–25 km from explosion point 2, a sharp distortion of the wave field is observed. At less than these distances, refracted waves from the crystalline basement are recorded (from explosion point 6 in the first, and from explosion point 2 in the second arrivals).

The nature of the change in the wave field is the result of a series of diffracted waves of fairly high intensity being recorded at the distances indicated. Their appearance in this interval of observations with amplitudes commensurable with those of the main waves, cannot be explained within the framework of the fault zones. Evidently they are not connected with faults, but are due to the presence of a large inhomogeneity in the upper layers of the crust.

The construction of a seismic section along the longitudinal profile on the basis of a system of travel–time curves was carried out using generally accepted techniques for refracted waves (the t_0 and time fields method). After a preliminary section had been constructed, the precise position of the boundaries was worked out by solving simple problems with the aid of radiation theory.

The final section is given in Fig. 3. Above all it should be noted that accuracy in constructing seismic boundaries depends on the precision with which velocities in the overlying media can be determined. As a rule, with fairly detailed systems of observation, boundary velocities in the layers can be measured with a high degree of precision (3–5%). The methods used for determining the average velocities in the overlying rock units (Kondrat'yev's, Pavlenkova's and others' methods) produce errors of 15–20%, which reduces the accuracy with which one can plot boundaries.

We shall describe the section from a closer examination of the profile orientated from north-west to south-east. As has already been pointed out, four discontinuities were constructed on the section. The uppermost is shown only tentatively, since data on it were obtained only near the explosion points. Moreover, in the upper part of the section the layers are found to be extremely heterogeneous, with numerous intercalations of lava flows and pyroclastics.

The second discontinuity is attributed to the top of the Tertiary (Neogene) deposits and occurs at depths of from 200 m on the eastern end of the profile to 1200–1300 m beneath the cones of the Northern

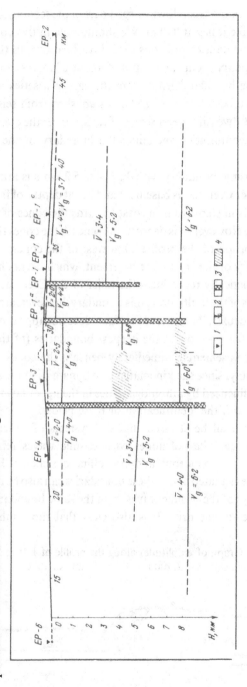

Fig. 3. Seismic section across areal zone of volcanism. 1, explosion points; 2, seismic boundaries and velocities at them; 3, fractures (from seismic data); 4, position of inhomogeneities (from seismic data). The line of small dashes shows the position of Cone II of the Northern Breakthrough.

Breakthrough (in the most downwarped part of the section); at the western end of the profile it lies at 800 m. We should note in this connection that according to the data of previous GSZ investigations, in the Kozyrevka basin this boundary occurs at a depth of about 2 km. As can be seen from the section, the boundary drops from the east in a series of steps along faults that have been fairly accurately established from seismic data. Of these, at least two are deep-seated faults, since they also affect the underlying discontinuities, including the boundary of the consolidated basement.

The next seismic boundary with $V_g = 4.8$–5.2 km/s is attributed to the top of the Upper Cretaceous basement and lies at depths of from 3 to 5 km. It also descends in steps from the east towards the cones of the Northern Breakthrough. However, it is deepest not beneath the cones themselves but in the fault portion of the profile 3 km west of the cones.

The boundary of the crystalline basement, which occurs here at depths of 7–8 km, practically reproduces the relief of the Cretaceous basement. The amplitudes of the faults along this boundary, however, are considerably smaller. In particular, the easternmost fault does not appear on it. It should be noted that the position of the deepest boundaries (of the Cretaceous and crystalline basements) immediately beneath the cones is determined much less reliably, since the kinematics and dynamics of the seismic waves are strongly influenced by inhomogeneities in the upper layers of the crust. We shall deal with the identification of these below.

It must above all be stressed that the systems of observations used enabled us to scan the volcanoes with seismic waves refracted at the boundaries of the Cretaceous and crystalline basements. Evidently, the inhomogeneities situated above these boundaries can also be revealed using transmitted rays of the seismic waves from the same boundaries within the confines of the seismic drift. It is also clear that those inhomogeneities

Fig. 4. Graph of amplitudes along the profile of 1/1975.
1, seismic stations; 2, data of 1975; 3, data of 1976.

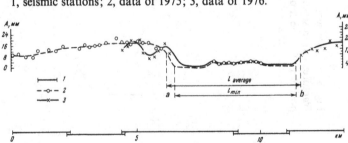

which are situated above the Cretaceous basement and did not fall within the scheme of the transmitted rays could be missed and remain unrevealed by such a method.

Let us consider the identification of the inhomogeneity obtained from the results of the studies in 1975 and from additional data in 1976. As has been pointed out, a zone of seismic shadow was discovered on profile 1/1975 during the study of the northern cones by shooting from point 1. This is confirmed by the 1976 data and we have also succeeded in distinguishing its southern limit. Fig. 4 plots the amplitude graph of this wave. The seismic shadow is clearly visible, caused by an inhomogeneity with sharply reduced viscous–elastic parameters in the propagation path of the seismic wave refracted at the boundary of the crystalline basement. The nature of the attenuation of the seismic wave in the shadow zone (oscillation on the boundaries of the shadow, increased frequencies in this zone, appearance of a weak wave in the central part of the shadow) enables us confidently to describe it as diffractional. This was the first time experimental data had been obtained on an active volcano that permitted the nature of the attenuation of seismic waves in the shadow zone to be established with certainty. The geometrical shadow zone discovered was 4.8 km wide, which suggests that the dimensions of the inhomogeneity were between 2.4 km and 4.8 km. A similar picture was obtained on another shooting profile (profile 5/1975) from explosion point No. 2 (see Fig. 1). The combination of the data from these two profiles enables us to delineate the inhomogeneity, which is situated in the plane of the crystalline basement at a depth of ≈ 8 km. Estimating the dimensions of the inhomogeneity from those of the geometrical shadow, we obtain a body extended in a roughly north–south direction and situated beneath the cones

Fig. 5. Graph of amplitudes along the profile of 1/1976 from explosion point 7.
1, seismic stations; 2, data of 1975.

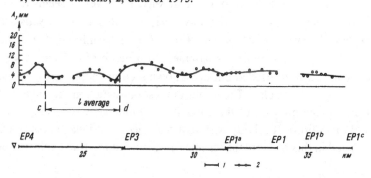

of the Northern Breakthrough. Its most probable position is shown in Fig. 3 (in section) and Fig. 1 (in plan).

The results of the seismic shooting from observations of profile 1/1976 from explosion point 7 are well represented in a montage of the seismograms, where the shadow zone of the seismic waves refracted at the boundaries of the Cretaceous and crystalline basements can be clearly identified. The amplitude graphs emphasise the diffractional aspect of the attenuation of the seismic waves in the shadow zone (Fig. 5). The width of the geometrical shadow zone revealed on this profile is 3.2 km. The existence of an inhomogeneity with reduced viscous–elastic parameters in the propagation path of the seismic waves is borne out by data obtained on the same profile from explosion point No. 8 (see Fig. 1). For practical purposes, no seismic readings were obtained from this point anywhere in the 48-channel seismostation, whereas intense readings of seismic waves from this explosion were obtained on the array of another station situated further along this profile.

Thus on profile 1/1976, both longitudinally (from explosion points 2 and 6) and non-longitudinally (from explosion points 7 and 8) anomalies have been established in the propagation of seismic waves refracted at the boundaries of the Cretaceous and crystalline basements. An analysis of the dynamic features of these waves shows that they can be well explained in terms of the diffraction of seismic waves at an inhomogeneity with reduced viscous–elastic parameters (Averko & Balesta, 1970; Farberov, 1974; Utnasin *et al.*, 1974; and others). Evidently, in each concrete case it is possible to select a particular inhomogeneity causing the diffraction that could explain the wave field observed. Thus, in order to explain the diffractional aspect of the seismic waves recorded from explosion point 7, it is sufficient to place in their propagation path a body that will take in the shadow zone (either on a discontinuity, or above it in places where the seismic rays are cut off). The anomalies obtained from the other explosion points may be explained in a similar way. However, it seems to us that all the characteristics of the wave field revealed may be satisfactorily explained if we assume the existence of a single inhomogeneity on the boundary of the Cretaceous basement at a depth of 4–5 km. A series of ray constructions carried out in a known section enable us to outline this inhomogeneity in plan. The results of these constructions are given in Figs. 1 and 3 (the position of the anomalous inhomogeneity is shown in plan and in section). If a body with reduced viscous and elastic parameters is situated in this fashion, the seismic waves spreading from explosion points

7 and 8 will indeed undergo diffraction at the cut-off points of the seismic rays (taking into account seismic drift). An anomalous picture will also be observed on the longitudinal profile for a wave refracted at the boundary of the crystalline basement, as in fact can be seen on the travel–time curves obtained from explosion points 2 and 6 (see Fig. 2). To satisfy the conditions indicated, the dimensions of the present anomalous inhomogeneity must be no less than 3×5 km. Judging by the diffraction picture obtained on the longitudinal profile 1/1976 from explosion point 6, the diffracting body cannot be very thick (diffraction waves are recorded only at a certain interval from the explosion point, after which a 'normal' refracted wave from the crystalline basement is observed). On this basis the dimensions of an inhomogeneity situated on the boundary of the Upper Cretaceous basement (at a depth of 4–5 km) and satisfying the wave picture obtained from all the explosion points, are estimated as 3×5 km. Evidently these may be the minimal dimensions, since (*a*) it is difficult to evaluate the true dimensions owing to diffraction phenomena, and (*b*) the body may extend considerably further south and south-west as the observations were directed non-longitudinally and thus do not permit definition of the southern limit of the inhomogeneity (this also applies to its northern limit).

Thus from the results of the various seismic investigations carried out in 1975–76 in the region of the New Tolbachik Volcanoes, we can draw the following principal conclusions.

1. With the aid of KMPV continuous longitudinal seismic profiling, the structure of the upper portion of the crust in the Tolbachik areal zone of volcanism has been established, including the top of the consolidated basement. According to these seismic data, the New Tolbachik Volcanoes are situated in a complex downwarp structure, bounded by step-like faults. The seismic discontinuities are attributed to the surface of the Neogene deposits, that of the Upper Cretaceous basement, and that of the crystalline basement respectively, and occur at depths of 200–1300 m; 3–5 km; and 7–8 km.

2. Seismic shooting methods have identified an inhomogeneity within the crystalline basement at a depth of 7–8 km. It is situated beneath the Northern cones and oriented in a more or less north–south direction. Evidently it may be identified with a magmatic intrusion from deep within the Earth.

3. The complex of seismic data which include a system of longitudinal and non-longitudinal observations, has enabled us to identify an

326   *Balesta* et al.

inhomogeneity with reduced viscous–elastic parameters, situated on the boundary of the Upper Cretaceous basement at a depth of 4–5 km. Its dimensions may be put at 3 × 5 km. This inhomogeneity, situated higher in the section than the first, can probably be interpreted as a peripheral magma chamber or sill-like intrusion (Fedotov *et al.*, 1976), feeding the northern group of cones of the GTFE.

4. Seismic shooting has shown that directly beneath the Southern Breakthrough, between discontinuities attributable to the top of the Upper Cretaceous and crystalline basement, no inhomogeneities greater than a wave length of 300–500 m have been discovered by our present methods.

5. The seismic shooting shows that the inhomogeneity discovered on the boundary of the crystalline basement is limited in the south and stretches northwards only as far as Cone III of the Northern Breakthrough. This indicates that there is evidently no feeder link between the Southern and Northern Breakthroughs, or Ploskiy Tolbachik and the New Tolbachik Volcanoes, at this depth.

In conclusion, the authors sincerely thank S. A. Fedotov for valuable discussions of the results of their investigations.

References

Averko, Ye. M. & Balesta, S. T. 1970. Obrabotka rezul'tatov seysmicheskogo modelirovaniya ochaga vulkana. (Evaluating the results of seismic modelling of a volcano chamber.) *Geologiya i geofizika*, No. 8, 32–44.
Farberov, A. I. 1974. *Magmaticheskiye ochagi vulkanov Vostochnoy Kamchatki po seysmologicheskim dannym.* (*Magma chambers of the volcanoes of East Kamchatka from seismological data.*) M., Nauka, 88 pp.
Fedotov, S. A., Gorel'chik, V. I. & Stepanov, V. V. 1976. Seysmologicheskiye dannyye o magmaticheskikh ochagakh, mekhanizme i razvitii bazal'tovogo treshchinnogo Tolbachinskogo izverzheniya v 1975 g. na Kamchatke. (Seismological data on the magma chambers, mechanism and development of the basalt Tolbachik fissure eruption of 1975 in Kamchatka.) *Dokl. AN SSSR*, **228**(6), 1407–10.
Utnasin, V. K., Anosov, G. I., Balesta, S. T., Markhinin, Ye. K. & Fedorchenko, V. I. 1974. Glubinnoye stroyeniye Klyuchevskoy gruppy vulkanov i problema magmaticheskikh ochagov. (Deep structure of the Klyuchi group of volcanoes and the problem of magma chambers.) *Sovetskaya Geologiya*, No. 2, 38–54.

Study of volcanic tremor in the area of the Tolbachik eruption, using seismic instrumentation

A. I. Farberov and I. V. Garetovskaya

Introduction

Many years' study of volcanic tremor in different active volcanoes of the world have shown that it is characterised by a broad band of vibration frequencies – from ≈ 0.15 to ≈ 50 kHz (Sassa, 1935, 1936; Finch, 1949; Shimozuru et al., 1966; Tokarev, 1966; Farberov & Balesta, 1966; Schick & Riuscetti, 1973; and others). Tremor with periods from ≈ 0.3–0.5 s to 2–3 s has been studied in some detail, as it can be comparatively easily recorded with standard seismological instruments over a wide range of distances. Short period volcanic tremor with $f \gtrsim 5$ Hz has been studied in less detail, although it has been identified on many active volcanoes – Nyiragongo, Aso, Kilauea and others. The main causes of this situation are the low intensity of the high frequency seismic signals, their relatively high attenuation with distance, and the (as a rule) poor sensitivity of the seismological instruments used. Nonetheless, high frequency volcanic tremor is of considerable interest, since it is connected with the process of the generation of body waves. The question of the type of waves, together with the problem of the depth and dimensions of the region producing the vibrations, is extremely important for an understanding of the mechanism of the sources of volcanic tremor.

Instrumentation, methods, experimental data

During the Tolbachik eruption, for the first time in volcanological practice, standard seismic instrumentation was used for the study of short period volcanic tremor. Seismic vibrations were recorded by multi-channel stations of the Poisk type, which have a high sensitivity – the amplification of rate of displacement was $\approx 10^5$ in the frequency field ≈ 5–50 Hz. This instrumentation enabled us to concentrate on the task of studying the

327

arrival direction of tremor waves, and their propagation velocities. The seismic vibrations were received on angular profile installations, which are a kind of stationary antennae. Vertical SV2-05 seismic detectors were used, with a natural frequency of 5 Hz and at steps of 100 m or 15 m along the profile. For the various angular arrays the angle between the profiles was 60, 78 or 90°, while the length of the profiles was 700–1400 m spaced at 100 m, or 165 m with steps of 15 m. Since the seismic stations were not designed for uninterrupted registration, recording was carried out in sessions lasting from a few tens of seconds to one and a half to two minutes, as a rule daily. Work on the recordings was ancillary and carried out only in periods when seismic studies were not going on in the region of the eruption.

Nevertheless, fairly prolonged series of observations at several sites enabled us to obtain information not only about the approach directions and velocities of the tremor waves, but also about the change in the basic characteristics of the tremor at different times in the eruption. In all 236 seismograms were recorded, 144 in 1975, and 92 in 1976. These recorded several types of vibrations: atmospheric waves, propagated at the detection site with the speed of sound, seismic waves from earthquakes (in particular, some that could be felt) and volcanic tremor. The nature of some of the vibrations registered has still to be deciphered. Some of the seismograms are unreadable because of the superimposition on the recording of intense oscillations connected with powerful atmospheric vibrations of sonic and infrasonic frequency. The present study examines data from the manual processing of seismic material for one type of signal – relatively low frequency volcanic tremor. This consists of continuous vibrations with a variable amplitude and a frequency of 5–10 Hz. It has been found on the overwhelming majority of seismograms and is characterised by a series of dynamic and kinematic features that are fairly persistent in time. In some seismograms, and on recordings throughout the period in question, these vibrations are as a rule characterised by the way their amplitudes stand out against a background of higher frequency tremor with $f \gtrsim 12$–15 Hz. Pulses of increased amplitude, consisting of a small number of phases, are registered at intervals from a few fractions of a second to several seconds. In some of these pulses fairly distinct, mostly straight or slightly curved synphasal axes can be distinguished. In cases not complicated by oscillation interference, when the synphasal axes can be traced simultaneously on both profiles of an angular array (see Fig. 1), the approach directions of the waves, and their apparent velocities V^* along these directions, were

Fig. 1. Example of recording of a pulse of short-period volcanic tremor, and of sound wave from a volcanic explosion. Angular array in the Gora Vysokaya area, distance to the Southern Cone 7 km.

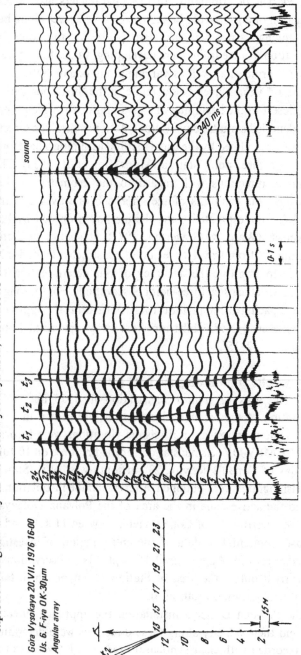

calculated. The determination of these parameters was preceded by the construction of correlatedly linked travel–time curves along both profiles of the angular array for successive phases of the wave train under consideration. As the trains were developing, values for V^* were constant along both profiles, as well as varying along one or both profiles.†

All the V^* values given below relate to the 'first appearances', i.e. the phases closest to the beginning of a given pulse, for which synphasal axes can be identified simultaneously along both profiles. The directions to the source were calculated by the usual method assuming a flat front for waves arriving at a three-point station (Monakhov *et al.*, 1959). The errors in determining directions with regard to the geometry of the receiving installations, the azimuths of the approach of waves and the absolute values V^*, were estimated using the method proposed by P. Ward and S. Gregersen (Ward & Gregersen, 1973). It should be noted that by calculating V^* from travel–time curves constructed from 8–12 points, we were able to improve the accuracy of our determination of direction compared with the usual three-point observations. For the long arrays with a detector spacing of 100 m, the errors in measuring the approach azimuths varied between ± 2–$6°$, whilst for short arrays spaced at 15 m they were as much as ± 12–$15°$.

Results of processing the 1975 data

The sites of tremor registration in 1975 and 1976 and the configuration of the receiving installations are shown in Fig. 2. In the course of the eruption natural seismic vibrations were recorded using a Poisk 24-KMPV-OV seismic station for the first time on 10 July 1975, the fourth day of Cone I's activity, 23 km south-west of it (Station 1). Here three seismograms were obtained, after which the installation was moved closer to the active cone in the area of the Polyana Dolgaya landmark 16 km west-north-west of Cone I. Here, between 11 July and 19 August, the most representative data of the entire period of investigation were obtained. From 29 August until 15 September observations were once again carried out in the area of Station 1, but on an installation of a different configuration (Station 3).

At the Polyana Dolgaya installation the approach directions of the waves and their velocity along these directions were determined for 190 pulses recorded with clear synphasal axes. The V^* values obtained over

† An analysis of the causes of this phenomenon as evaluated by computer will be given in a separate publication.

the period in question vary within considerable limits, comprising no less than 3.5–4.0 km/s (see Fig. 3(*a*)). Fig. 3(*b*) shows the variation in time of the approach at this installation. The azimuths – $\alpha°$ in Fig. 3(*a*) and (*b*) were calculated relative to the longitudinal 'shoulder' of the installation, which extends approximately east–west in the direction of the Northern Breakthrough. The number of measurements in each sector of the azimuths is not indicated because of the short duration of the recording sessions, their varying number and duration on different days, and the differences in resolution of the useful waves on the various seismograms, facts that mean we cannot make quantitative correlations here.

It can clearly be seen that in the initial stage of the eruption, when as yet only Cone I was active, tremor waves arrived from a broad sector of azimuths that includes both the Northern and Southern Breakthrough regions. In a few days even more southerly directions of approach were recorded. The picture remained essentially unchanged when Cones II and III became active. The most persistent 'generator' of tremor over the period in question was the azimuthal sector that includes Gora 1004 and Gora Vysokaya – waves emanating from here were recorded on every day of the observations. On the other hand, waves were recorded by no means every day from the sector where the cones of the Northern Breakthrough are situated. It is important to emphasise that the waves we examined were

Fig. 2. Arrangement of angular profile arrays for recording tremor in 1975 and 1976. Asterisks mark active cones; N, Northern Breakthrough; S, Southern Breakthrough.

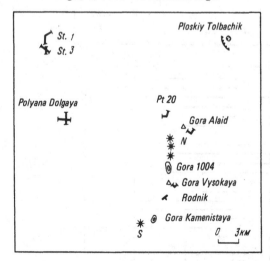

Fig. 3. (*a*) Dependence of the apparent velocity of tremor waves on direction of approach to the Polyana Dolgaya instrument. (*b*) Alteration with time of the direction of approach of tremor waves to the Polyana Dolgaya instrument.
Here and subsequently: N, Northern Breakthrough; S, Southern Breakthrough; 1, Ploskiy Tolbachik; 2. Gora Alaid; 3, Gora 1004; 4, Gora Vysokaya; 5, Gora Kamenistaya.

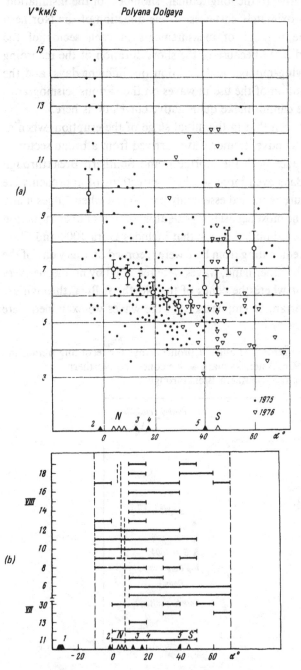

recorded both during surface explosive and effusive activity and during periods of inactivity on 9 August at 10.59, 12.00 and 18.00 hours when Cone I had ceased activity and Cone II had not yet formed, and on 10 August at 18.00 in a five-hour pause (14.30–19.30) in the activity of Cone II.

The significant changes in the average V^* values, depending on the approach directions of the waves to the Polyana Dolgaya installation (Fig. 3(a)), may indicate that in the northern and southern marginal portions of the zone where tremor is generated, the sources are at greater depth. Without outlining in plan the region where vibrations are generated, it is difficult to evaluate the range of depths this region affects. It is natural to assume that the generation of tremor was confined to that cinder cone area where the eruption of the Northern and then the Southern Breakthrough originated (epicentral distances for this installation: 15–20 km). If this is so, then $V^* = 3.5–5.2$ km/s corresponds to velocities of longitudinal waves refracted at the boundary of the Cretaceous basement and at a series of higher-placed boundaries (see data from the velocity profile in Balesta *et al.* (this volume)). Their sources may be situated in the interval of depths below the Earth's surface and above the seismic boundary with $V_r = 4.8–5.2$ km/s. Waves with greater apparent velocity values should be regarded in this case as direct longitudinal waves from deeper sources situated within the Cretaceous and crystalline basements.

Fig. 4 shows how V^* depends on the approach direction of the waves

Fig. 4. Dependence of apparent velocities of tremor waves on direction of approach to instrument 3.

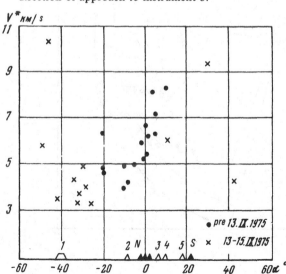

at installation 3, which operated from 29 August to 15 September, i.e. right up until Cone II ceased activity. The azimuths $\alpha°$ in Fig. 4 were calculated relative to the longitudinal 'shoulder' of the installation, which extended towards the Northern Breakthrough. As with the Polyana Dolgaya installation, a considerable range of V^* variation is observed, whilst the apparent velocity values obtained are not less than ≈ 3.5 km/s. The sector from which waves approached this array widened northwards considerably, compared with the Polyana Dolgaya installation (see Fig. 3), and from 13 September it included the area of Ploskiy Tolbachik volcano. Lower V^* values were recorded from this area than from Gora Vysokaya. The most stable 'generator' of tremor over the period when installation No. 3 was operating was the sector that includes the region of the Northern Breakthrough – from Gora Alaid to Gora Vysokaya.

The one-sided positioning of the installations vis-à-vis the active volcanic zone, and the absence of simultaneous observations at several points, meant that we were unable to outline the region where tremor was being generated. At the same time, the fact that three installations operating at

Fig. 5. Sectors of approach of tremor waves to the angular profile array in July–September 1975.

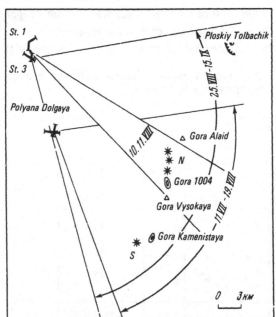

different times give partially overlapping sectors of wave approach direction may indicate that at least part of the area where these vibrations were generated had a stable spatial position from July to the first half of September 1975 (Fig. 5). It should be stressed that although the predominant directions of wave approach were from the region of the eruption, in several cases directions were obtained that differ sharply from the others. Moreover, for a number of waves curvilinear synphasal axes were observed, symmetrical in relation to the central portion of one of the profiles and with finite V^* values along the other; this could be caused by the spherical nature of the wave front from a near source. A few pulses of vibrations are characterised by vertical angles of emergence, since $V^* \approx \infty$ along both profiles of an angular array. All this strongly suggests that, simultaneous with volcanic tremor pulses, some pulses of vibrations with the same predominant frequencies are being recorded from the micro-earthquakes situated at various depths and at various azimuths from the angular arrays.

Results of processing the 1976 data

In 1976, whilst the Southern Cone was active, observations of volcanic tremor using seismic instrumentation were repeated at the Polyana Dolgaya array from 4 to 22 July. As can be seen from Fig. 3(a), the sector of the approach directions of the tremor waves had narrowed by approximately half (up to 40°) compared with 1975. The apparent velocity values were on average lower, and the significant scatter of V^* was retained. The level of the signal fell by two to three times compared

Fig. 6. Dependence of apparent velocities of tremor waves on direction of approach of the waves to the angular profile array in July–September 1976.

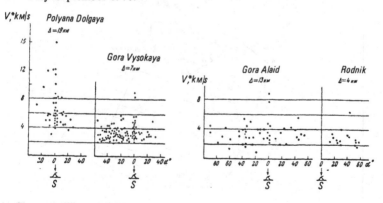

with July–August 1975, although the difference in the distances from this array to the cones of the Northern and Southern Breakthroughs does not exceed 2–3 km.

After repeated observations at Polyana Dolgaya, recordings were made at a number of other points closer to the Southern Cone (Point 20, Gora Alaid, Gora Vysokaya and Rodnik; see Fig. 2). The apparent velocities of the tremor waves proved to be practically independent of the approach directions of waves for all the observation points (Fig. 6). Moreover, at

Fig. 7. Diagram showing area of tremor generation from observations in July–September 1976. Explanation in text.

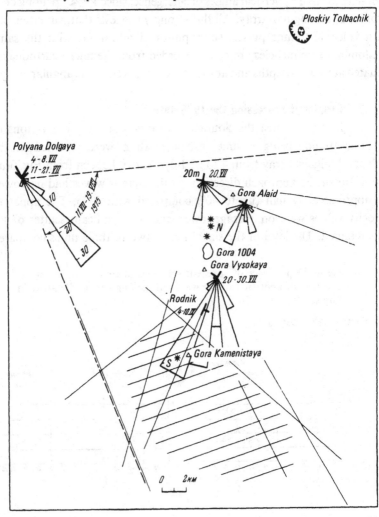

points to the north-east and north of the Southern Cone the range of variation in velocities was practically constant (2–6 km/s). Only occasional jumps in values are noted in the direction directly towards the Southern Cone. To the north-west of the cone, in the region of Polyana Dolgaya, velocities are significantly higher, 3–12 km/s.

In order to determine from these data the range of depths involved in the region of tremor generation, the position of the region where vibrations are generated was evaluated in plan. Fig. 7 shows, in the form of vector diagrams, the distribution of the various measurements of approach directions of tremor waves to the observation points Polyana Dolgaya, Point 20, Gora Alaid and Gora Vysokaya. Taking into account the precision of azimuth determination, averaging was carried out at 15° intervals. The length of each 'petal' corresponds to the number of measurements in the relevant sector. The scale for all the diagrams is the same and is shown for the angular array at Polyana Dolgaya. The dates of observations are indicated beside the arrays. Maxima directed at the region of the Southern Cone are identified. There are also other sectors of approach that do not coincide with the direction to the Southern Cone. The data obtained at Gora Vysokaya as well as at Gora Alaid and Rodnik indicate that the region of tremor extends a considerable distance to the south-east of the Southern Cone. Two lateral 'petals' – the eastern for the array at Point 20 and the western for the array near Gora Alaid, are evidently connected with lateral refraction of the tremor waves, which skirt round an inhomogeneity beneath the Northern Breakthrough region as they approach the station. They were not taken into account in drawing the outline. From data on the propagation velocities, angles of approach and amplitude of signals, an estimate has been obtained of what are evidently the limiting dimensions of the tremor generating region. The north-western boundary of this region passes no further than 2–3 km from the Southern Cone, and the north-eastern boundary passes to the south of Rodnik station, i.e. 2–3 km from the Southern Cone. The zone extends from the north-west to the south-east over a distance of not less than 7–10 km. The south-western boundary turned out to coincide with the limit of tremor obtained in 1975. It is probable that this boundary passes further to the north-east, but in the absence of observations south of Gora Kamenistaya we are unable to narrow down the source region of the vibrations in this direction.

If tremor is generated within the shaded region, then at the distances at which observations were carried out a proportion of the velocity

338   *Farberov and Garetovskaya*

readings obtained corresponds to sources in a range of depths from 0 to a boundary with velocities 4.0–4.4 km/s, i.e. a depth interval not exceeding 1–1.5 km. However, higher velocity values have also been observed and require placing some of the tremor sources at depths of at least the 4.8–5.2 km/s velocity boundary, i.e. 3–5 km. The most probable interpretation at the present stage of data analysis is that vibrations are being generated throughout this region at a fairly even depth. The data available enable us, moreover, to put forward the hypothesis that in a local sector immediately beneath the region of the Southern Cone tremor is also proceeding from greater depths that include the crystalline basement, i.e. from not less than 6–8 km.

In 1976, as in 1975, the tremor was recorded in differing regimes of explosive activity. The longest series of observations was obtained from the array at Gora Vysokaya, in the period from 20 to 30 July 1976 – 26 recordings from 1–4 sessions a day. It can be clearly seen from Fig. 8 that relatively low-frequency tremor (5–10 Hz) was recorded in all the eruption regimes of the Southern Cone during explosive activity, in the pauses between explosions, and in periods of quiescence and ash discharge. Sound

Fig. 8. Characteristics of the types of seismic vibrations registered during various explosive regimes of the activity of the Southern Cone. 1, low frequency; 2, sound; 3, high frequency; 4, explosion; 5, lull; 6, pause; 7, ash.

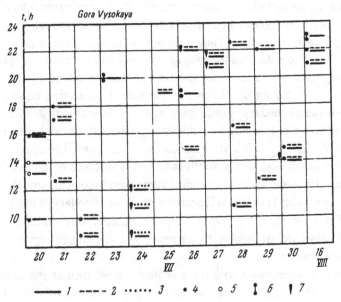

waves from volcanic explosions were fairly frequently recorded (an example is given in Fig. 1). Moreover, throughout the period under consideration and especially during explosive activity, tremor waves arrived at the array on Gora Vysokaya from a fairly wide azimuthal sector, which by no means always included the Southern Cone itself (Fig. 9).

Discussion of results, conclusions

The material presented here shows that with the aid of seismic instrumentation it is possible during an eruption to record short period volcanic tremor at considerable distances from the volcano. Judging from the combination of its dynamic and kinematic features, the volcanic tremor examined in this paper should be classed as body waves. The question of the proportion of transverse waves among the vibrations recorded remains open, since the observations were conducted only with vertical instruments. If not only longitudinal but transverse waves were recorded, we should have to postulate a greater depth and/or 'compactness' along the vertical section of the region generating the tremor. The complex interference

Fig. 9. Alteration with time, from beginning to end, of a recording of approach directions that follow tremor. Dashes indicate moment of arrival of sound waves from volcanic explosions.

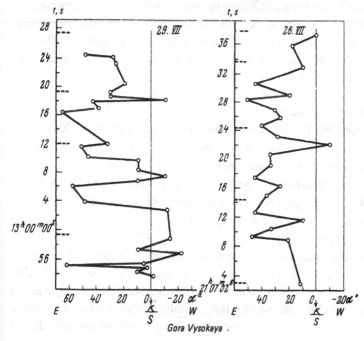

nature of seismic readings in the frequency range 5–10 kHz is most probably caused by the formation of a large number of tremor pulses coming from various parts of this region and from different depths. From the initial stage of the eruption onwards, short period tremor is generated in an extended zone, a sector that includes the Northern and Southern Breakthroughs. This agrees with the data of seismological observations (see Fedotov, Gorel'chik *et al.*, this volume), which show that earthquakes of the first swarm, preceding the beginning of the eruption, 'ripped open' the axial part of the zone of cinder cones in the region stretching from Gora Alaid to Gora Kamenistaya.

The type of vibration considered here is not directly connected with the nature of the surface volcanic activity and is doubtless caused by deeper processes. A discussion of this question would be beyond the scope of the present paper. We can only put forward the most general suggestion that one of the causes of the vibrations was the pulsation of a magmatic pressure that was initiated and became active in a system of fissures within the crust; and that this happened before and during the eruption of the magma responsible over a considerable part of the cinder cones zone extending north-north-east to south-south-west for 15–20 km. It is therefore particularly interesting that the estimates of the length and general orientation of the zone generating the tremor (north-west to south-east in the area of the Southern Cone – see Fig. 6) should correspond quite closely with those of the orientation and length of the fissure (or system of fissures) feeding this cone, according to geodetic and tilt measurements (see Fedotov, Enman *et al.*, this volume).

The analysis carried out of the data obtained enables us to formulate a number of conclusions.

1. During the Tolbachik eruption seismic survey instrumentation was used to register short period volcanic tremor with frequency 5–10 Hz, consisting of body waves and being of deep-seated origin.

2. During the activity of both the Northern (1975) and the Southern (1976) Breakthroughs, this tremor was generated within zones stretching a considerable distance horizontally, and it was not directly connected with surface eruptive activity.

3. The studies carried out confirmed data obtained earlier (Farberov & Balesta, 1966) concerning the presence in volcanic tremor of vibrations with a frequency of $\gtrsim 6$ Hz at distances of more than 10 km from an active volcano. They also demonstrated the possibilities of using spatial correlative systems of the seismic survey type for studying seismic phenomena during eruptions.

References

Farberov, A. I. & Balesta, S. T. 1966. Ob issledovanii vulkanicheskogo drozhaniya. (Studying volcanic tremor.) *Byul. vulkanol. stantsiy*, No. 40, 45–60.

Finch, R. H. 1949. Volcanic tremor (part 1). *Bull. Seismol. Soc. Am.*, **39**(2), 73–8.

Monakhov, F. I., Pasechnik, I. P. & Shebalin, N. V. 1959. *Seysmicheskiye i mikroseysmicheskiye nablyudeniya na sovetskikh stantsiyakh v period MGG*. (*Seismic and microseismic observations at Soviet stations in the International Geophysical Year*.) M., Izd-vo AN SSSR, 39 pp.

Sassa, K. 1935. Microtremors and eruption earthquakes (1). *Mem. Coll. Sci. Kyoto Imp. Univ.*, Ser. A, **18**, 255–93.

Sassa, K. 1936. Microseismometric study on eruption of the volcano Asso (2). *Mem. Coll. Sci. Kyoto Imp. Univ.*, Ser. A, **19**, 11–56.

Schick, R. & Riuscetti, M. 1973. An analysis of volcanic tremors at South Italian volcanoes. *Zeitschrift für Geophysik*, **39**, Part 2, 247–62.

Shimozuru, D., Kamo, K. & Kinoshita, W. T. 1966. Volcanic tremor of Kilauea volcano, Hawaii, during July–December 1963. *Bull. Earthqu. Res. Inst. Tokyo Univ.*, **44**, Part 3, 1093–1134.

Tokarev, P. I. 1966. *Izverzheniya i seysmicheskiy rezhim vulkanov Klyuchevskoy gruppy.* (*Eruptions and the seismic regime of volcanoes of the Klyuchi group.*) M., Nauka, 118 pp.

Ward, P. L. & Gregersen, S. 1973. Comparison of earthquake locations determined with data from a network of stations and small tripartite arrays on Kilauea volcano – Hawaii. *Bull. Seismol. Soc. Am.*, **63**, 679–711.